Lecture Notes in Mechanical Engineering

Lecture Notes in Mechanical Engineering (LNME) publishes the latest developments in Mechanical Engineering—quickly, informally and with high quality. Original research reported in proceedings and post-proceedings represents the core of LNME. Volumes published in LNME embrace all aspects, subfields and new challenges of mechanical engineering. Topics in the series include:

- Engineering Design
- Machinery and Machine Elements
- Mechanical Structures and Stress Analysis
- Automotive Engineering
- Engine Technology
- Aerospace Technology and Astronautics
- Nanotechnology and Microengineering
- Control, Robotics, Mechatronics
- MEMS
- Theoretical and Applied Mechanics
- Dynamical Systems, Control
- Fluid Mechanics
- Engineering Thermodynamics, Heat and Mass Transfer
- Manufacturing
- Precision Engineering, Instrumentation, Measurement
- Materials Engineering
- Tribology and Surface Technology

To submit a proposal or request further information, please contact the Springer Editor of your location:

China: Ms. Ella Zhang at ella.zhang@springer.com
India: Priya Vyas at priya.vyas@springer.com
Rest of Asia, Australia, New Zealand: Swati Meherishi at swati.meherishi@springer.com
All other countries: Dr. Leontina Di Cecco at Leontina.dicecco@springer.com

To submit a proposal for a monograph, please check our Springer Tracts in Mechanical Engineering at https://link.springer.com/bookseries/11693 or contact Leontina.dicecco@springer.com

Indexed by SCOPUS. All books published in the series are submitted for consideration in Web of Science.

More information about this series at https://link.springer.com/bookseries/11236

Ivan Tolj · M. V. Reddy · Achmad Syaifudin
Editors

Recent Advances in Mechanical Engineering

Select Proceedings of ICOME 2021

Springer

Editors
Ivan Tolj
Faculty of Electrical Engineering, Mechanical Engineering and Naval Architecture
University of Split
Split, Splitsko-dalmatinska, Croatia

M. V. Reddy
Nouveau Monde Graphite (New Graphite World)
Hydro-Québec
Montreal, QC, Canada

Achmad Syaifudin
Faculty of Industrial Technology and Systems Engineering
Institut Teknologi Sepuluh Nopember
Surabaya, Indonesia

ISSN 2195-4356 ISSN 2195-4364 (electronic)
Lecture Notes in Mechanical Engineering
ISBN 978-981-19-0866-8 ISBN 978-981-19-0867-5 (eBook)
https://doi.org/10.1007/978-981-19-0867-5

This Springer imprint is published by the registered company Springer Nature Singapore Pte Ltd.
The registered company address is: 152 Beach Road, #21-01/04 Gateway East, Singapore 189721, Singapore

Contents

About the Editors

Dr. Ivan Tolj is currently an Assistant professor at FESB—Faculty of Electrical Engineering, Mechanical Engineering and Naval Architecture. He obtained Doctor of Philosophy (Ph.D.) and Master of Engineering (M.Eng.) from University of Split, Croatia. His major areas of research interests include Thermodynamics, PEM fuel cells, hydrogen powered vehicles, MH hydrogen storage, and Renewable energy sources in Mechanical Engineering and Naval Architecture. He has published 63 papers papers in respected international journals. He received Faculty Award in University of Split in 2005, as excellent results achieved during study, and also Recognition as the best students of FESB in 2003. Currently, he is an Assistant Subject Editor of the International Journal of Hydrogen Energy.

Dr. M. V. Reddy obtained his Ph.D. (2003) with the highest honours in the area of Materials Science and Engineering from the University of Bordeaux, ICMCB-CNRS/ENSCPB, France. From July 2003 to May 2019, he worked at the Department of Materials Science and Engineering, Chemistry and Physics, National University of Singapore (NUS). Dr. Reddy has published around 200 papers in various international journals. These have recently placed him within the top 2% highly cited researchers in Energy. Dr. Reddy is also serving as an editorial advisory board member in Materials Research Bulletin and Journal of Energy Storage (Elsevier) as well as several open access Journals (Materials, Energies and Molecules, MDPI) and the Regional editor: Nanoscience & Nanotechnology-ASIA.

Dr. Achmad Syaifudin Ph.D. is now currently an asistant professor at laboratory of Solid Mechanics, department of Mechanical Engineering, Institut Teknologi Sepuluh Nopember, Indonesia. After earning a doctorate from Hokkaido University, Japan, he has 2 research in progress that are included in the national research priorities, i.e. structural integrity of high-speed train structures, and development of PMMA-HA composite for a dental implant. He is a reviewer at several respected international journals (BioMedical Engineering Online, Journal of the Brazilian Society of Mechanical Sciences and Engineering, Scientific Reports, Journal of Mechanical Engineering and Sciences, and Journal of Mechanics). Besides, he is also an editorial board member for journal Rotasi published by Diponegoro University.

Cavitation Detection of Centrifugal Pumps Using SVM and Statistical Features

Berli Kamiel(✉), Taufiq Akbar, Sudarisman, and Krisdiyanto

Mechanical Engineering Department, Universitas Muhammadiyah Yogyakarta, 55183 Yogyakarta, Indonesia
berlikamiel@umy.ac.id

Abstract. Centrifugal pump is one type of pump that is widely used in the industry. Its mechanism which creates pressure change may cause cavitation. Generally, the cavitation increases noise and vibration level, if not properly maintained leads to catastrophic failure and total stop of the whole process. Therefore, it is needed to develop a method that can detect cavitation as early as possible. Recently, pattern recognition-based fault detection is gaining attention from many researchers due to its superior detection and classification performance. Support vector machine (SVM) is one of the pattern recognition techniques which requires statistical features as input for building the model. However, the selection of statistical features is arbitrary. In this study, ten statistical features are extracted from the time-domain vibration signal and selected using Relief Feature Selection. The selected features are used as input for two types of SVM, binary, and multi-class, to classify new vibration data. Each multi-class classification result is optimized by the Bayesian Optimization algorithm and Grid Search Method. The result shows that root mean square, standard deviation, variance, entropy, and standard error are several features that indicate the best plot. The feature selection process reveals that variance, root mean square, and standard error are the best feature to use for SVM classification. The binary SVM method shows the best plot on early cavitation with an accuracy of 99%. The Bayesian optimization algorithm with multi-class SVM is the best combination to classify all pump conditions with an overall accuracy of 100%.

Keywords: Centrifugal pump · Support vector machine · Relief feature selection · Bayesian optimization · Grid search method

1 Introduction

Centrifugal pump is one type of pump widely used in the industry such as chemical, oil, and gas company because of its simple mechanism and construction. Considering the importance of its role, the performance is salient to be maintained. The reduction of its performance is usually affected by faulty components. One of the most common causes of components' failure is cavitation. Cavitation occurs due to the evaporation of liquid flowing fluid which is caused by a pressure drop below the saturated vapour pressure.

I. Tolj et al. (eds.), *Recent Advances in Mechanical Engineering*, Lecture Notes in Mechanical Engineering, https://doi.org/10.1007/978-981-19-0867-5_1

The cavitation diagnosis is very important to be carried out at an early stage because a decrease in pump capacity due to cavitation can disrupt production activities. This will certainly impact the level of productivity in an industry so that an effective method is needed in detecting early cavitation in centrifugal pumps.

Early cavitation detection on the centrifugal pump was examined and tested by several researchers in various methods. Al-Hashmi et al. [1] utilized spectrum analysis from vibration signals. On the other hand, Al-Obaidi [2] examined the use of statistical features in the time domain to detect faults in centrifugal pumps. Another approach was introduced by Nasiri et al. [3] which uses Artificial Neural Network (ANN) based vibration signal for detecting cavitation in a centrifugal pump. Sakthivel et al. [4] applied the Decision Tree algorithm in detecting cavitation in a centrifugal pump. Farokhzad [5] detected several faults of the pump using the Adaptive Network Fuzzy Inference System (ANFIS); however, the indication of cavitation at the initial stage was difficult to observe.

The researchers then try to use one of the pattern recognition-based methods, called support vector machine (SVM). This method is considered as the method which enables to classify data and provide information with a high level of accuracy. Ebrahimi and Javidan [6] investigated faults in centrifugal pumps using vibration signal analysis and the SVM method. Vibration signals were decomposed in three-level using wavelet transform, and descriptive statistical features were extracted from detail and approximation coefficients of the wavelet. Saberi et al. [7] found that the SVM method had the advantage of having a kernel function that could distinguish normal conditions and high noise on the centrifugal pump. Sakthivel et al. [8] compared the use of Proximal Support Vector Machine (PSVM), Gene Expression Programming (GEP), Wavelet-GEP, and SVM, which showed that SVM was the method with the highest accuracy.

The use of SVM in detecting early cavitation requires the help of a kernel function. The selection of the right kernel function affects the detection results. The recommended kernel function is Gaussian Radial Basis Function (RBF) because it can classify non-linear data groups [9]. In addition, the statistical features in the time domain used can also affect SVM performance. Rapur and Tiwari [10] proposed that the statistical parameters mean, standard deviation, and entropy because they give excellent accuracy. On the other hand, Elangovan et al. [11] proved that the promising statistical parameter is the standard error and minimum value.

However, there is no standard procedure in determining the use of statistical parameters, and the best classification algorithm for detecting cavitation in a centrifugal pump is the SVM. In this study statistical parameters are extracted from the time domain instead of the frequency domain since the spectrum from the vibration of fluid interaction in centrifugal pumps is dominated by noise and random high-frequency vibration which has less meaningful information. Further research and development of the SVM-based method are needed to get better accuracy. Hence this paper aims to find the best combination of the statistical features, and optimization algorithm in SVM to detect early cavitation and to classify several levels of cavitation in the centrifugal pump.

2 Support Vector Machine

SVM is one method that classifies data based on pattern recognition. Pattern recognition works with separating data into several groups or classes [12]. This method can be classified as part of an artificial intelligence system built for decision-making. Input data is used in various, such as numbers, images, sounds, or a signal wave. This method is very popular in the field of statistics. Up to now, the implementation of the analysis vibration signal method based on pattern recognition continues to develop. This affects the increasing number of new methods based on pattern recognition. The appearance of this new method indicates that the level of popularity in the future will be even better.

SVM is a method that is used for binary classification. Originally, SVM combined several sets of concepts in the field of pattern recognition. This method works by finding the best hyperplane that separates groups of data on a dimension perfectly into two classes. The pattern of the two classes assumed has been completely separated by a hyperplane in a dimension defined by Eq. (1).

$$\vec{w} \cdot \vec{x} + b = 0 \tag{1}$$

If $\vec{x_i}$ is in class -1, then as shown in Eq. (2).

$$\vec{w} \cdot \vec{x} + b \leq -1 \tag{2}$$

While the value $\vec{x_i}$ in class $+ 1$ is shown in Eq. (3).

$$\vec{w} \cdot \vec{x}_i + \geq +1 \tag{3}$$

Quadratic Programming (QP) problem is an effort to maximize the value of the distance between the hyperplane and its closest point by finding its minimum point like Eq. (4).

$$\min_{\vec{w}} \tau(w) = \frac{1}{2} w^2 \tag{4}$$

$$y_i(\vec{x}_i \cdot w + b) - 1 \geq 0, \forall_i \tag{5}$$

The problem in Eqs. (4) and (5) can be solved by the Lagrange Multiplier (α_i) technique, as in the following Eq. (6).

$$L(\vec{w}, b, \alpha) = \frac{1}{2}\vec{w}^2 - \sum_{i=1}^{l} \alpha_i \left(y_i \left(\left(\vec{x_i} . \vec{w} + b \right) - 1 \right) \right); \ (i = 1, 2, \ldots, l) \tag{6}$$

Furthermore, Eq. (4) is optimized by maximizing problem that only contained value (αi), as in the following Eq. (7).

$$\sum_{i=1}^{1} \alpha_i - \frac{1}{2} \sum_{i,j=1}^{1} \alpha_i \alpha_j y_i y_j x_i x_j \tag{7}$$

So, Eq. (8) is obtained:

$$\alpha_i \geq 0(i = 1, 2, \ldots, l) \sum_{i=1}^{l} \alpha_i y_i = 0 \quad (8)$$

Equation (8) shows that the value (α_i) is mostly positive, and the support vector correlates with this positive value (α_i).

Figure 1 shows several patterns of two classes (the pattern in class A is marked with a red box symbol and the pattern in class B is marked with a blue circle symbol) which will further be processed to search for the best hyperplane by giving rise to several discrimination boundaries (alternative split lines). In determining the best hyperplane, it can be done by finding the maximum point and measuring the margin. Figure 2 shows the best hyperplane located in the middle between the two classes, and the pattern near the hyperplane is Support Vector.

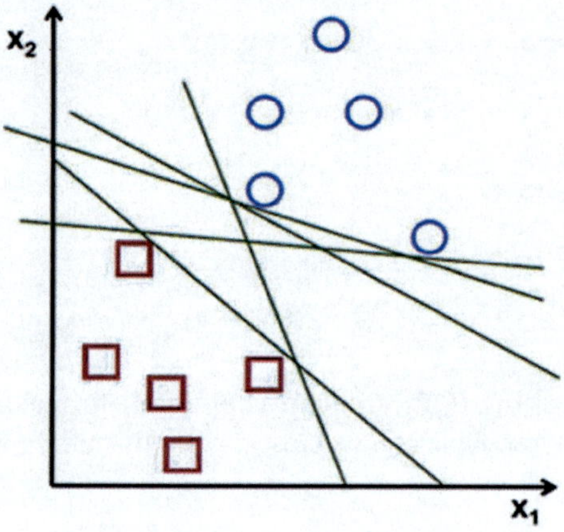

Fig. 1. Search for optimal hyperplane

3 Method

The data acquisition process was taken by recording 500 files of data from vibration signal for every four variations of pump conditions, including normal (0% valve blockage), level 1 cavitation (0.25% valve blockage), level 2 cavitation (0.50% valve blockage), and level 3 cavitation (0.75% valve blockage). The recording time was 10 s per file and paused for 2 s between recordings. The pump rotational speed was regulated at 2850 RPM, and a sampling rate of 17,066 Hz was set to produce a stable change in each condition. The recording was carried out using the NI 9234 acquisition device with the chassis NI DAQ 9174, and an accelerometer mounted on the pump inlet. The data acquisition process was regulated using the NI MAX and Matlab R2017a. The test rig setting consists of main components such as a centrifugal pump, closed-loop pipe network, gauges valves, water tank, and flowmeter.

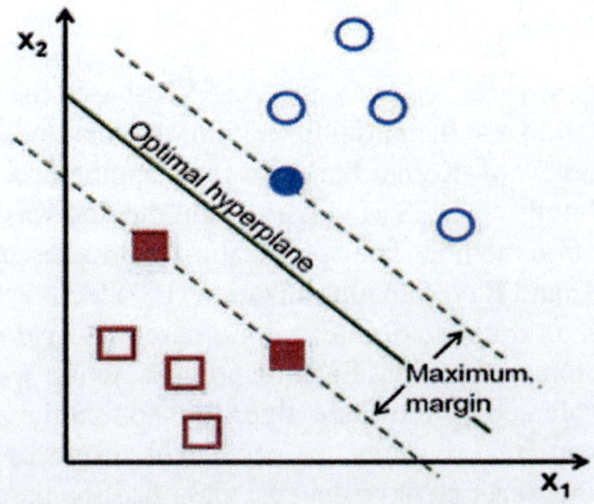

Fig. 2. Optimal separating hyperplane

3.1 Feature Extraction and Selection Process

The vibration signal was extracted into ten statistical features in the time domain as shown in Table 1. Each statistical feature was then plotted to show its characteristics to the distribution of data from the vibration signal of all conditions. All results of the extraction process were then selected. The feature selection process was done by using Relief Feature Selection. This method ranked the statistical features based on the weight of the information content. The feature selection process then produced the best data input for SVM classification.

Table 1. Statistical features in time domain

Statistical features	Formulas	Statistical features	Formulas
Root Mean Square (RMS)	$\sqrt{\frac{1}{n}\sum_{i=1}^{N}(x_i-\overline{x})^2}$	Crest factor	$\frac{\max(\lvert x(n)\rvert)}{\sigma}$
Standard Deviation (SD)	$\sqrt{\frac{\sum_{i=1}^{N}(x_i-\overline{x})^2}{N-1}}$	Mean	$\frac{1}{K}\sum_{k=1}^{K}x(k)$
Peak value	$\max\lvert x(N)\rvert$	Entropy	$\sum_{i=1}^{N}p(x_i)\log_{10}p(xi)$
Kurtosis	$\frac{\sum_{i=1}^{N}(x_i-\overline{x})^4}{(N-1)\sigma^4}$	Minimum value	$\min\lvert x(N)\rvert$
Variance	$\frac{(x_i-\overline{x})^2}{N-1}$	Standard Error (SE)	$\sqrt{\frac{1}{(n-2)}-\left[\sum(y-\overline{y})^2-\frac{[\sum(x-\overline{x})(y-\overline{y})]^2}{(x-x^2)^2}\right]}$

3.2 Classification Process

The classification stage was carried out using the SVM-based method. To determine the various levels of cavitation and the initial formation, two methods were applied, namely Binary and Multi-Class SVM. Kernel Function for mapping process used Radial Basis Function (RBF). The multi-class SVM classification method was carried out with three trials with an optimization method. The optimization method used at this stage was Grid Search Method (GSM) and Bayesian Optimization (BO) techniques.

GSM algorithm is an optimization technique based on grid search. Grid search is performed on each mapping function. Each mapping function with optimal results and not optimal will be evaluated by the GSM algorithm so that the number of evaluations produced in several mapping processes that occur. The results of optimization then are sorted according to functions that have the best classification parameters.

BO techniques were carried out to evaluate any mapping errors found. This technique was equipped with an acquisition function that was useful for determining which mapping function was not optimal. Then the mapping function was optimized and would be returned in the training process. The number of mapping functions evaluated according to the number of mapping error processes that occurred.

4 Result

Statistical features in time domain aim to find characteristics of data from vibration signals under normal conditions, level 1, 2, and 3 cavitation. Some features such as RMS, SD, variance, entropy, and SE show the separation of the pump conditions. However, for normal conditions and level 1 cavitation, they fail to separate the classes. Thus, the selection process is intended to select the best features as input for SVM classification.

The result of features selection using Relief Feature Selection can be seen in Fig. 3. The result of the selection shows that the variance has the highest weight meanwhile the crest factor is the lowest one. Three features with the highest scores were used as SVM input since they contained a weight of more than 0.0050. However, the SE is not used as input because it has the same value as SD.

The binary SVM classification was performed on four pump conditions. The two main stages carried out in this classification are the training and testing process. The training forms a classification model and generates a mapping function, while testing data evaluates the training model and determines the accuracy. The input data was separated using the cross-validation process to avoid overfitting. This process includes 800 samples for training and 200 samples for the testing process.

The steps taken in the multi-class SVM correspond to the binary SVM method. The method classifies the four pump conditions and shows the result of the spread of the pattern. The cross-validation process produces 1800 training data and 200 testing data, where all variations have the same amount of data in each set.

The classification was done by three experiments, those were multi-class SVM without optimization, using GSM optimization, and BO algorithm. From the experiments, the highest level of accuracy is shown by a combination of multi-class SVM and BO.

Based on the distribution of statistical parameters, RMS, SD, variance, entropy, and SE solely can separate four pump conditions properly with an exception for normal and

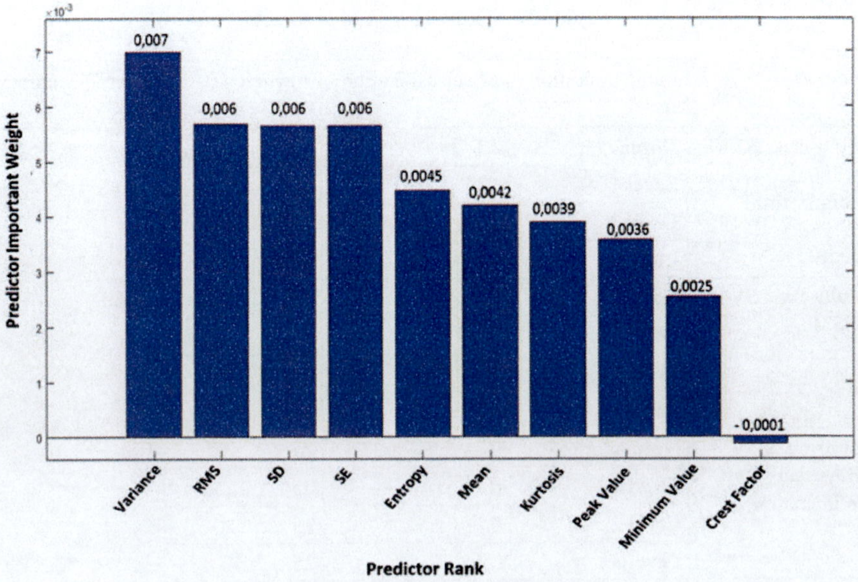

Fig. 3. Features selection result

level 1 cavitation classes which show less clustering effect. However, the results have a promising potential to be used as the input parameters for SVM.

Table 2 shows the success of SVM binary classification in detecting early cavitation phenomena, with an accuracy of 99%. The separation of the four classes is carried out using a combination of multi-class SVM and BO as depicted in Table 3 with a maximum accuracy of 100%. The obtained classification accuracy is a bit higher compared to that from [5] where they extracted statistical parameters from wavelet coefficients. The relief feature selection procedure is proven to be more effective to choose suitable statistical parameters than the wavelet-based parameters extraction.

Table 2. Binary SVM accuracy

No	Binary SVM	Accuracy (%)
1	Normal and Level 1 cavitation	99
2	Normal dan Level 2 cavitation	97
3	Normal dan Level 3 cavitation	100

The results of the study reveal that the SVM-based cavitation detection method show the early cavitation phenomenon. The combination of the SVM-BO algorithm is the most superior method to detect cavitation in several levels.

Table 3. Multi-Class SVM accuracy

Method	Pump condition	Misclassification	Accuracy (%)	Overall accuracy (%)
Multi-class SVM (without optimization)	Normal	23	13.50	58
	Level 1	13	18.50	
	Level 2	15	17.50	
	Level 3	33	8.50	
Multi-class SVM (GSM optimization)	Normal	0	25	99.50
	Level 1	0	25	
	Level 2	1	24.50	
	Level 3	0	25	
Multi-class SVM (Bayesian optimization)	Normal	0	25	100
	Level 1	0	25	
	Level 2	0	25	
	Level 3	0	25	

5 Conclusion

The study indicates that the character of each statistical feature in the time domain produces specific information to the vibration signal distribution. RMS, SD, and variance are the most suitable statistical features used as input for SVM classification. The SVM has been proven to detect early cavitation phenomena in the centrifugal pump. This is shown in the classification between normal and level 1 cavitation that has an accuracy of 99%. Development and optimization of SVM multi-class algorithms by using BO is the best combination method in detecting cavitation at several levels. The level of accuracy obtained with this combination is 100%.

References

1. Al-Hashmi S, Abdurrhman A, Nasir I, Elhaj M (2017) Spectrum analysis of vibration signals for cavitation monitoring. J Pure Appl Sci. Sebha University: 16 14–20
2. Al-Obaidi AR (2020) Detection of cavitation phenomenon within a centrifugal pump based on vibration analysis technique in both time and frequency domains. Exp Tech 44(3):329–347
3. Nasiri MR, Mahjoob MJ, Vahid-Alizadeh H (2011) Vibration signature analysis for detecting cavitation in centrifugal pumps using neural networks. IEEE Int Conf Mechatron 13–15
4. Sakthivel NR, Sugumaran V, Babudevasenapati S (2010) Vibration based fault diagnosis of monoblock centrifugal pump using decision tree. Expert Syst Appl 37(6):4040–4049
5. Farokhzad S (2013) Vibration based fault detection of centrifugal pump by fast fourier transform and adaptive neuro-fuzzy inference system. J Mech Eng Technol: 82–87
6. Ebrahimi E, Javidan M (2017) Vibration-based classification of centrifugal pumps using support vector machine and discrete wavelet transform. J Vibroengineering 19(4): 2586–2597

7. Saberi M, Azadeh A, Nourmohammadzadeh A, Pazhoheshfar P (2011) Comparing performance and robustness of SVM and ANN for fault diagnosis in a centrifugal pump. In:19th International Congress on Modelling and Simulation, Perth
8. Sakthivel NR, Binoy.B.Nair, Sugumaran V (2012) Soft computing approach to fault diagnosis of centrifugal pump. Appl Soft Comput 12(5): 1574–1581
9. Sakthivel NR, Saravanamurugan S, Nair BB, Elangovan M, Sugumaran V (2016) Effect of kernel function in support vector machine for the fault diagnosis of pump. J Eng Sci Technol 11(6): 826–838
10. Rapur JS, Tiwari R (2017) Experimental time-domain vibration-based fault diagnosis of centrifugal pumps using support vectorr machine. ASCE-ASME J Risk Uncert Eng Sys Part B Mech Eng 3(4)
11. Elangovan M, Sugumaran V, Ramachandran KI, Ravikumar S (2011) Effect of SVM kernel functions on classification of vibration signals of a single point cutting tool. Expert Syst Appl 38(12):15202–15207
12. Mavroforakis ME, Theodoridis S (2006) A geometric approach to support vector machine (SVM) classification. IEEE Trans Neural Netw17(3): 671–682

Design Optimization of Hooked Plate on the Automatic Coupler for High-Speed Train

Jean Mario Valentino [1,2](✉), Agus Sigit Pramono[3], and Achmad Syaifudin[3]

[1] Graduate Program of Mechanical Engineering, Institut Teknologi Sepuluh Nopember, Surabaya, Indonesia
jean.mario@bppt.go.id

[2] Research Center for Transportation Technology, National Research and Innovation Agency, Jakarta Pusat, Indonesia

[3] Department of Mechanical Engineering, Institut Teknologi Sepuluh Nopember, Surabaya, Indonesia

Abstract. Weight reduction is a major issue in designing high-speed train components to reduce train power consumption. In this study, an automatic coupler that is being developed for the high-speed train in Indonesia is selected for weight reduction by using topology optimization. The simulation model is referring to an automatic coupler model for light trains that had been developed previously. The main component of the automatic coupler selected for optimization is the hooked plate. Longitudinal train force analysis was carried out to obtain tensile and compressive forces on the coupler. Iteration of retaining every 10% is performed on topology optimization simulation. The new design is made with an over-designed form by giving excess material to the geometry. This study indicates that the mass reduction on the hooked plate is 24% form based on the von Mises stress distribution under high-speed train loading. In total, the new design contributes to reducing the mass of the automatic coupler by 4%.

Keywords: Automatic coupler · Topology optimization · High-speed train

1 Introduction

One of the main issues in high-speed trains is weight reduction and energy consumption. Reducing vehicle weight is the goal in the design of the high-speed train. The potential weight reduction of lightweight components, material substitution, and new technologies in the design of the train's structure can reduce power consumption, less track wear, and the ability to carry larger loads [1]. Another obstacle is the excessive weight on the train at high speed which can cause a high load on the braking system [2]. There are several methods to reduce vehicle weight, such as the use of lighter materials, minimizing component size, and reducing unwanted material from components [3].

Structural optimization is a method that can overcome the problem of reducing the weight of the vehicle. Structural optimization is a step in the design process to find the best design with certain constraints or constraints on solid objects [4]. Structural optimization is classified into three groups based on variables during optimization, i.e. sizing

I. Tolj et al. (eds.), *Recent Advances in Mechanical Engineering*, Lecture Notes in Mechanical Engineering, https://doi.org/10.1007/978-981-19-0867-5_2

optimization, shape optimization, and topology optimization [5]. Sizing optimization is a type of structural optimization based on dimensional variables with respect to the main axes. Optimization is done only on the size of the length, width and thickness. Shape optimization is a type of structural optimization based on variable geometry through element nodal coordinates so that the resulting design results in a new curvature that has a volume smaller than the initial design volume. Topology optimization is a type of structural optimization that aims to determine the optimal design based on the stress distribution on a material in the design space. This method offers a design alternative with complex shapes that the designer may not have imagined intuitively. Topology optimization is generally on one component but can also be done on the assembly model [6]. Constraints in topology optimization, in general, are in manufacturing due to the geometry with complicated shapes. However, some studies can optimize the topology optimization model with metal additive manufacturing [7]. Advances in additive manufacturing technology include 3D printing, and the results of topology optimization can be produced.

Topology optimization as a material reduction method has been widely applied in various industries, one of which is the railway vehicle. Topology optimization for critical components such as bogies, especially for bolster components in bogies, has been researched [8]. Topology optimization on the carbody high-speed train has also been carried out [9]. One of the critical structural components besides the bogie and carbody on the train is the coupler. A coupler is an equipment that connects between trains and serves to connect and separate trains. The coupler transmits the load according to the maximum number of trains assembled without permanent deformation. In addition, the movement of the coupler is able to adjust the movement of the train according to the railroad being traversed, structurally sturdy, and has high compatibility [10]. Topology optimization on the coupler of light trains is also very useful to reduce its weight, without omitting the main function [11]. This study aims to reduce the mass of the automatic coupler component using topology optimization to reduce weight while maintaining rigidity and remaining safe when operating.

2 Method

Currently, an industrial-scale automatic coupler prototype for light trains has been developed [12]. In this study, design, numerical simulations, prototyping to tensile and compression tests have been carried out [13]. The implemented methodology is depicted in Fig. 1, which proposed the dynamic safety factor as parameter to evaluate model after reducing process in topology analysis. The simulation model is derived from the design of light trains' coupler. The working principle of the automatic coupler is that the coupling link will connect to the hooked plate in the other coupler where the movement of the hooked plate is assisted by a locking device. The hooked plate is the main part of the connection between the two automatic couplers. The coupling link is the part that connects two hooked plates. The main components of the automatic coupler are covers, hooked plate, coupling link, locking device, as shown in Fig. 2. The masses of each component can be summarized in Table 1. The hooked plate component was chosen for topology optimization because it is the critical component of the coupler that has a significant mass compared to the other main components. The hooked plate has a mass of 8.2 kg.

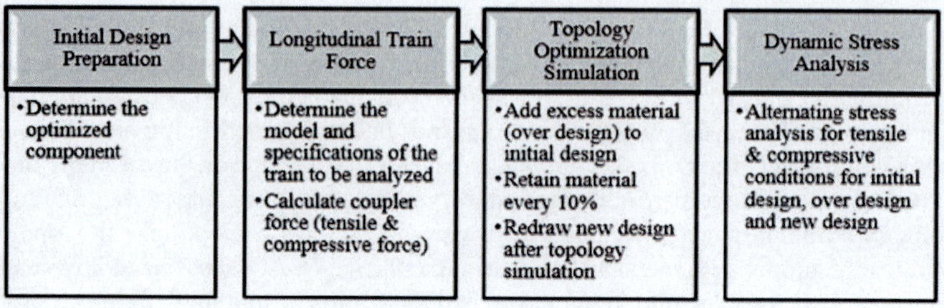

Fig. 1. Optimization procedure

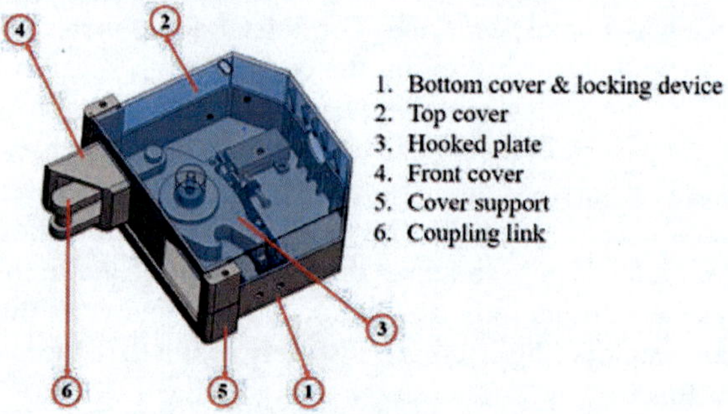

Fig. 2. Automatic coupler component

Table 1. Automatic coupler component mass

Component	Mass (kg)	Quantity	Total mass (kg)
Bottom cover and locking device	9.32	1	9.32
Top cover	8.3	1	8.3
Hooked plate	8.2	1	8.2
Front cover	5.41	1	5.41
Cover support	1.19	4	4.76
Coupling link	3.31	1	3.31

Longitudinal force analysis is carried out with the aim of knowing the maximum force that subjected to the coupler [14]. This force is used as loading in the simulation of tensile and compressive stress that occurs in the coupling process. The configuration used in this study is a trainset consisting of 3 trains [15], namely Motor Cab (MC1), Trailer (T), Motor Cab (MC2), as shown in Fig. 3. The technical specifications for the high-speed train are as follows: the operational speed is 250 km/h, the radius of the train wheel is 460 mm, the acceleration is 0.5 m/s^2, the deceleration is 0.5 m/s^2, the height of the train body is 4050 mm, the width of the body of the train is 3360 mm, and the axle load is 14 tons with the passenger loading conditions in crush conditions. The total number of passengers in crush conditions is 225 pax, with the mass of each passenger is 70 kg. In this study, the train runs on a straight track without a gradient.

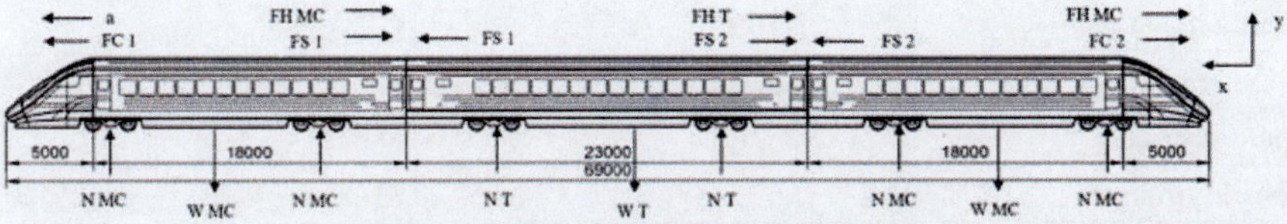

Fig. 3. High-speed train configuration

Based on the free-body diagram, the resistance force that caused by rolling and drag resistance can be formulated by means of Eqs. (1) and (2), respectively. F_H is the drag force with C_{rr} being the rolling resistance coefficient. N is the normal force, and r is the radius of the wheel. F_D is the drag force, C_D is the drag coefficient, and ρ is the density of air in kg/m^3. A is the area in front of the train interacting with the air, and V is the train speed in m/s. The coupler force F_C is described as the force received by the coupler when the train is running. The magnitude of the coupler force is obtained by a tensile force and a compressive force of 296.65 kN.

$$F_H = \frac{C_{rr} \times N}{r} \tag{1}$$

$$F_D = C_D \times \rho \times A \times \frac{V^2}{2} \tag{2}$$

The objective of topology optimization is to obtain the best stiffness to weight ratio. The design variable in topology optimization is the density of elements. Constraints in topology optimization are areas subjected to force, fixture parts, and parts that have been optimized for stress concentration. Manufacturing controls include a preserved region to keep the boundaries of the region from being optimized, thickness control to limit the thickness of the components, de-mold direction to maintain the casting boundaries to ensure easier shapes to produce, and a symmetry plane to keep the geometry symmetrical with the same shape with the other side.

The optimization method used in this simulation is to add excess material (over design) to the selected initial design to eliminate stress concentrations that may occur while maintaining the constraint area and contact point with other components, as shown in Fig. 4. The excess material is added over the space that allows for the movement of the hooked plate. This strategy has the advantage to improve the initial model that had been developed without topology analysis. Tensile conditions are the opposite of compressive conditions. The force is applied to area A with the load value that has been obtained from the longitudinal analysis of the train forces in the negative y-axis direction. Fixed supports are placed in areas B, C, and D. The material used in the hooked plate component is AISI 1045 steel with the isotropic elastic material model as shown in Table 2.

Table 2. Material properties used for AISI 1045

Properties	Value	Dimension
Density	7.85	g/mm^3
Young modulus	205	GPa
Poisson's ratio	0.29	
Yield strength	530	MPa
Tensile strength	625	MPa

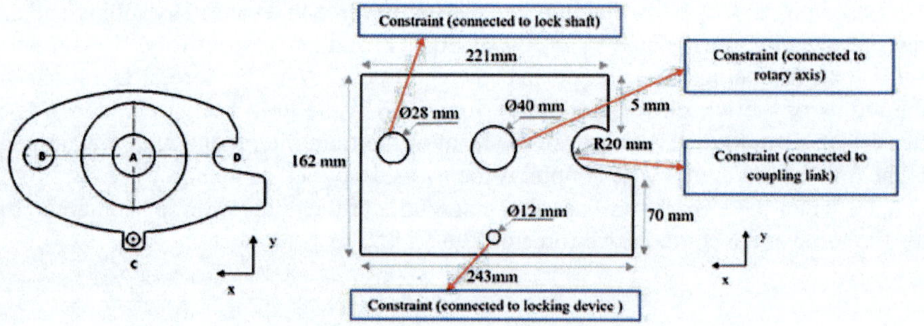

Fig. 4. Initial design (left) and over design with constraint (right)

Coupler is one of the train's main components that experience repeated loads. The repeated loads are coming from the acceleration condition that is derived as the tensile load and the deceleration condition that is derived as compression load. This stage is carried out to verify the design of the topology simulation results. To get the stress amplitude (σ_a) use the formula:

$$\sigma_a = \frac{\sigma_{max} - \sigma_{min}}{2} \tag{3}$$

To get the *mean stress* (σ_m) use the formula:

$$\sigma_m = \frac{\sigma_{max} + \sigma_{min}}{2} \tag{4}$$

By referring to Soderberg's average stress correction theory, to calculate the equivalent alternating stress $\left(\sigma_{a_{eq}}\right)$ the following formula is used:

$$\sigma_{a_{eq}} = \sigma_a \frac{S_y}{S_y - s_m} \tag{5}$$

In redesigning according to topology results, it is necessary to determine the safety factor to evaluate the safety of machine elements with minimum dimensions. The factor of safety is defined by the ratio of the yield stress of the material. The failure criteria for the construction is the maximum stress that occurs at a maximum of 75% from the yield stress of the material [16].

3 Results and Discussion

In the topological simulation, the material is retained every 10 to 50% as in Fig. 5. Material reduction of more than 50% is not acceptable because some constrained part is missing. After getting the geometric shape with 50% retain, it is redrawn according to the geometric shape. The results of the topology cannot be directly converted into a solid part file. A smooth line modification is applied to the optimized model; thus, the results of the re-drawing may not be similar.

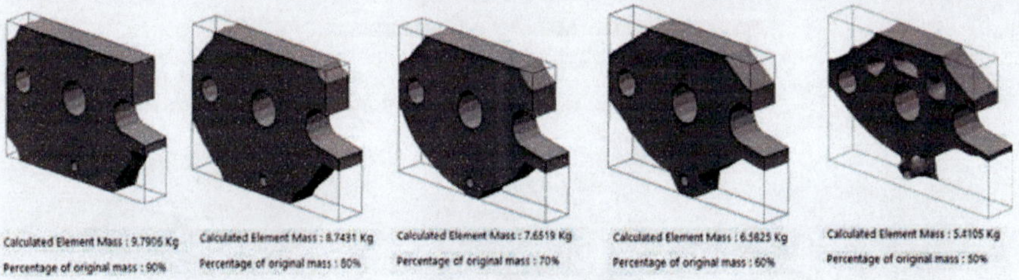

Fig. 5. Topology result

Based on the design carried out, the initial design became an "over" design, there was an increase in mass of 31% (8.2 to 10.82 kg). After topology optimization, there was a mass reduction of up to 50% (5.41 kg). New design drawings pay attention to aesthetics and ease of manufacture, so the new design has a 14% increase in mass over topological optimization designs. Compared to the original design, the new design has a mass reduction of 24% as in Fig. 6.

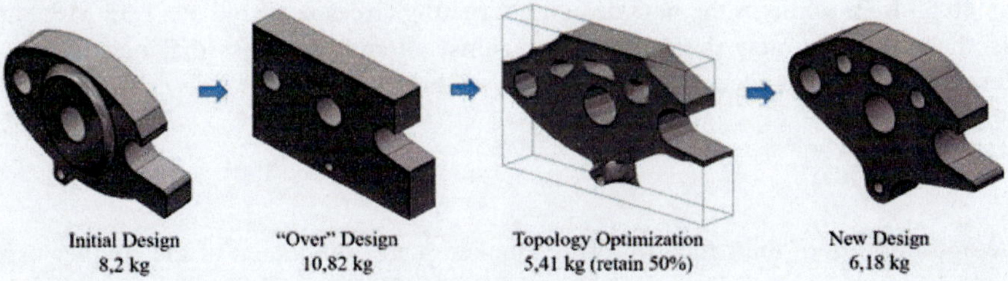

Fig. 6. Design comparison

The results of the static analysis for tensile and compressive conditions show that the stresses in the over design are reduced by about 29% for tensile stresses and 13% for compressive stresses with the same location compared to the initial design as in Fig. 7. The comparison of over design with the new design is 35% for tensile stress and 0.6% for compressive stress with the same max stress location. The comparison of the initial design with the new design is 4.7% for tensile stresses and 13% for compressive stresses with the same maximum stress location. The factor of safety has increased from 1.8 in the initial design to 2.07 in the new design.

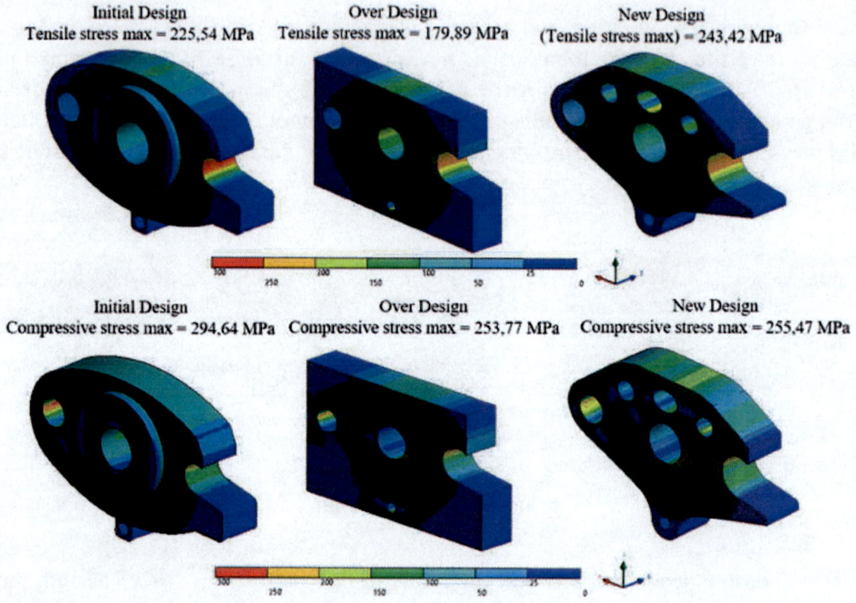

Fig. 7. Tensile and compression simulation

In the hooked plate component, there is loading fully reversed in accordance with operational conditions, namely acceleration and deceleration when braking occurs. According to the initial design, as shown in Table 3, alternating stress that occurred was 40.65 MPa while in the new design alternating stress occurred at 11.38 MPa, and this shows that the new design is safer against alternating stress that occurs due to alternating loads that occur in the coupler.

4 Conclusion

The optimization of the structure for the hooked plate component of the coupler using topology optimization is investigated in this paper. Based on the optimization results, it was found that the mass reduction in the initial hooked plate design was 2.02 kg or 24% lighter than the initial design. In total, the new design contributes to reducing the mass of the automatic coupler by 4% with a maximum of 50% retained. This shows that this topology optimization method looks for the optimal distribution in a predetermined

Table 3. Tensile and compression analysis

Properties	Initial design (MPa)	Over design (MPa)	New design (MPa)
σ tensile loading	255.54	179.89	243.42
σ compression loading	294.64	253.77	255.47
σ amplitude	19.55	36.94	6.02
σ mean	275.09	216.83	249.44
σ alternating	40.65	62.52	11.38
Factor of safety	1.80	2.09	2.07

design space. Tensile stress, compression stress, and alternating stress have decreased in the new design, as a result, the safety factor increases. Nevertheless, this study only considers the fluctuating load originating from the maximum and minimum loading modes. In actual conditions, each loading mode yields its alternating stresses, which requires Palmgreen Miner's theory to accumulate the lifetime. Therefore, dynamic analysis based on historical loading data should be implemented in the future to obtain more realistic loading phenomena.

Acknowledgements. The authors gratefully acknowledge the financial support provided by the Directorate of Research and Community Service Institut Teknologi Sepuluh Nopember, Indonesia, with funding agreement letter number 1251/PKS/ITS/2021.

References

1. De la Guerra E (2018) Lightweight primary structures for high-speed railway carbodies. In: International congress high-speed rail technology: long term impacts. pp 1–12
2. Lei Y (2020) Lightweight design of high-speed train under the development of new materials. In: ICVMEE. pp 217–221
3. Tyflopoulos E, Lien M, Steinert M (2021) Optimization of brake calipers using topology optimization for additive manufacturing. Appl Sci 11(4):1–21
4. Gao J, Xiao M, Zhang Y, Gao L (2020) A comprehensive review of isogeometric topology optimization: methods, applications and prospects. Chin J Mech Eng (Engl Edn) 33(1)
5. Al Ali M, Takezawa A, Kitamura M (2019) Comparative study of stress minimization using topology optimization and morphing based shape optimization
6. Sha L, Lin A, Zhao X, Kuang S (2020) A topology optimization method of robot lightweight design based on the finite element model of assembly and its applications. Sci Prog 103(3):1–16
7. Bandyopadhyay A, Traxel K (2018) Invited review article: metal-additive manufacturing—modeling strategies for application-optimized designs. Addit Manuf 22:758–774
8. Srivastava PK, Shukla S (2020) Reducing weight of freight bogie bolster using topology optimization. Rev Geintec 11(3):324–339
9. Mrzygłód M, Kuczek T (2014) Uniform crashworthiness optimization of car body for high-speed trains. Struct Multidiscip Optim 49(2):327–336

10. Massa A (2012) Numerical study of the noninertial systems: application to train coupler systems 68:215–233
11. Johnsen S (2013) Structural topology optimization: basic theory, methods and applications
12. Syaifudin A, Ariatedja JB, Wasiwitono U (2019) Prototyping coupler skala industri untuk LRT Indonesia. Surabaya
13. B2TKS BPPT (2019) Pengujian tarik dan tekan statis purwarupa coupler kereta. Tangerang Selatan
14. Wu Q, Spiryagin M, Cole C (2016) Longitudinal train dynamics: an overview. Veh Syst Dyn 54(12):1688–1714
15. PTSPT BPPT (2020) Persyaratan teknis dan spesifikasi teknis kereta cepat (high speed train) Indonesia. Jakarta
16. Menteri Perhubungan RI (2019) Peraturan Menteri Perhubungan Republik Indonesia No 69 Tahun 2019

Numerical Study on the Stress Distribution Analysis of Two-Planar DKDT Tubular Joint Under Variation of Axial Loading Conditions

Irma Noviyanti and Rudi Walujo Prastianto(✉)

Ocean Engineering Department, Institut Teknologi Sepuluh Nopember, Surabaya, Indonesia
rudiwp@oe.its.ac.id

Abstract. A multiplanar tubular joint is the most common tubular joint type of jacket offshore platforms. Nevertheless, previous studies subjected to this topic were relatively limited. The chosen tubular joint analyzed in this paper is obtained from an existing minimum jacket platform and numerically analyzed by using FE model. This research has the objective to obtain stress distribution along the brace-chord intersection lines of a two-planar DKDT welded tubular joint under three modes of axial loading: tension, compression, and combination of both. The stress distribution will be evaluated at the weld toe of both on the chord and the brace sides. The result shows that stresses occurred on the chord side are greater than on the brace side. Though the stress distribution trend for both side of the weld toe is close to similar, there are several shifting on the exact location. Hotspot stress from the tension axial loading case has the greatest value followed by compression axial loading and combined axial loading cases.

Keywords: DKDT multiplanar tubular joint · Finite element analysis · Local stress distribution

1 Introduction

The oil and gas site in Indonesia which depends on the conventional fields is decreasing averagely by 14.5% between 2014 and 2019 and is steadily shifting to operate in the non-conventional area (a marginal field) [1]. A marginal field has the criteria of poor economic performance, difficult development, and is only in the range of 5–20 million barrels reserves [2]. Thus, several adjustments were undertaken to make production feasible such as the need to build a more minimal version of the conventional jacket structure.

A study related to minimum jacket platforms has been conducted before which focused on the fatigue life comparison between the minimum and conventional jacket. It is revealed that the minimum jacket platform has a much lower fatigue life compared to the conventional one. However, the stress concentration factor (SCF) used for calculating the fatigue life was obtained from the stresses through global analysis and using the empirical SCF equation which is supposedly limited to the uniplanar joint [3, 4]. Hence,

I. Tolj et al. (eds.), *Recent Advances in Mechanical Engineering*, Lecture Notes in Mechanical Engineering, https://doi.org/10.1007/978-981-19-0867-5_3

a better method to analyze the SCF value of the tubular joint is required, which are through empirical testing and finite element models [5]. Generally, the finite element (FE) method is the common analysis method used since it took a lower cost. Various researches using FE analysis have been conducted to investigate the characteristic of the tubular joint in terms of stress distribution and SCF. Liu et al. [6] executed an experimental and numerical study to assess tubular joint behavior subjected to transverse impact loading.

The multiplanar tubular joint is well known as the most prevalent type of tubular joint used in jacket structures. Prastianto et al. [7–9] examined the way multiplanarity affects the behavior of a jacket platform's tubular joint. The result shows that multiplanarity has a substantial effect, particularly on the hotspot stress (HSS) and SCF value.

Each tubular joint of every jacket platform is unique that a specific study is essential to conduct in various conditions. A previous study was conducted to attain stress distribution in a DKDT tubular joint of a minimal jacket platform. Stress distributions were investigated for three different loading types which are axial loading, in-plane bending moment, and out-of-plane bending moment. The analysis concluded that axial loading has the most impact on the generated stresses of the tubular joint [10]. Thus, this present paper will conduct a further investigation that will cover a stress distribution analysis on the tubular joint under various axial loading conditions. The axial loading will differ in three modes, which are tension axial loading, compression axial loading, and combination axial loading.

2 Method of Study

2.1 Tubular Joint Geometry Modeling

Tubular joint modeling is performed in Salome Meca software using the Geometry module [11]. The welding geometry and size of the tubular joint was modeled and referring to AWS codes [12]. The 3D solid element is used because it gives more representative stresses output compared to shell elements [13]. Moreover, the tubular joint geometrical properties and loading magnitude are presented in Table 1. OD and Wt stands for outer

Table 1. Tubular joint properties

Element	Length (m)	OD (m)	Wt (m)	Axial Loading (kN)
CHORD 1	11.67	1.34	0.06	–
CHORD 2	16.00	1.34	0.06	–
BRACE 1	7.15	0.60	0.02	223.081
BRACE 2	8.90	0.60	0.02	340.656
BRACE 3	13.10	0.76	0.03	−1,049.596
BRACE 4	13.16	0.71	0.03	221.542
BRACE 5	13.85	0.71	0.03	−586.596
BRACE 6	12.25	0.71	0.03	−106.988

diameter and wall thickness of the tubular members, respectively. Material properties of the joint are modulus of elasticity of 2×10^5 MPa, yield stress of 250 MPa, and Poisson's ratio of 0.26.

Magnitude of the axial loads applied on each brace of the tubular joint is obtained from a global stress analysis as described in the previous work [10]. The modeling result of the DKDT joint with the angle among the braces is displayed in Fig. 1.

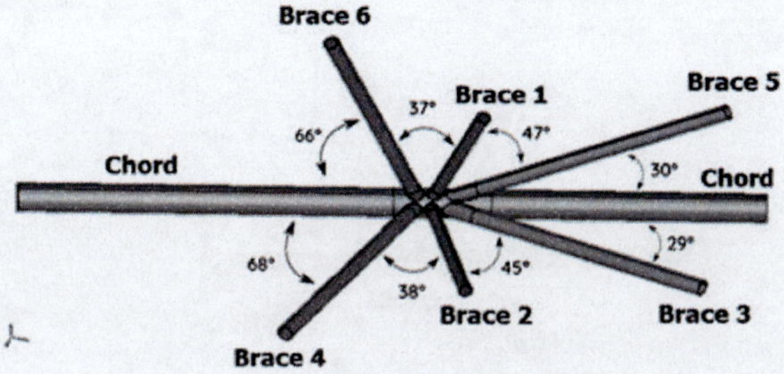

Fig. 1. Model of the DKDT tubular joint

2.2 FE Modeling

The finite element analysis is performed by Salome Meca software [14]. For the tubular joint in offshore structures, the end-fixity of each chord is ranging from almost pinned to almost fixed with generally being fixed [15]. Therefore, the end fixity taken in this study is assumed to be fixed supported for the chord ends. Three variations of the axial loading are considered in this study which are combination axial loading (Case 1 as the basic case), tension axial loading (Case 2), and compression axial loading (Case 3).

In Case 1, tension axial load is applied to the braces 3, 5, and 6 in combination with compression axial load on the braces 1, 2, and 4. For Case 2, tension axial load applied to all braces, and in Case 3, compression axial load applied to all braces. Figure 2 illustrates the boundary and loading conditions applied to the joint model.

A meshing sensitivity analysis was performed to ensure that the results of the FE analysis were not influenced by the poor quality or size of the mesh [16]. Meanwhile, the meshing is made smaller (fine mesh) on regions around the weld zone and gradually coarser for other area [8]. The meshing result of the model is shown in Fig. 3, and the result of the meshing sensitivity analysis is plotted in Fig. 4.

The graph of Fig. 4 indicated that the generated stress within the model is nearly stationary from the sixth trial of meshing and the subsequent meshing with the error is below 5%. Hence, it can be concluded that the sixth and subsequent meshing have good quality by mean that the output is unaffected by the number of elements. Therefore, the tubular joint model with 524,043 meshing elements is used for further assessment.

For the analysis, stresses along the weld toe of the model are derived into two sides separately: on the chord side and brace side as indicated in Fig. 5. Then the stress distributions will be evaluated along the length of the brace-chord intersection line within 4 regions as can be seen in Fig. 6.

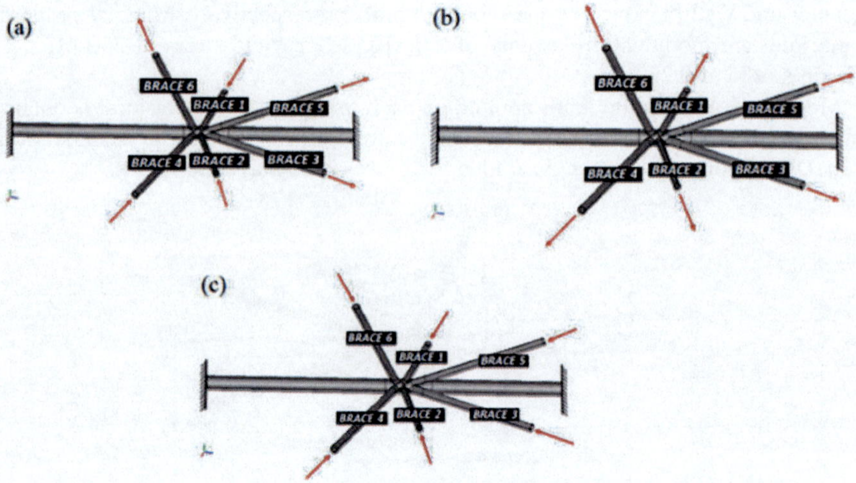

Fig. 2. Boundary and loading conditions of the tubular joint model

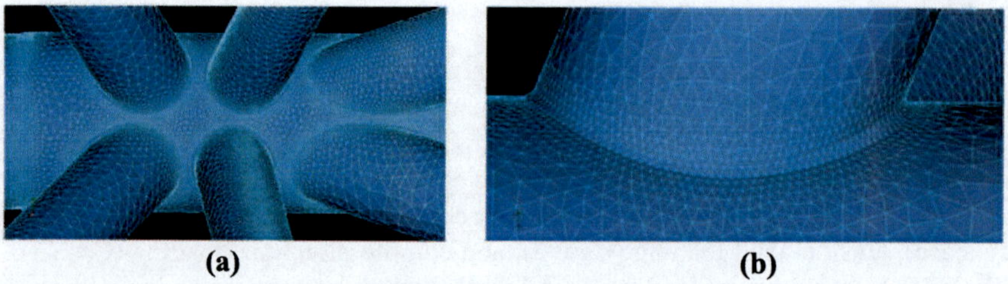

Fig. 3. **a** Meshing of the tubular joint model, **b** Detail meshing in the weld region

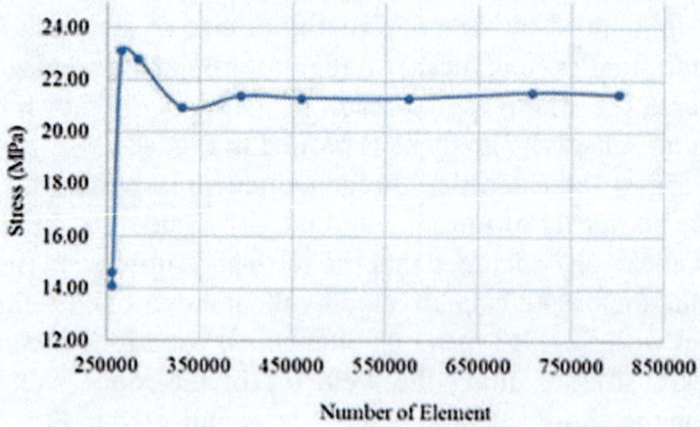

Fig. 4. Result of the meshing sensitivity analysis

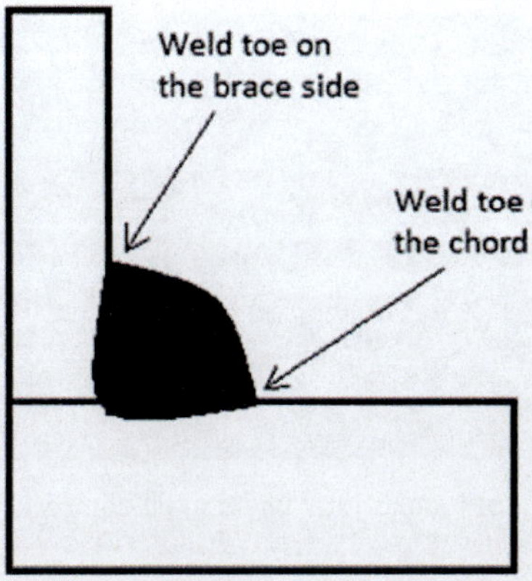

Fig. 5. Stress distribution extraction point

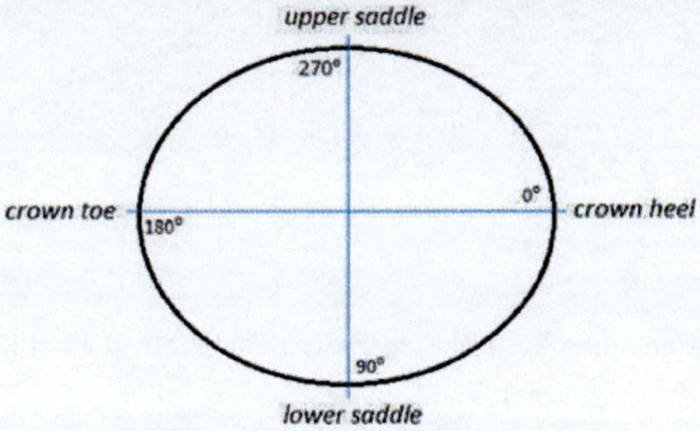

Fig. 6. Stress distribution regions

3 Results and Discussion

The result of the FE stress analysis on the tubular joint subjected to all three cases of combined axial loading, tension axial loading, and compression axial loading can be displayed as stress distribution in the form of different color over the surface of the joint. For example, Fig. 7 shows the stress distribution of the tubular joint under axial compression loading.

Meanwhile, the stress distribution at weld toes along each of the brace-chord intersection line for all three loading cases studied are graphically presented in Figs. 8, 9 and 10. Hotspot stresses for both in the chord and brace sides are identified in the figures.

3.1 Stress Distribution Subjected to Combined Axial Loading

For Case 1 of combined axial loading both the hot spot stress (HSS) in the chord side and brace side occurred at same brace that is Brace 3 with the value of 29.83 MPa and 27.24 MPa, respectively. The HSS on the chord and the brace sides occurred on a position of 202.5 deg and at about of 180 deg (crown toe point), respectively. However, a bit shifting occurred for the HSS at the brace side.

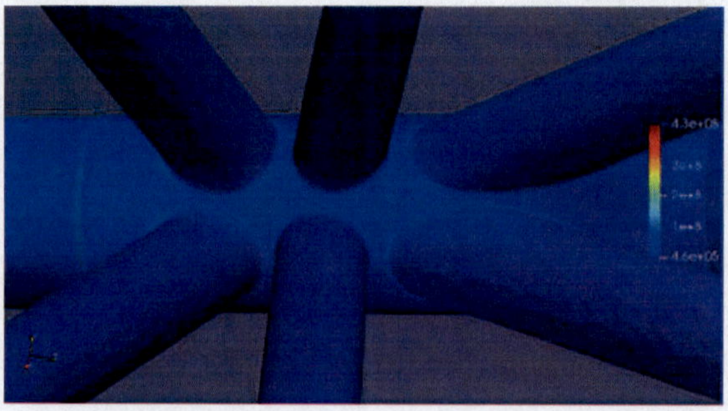

Fig. 7. FE result of tubular joint under axial compression loading

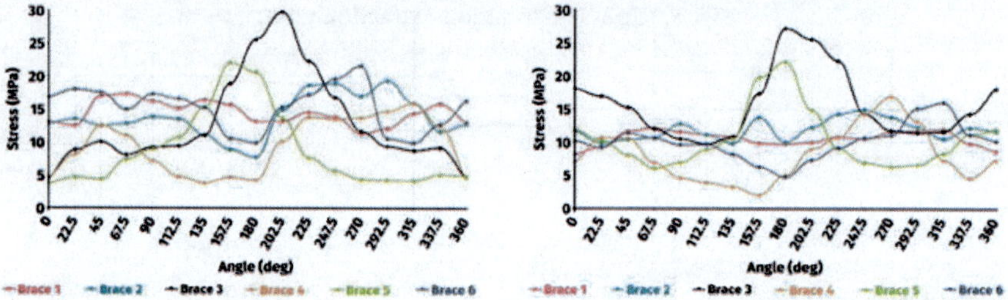

(a) (b)

Fig. 8. Stress distribution along the weld toe under the combined axial loading: **a** on the chord side, **b** on the brace side

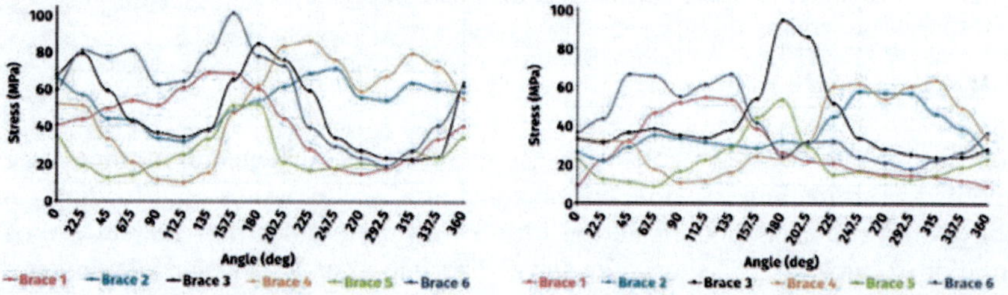

(a) (b)

Fig. 9. Stress distribution along the weld toe under tension axial loading: **a** on the chord side, **b** on the brace side

3.2 Stress Distribution Subjected to Tension Axial Loading

In this Case 2, the two HSS occurred at different braces. Figure 9 shows HSS in the chord side occurred on the Brace 6 at 102 MPa, whereas for the brace side arose on the Brace 3 at 95.1 MPa. Also, the both HSS occurred at different point along the brace-chord intersection line which are at 157.5 deg for the HSS in the chord side and at 180 deg (crown toe point) for the HSS in the brace side, respectively.

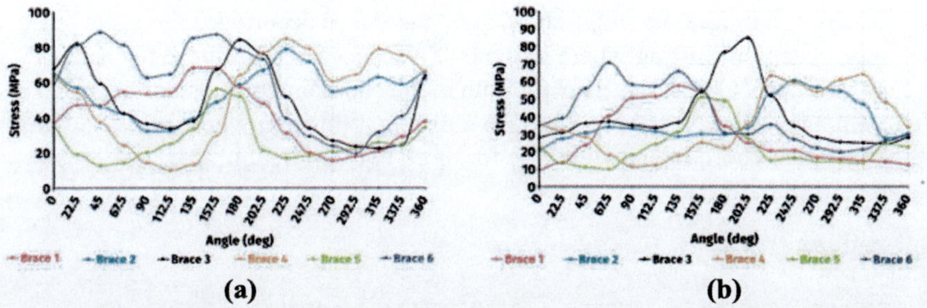

Fig. 10. Stress distribution along the weld toe under compression axial loading: **a** on the chord side, **b** on the brace side

3.3 Stress Distribution Subjected to Compression Axial Loading

Meanwhile, in this Case 3 the two HSS also occurred at different braces. The HSS in the chord's side occurred on the Brace 6 at 88.4 MPa, while in the brace side, the HSS occurred on the Brace 3 at 85 MPa. Moreover, in this loading case there are 4 peaks of large stresses in the range of 80–90 MPa appeared at chord' side of the Brace 3, Brace 4, and Brace 6. This situation indicated that the chord wall of the joint experienced more severe stresses than that experienced by the brace's wall.

The HSS in the chord's side of the Brace 6 occurred at a position of 45 deg, while the second large stress occurred at a point of about 150 deg. Then at the Brace 3 and Brace 4 the HHS on the chord side is located at a point of 180 deg (crown toe point) and 225 deg, respectively. On the other hand, the HSS in the brace's side at the Brace 3 occurred at a point of about 200 deg.

4 Conclusion

The stress analysis of the two-planar DKDT tubular joint in the present paper was conducted based on Finite Element Method (FEM) by using an open-source Salome Meca software. The tubular joint was modeled by using 3D solid element with fine meshing under three different axial loading conditions. The objective of the study is to examine the stress distribution at the weld toes along the chord-brace intersection line evaluated for two different sides which are the chord and brace sides.

Through the analysis, it could be concluded that for all loading cases, the HSS at the chord's side have higher values compared to the HSS occurred at the brace's side. Furthermore, the tubular joint subjected with combined axial loading has the least value of the HSS, followed by compression axial loading case and the tension axial loading case.

The HSS due to the compression axial loading (Case 3) was 3 times greater than from the case of combined axial loading (Case 1), whereas the HSS caused by the tension axial loading (Case 2) is 3.5 times greater than Case 1.

Lastly, the HSS on the chord and brace sides were not always occurred at the same brace and the HSS position along the chord-brace intersection line as well. Even though for Case 1 the HHS for both in the chord and brace sides occurred at the same brace, but there was a position shifting where in the HSS was occurred for those two sides.

For Case 2 and Case 3 the HHS for both in the chord and brace sides occurred at the different brace. Moreover, there was also a different position where the HSS occurred within the brace-chord intersection line for those two different braces.

References

1. SKK Migas (2019) Laporan Tahunan 2019 SKK Migas. SKK Migas, Jakarta
2. Zhao Z, Li Y, Tang Y, Ji X (2019) Conceptual design and numerical analysis of a novel platform for marginal oilfields development. Ocean Engineering 187
3. Giffary, Muhammad Nabil (2020) Fatigue life comparison of modified conventional 3 legged jacket structure. Institut teknologi sepuluh nopember, Surabaya
4. Efthymiou MD, Durkin S (1985) Stress concentrations in T/Y and GAP/overlap K-joints. In: Proceedings of the 4th international conference on behaviour of offshore structures, pp 429. Delft
5. Mirtaheri M, Zakeri HA, Alanjari P, Assareh MA (2009) Effect of joint flexibility on overall behavior of jacket type offshore platforms. American J. of Eng Appl Sci. 2(1):25–30
6. Liu K, Liu B, Wang Z, Wang G. Guedes Soares C (2018) An experimental and numerical study on the behaviour of tubular components and T-joints subjected to transverse impact loading, International journal of impact engineering 120: 16–30
7. Prastianto RW, Hadiwidodo Y, Widhestomo SW, Yazhahir R (2018) Stress distribution along the weld toes of tubular KT and KDT joints under balance axial loads and in-plane-bending moments. In: ISOCEEN, scitepress, lda, pp 170–175
8. Prastianto RW, Hadiwidodo YS, Fuadi IF (2017) Stress concentration factor of a two-planar double kt tubular joint due to in-plane bending loading in steel offshore structures. In: ISOCEEN, MATEC Web Conf, vol 177, EDP Science, pp 1–13
9. Prastianto RW, Hadiwidodo YS, Fuadi IF, Naibaho M.B.Y.P, Rahardianto P.N (2018) Stress concentration factor distribution of inclined brace in multiplanar offshore tubular double kt joints. In: IOP Conf Ser Earth Environ Sci. 162, IOP Publishing
10. Noviyanti Irma, Prastianto RW, Murdjito (2021) Stress distribution analysis of DKDT tubular joint in a minimum offshore structure, Rekayasa 14(2)
11. Aubry JP (2013) Beginning with code_aster. Framasoft, Paris
12. American Welding Society (1981) Structural welding code - steel. vol. 552, american national standards institute
13. Ahmadi H, Ziaei Nejad A (2017) Geometrical effects on the local joint flexibility of two-planar tubular DK-joints in jacket substructure of offshore wind turbines under OPB loading. Thin-Walled Struct 114, pp 122–133
14. Sellenet N (2016) Operator CALC_CHAMP Code-Aster. Electricite de France, Paris
15. Ahmadi H, Zavvar E (2015) Stress concentration factors induced by out-of-plane bending loads in ring-stiffened tubular KT-joints of jacket structures. Thin Walled Struct 91:82–95
16. Ahmadi H, Niaki MA (2019) Effects of geometrical parameters on the degree of bending (DoB) in two-planar tubular DT-joints of offshore jacket structures subjected to axial and bending loads. Mar Struct 64:229–245

The Influence of the Buttering Layer on the Creep Properties of Ferritic-Martensitic T91 and Austenitic Stainless Steel 347H Dissimilar Welding

Andik Santoso[1,2], Teguh Widjajanto[1], Fahmi Mubarok[2], and Suwarno[2](✉)

[1] PT. Pembangkitan Jawa Bali, Surabaya, Indonesia
[2] Department of Mechanical Engineering, Institut Teknologi Sepuluh Nopember, Surabaya 60111, Indonesia
warno@me.its.ac.id

Abstract. In the new supercritical boilers, a combination of tubing is used to achieve mechanical integrity at the most economical cost. The boiler uses dissimilar metal welding (DMW) to obtain the joint. The DMW process poses several challenges that are caused by differences in the physical properties of the base metals. In the present work, the effect of the type of filler metals (Inconel ERNiCr-3 and ERNiCrMo-3) and a buttering process on the creep properties of DMW between ferritic steel T91 and austenitic stainless steel 347H characteristics are presented. Creep testing results show that ERNiCr-3 produced slightly better creep strength compared with ERNiCrMo-3. Nevertheless, the buttering layer has a significant impact on the creep properties. The data was analyzed using the Weld Strength Reduction Factor using time initiation of tertiary creep (WTRF) to material capability. WTRF of direct Welding with ERNiCr-3 and ERNiCrMo-3 both are 0.32 and at 600 °C application. WTRF of ERNiCrMo-3 is 0.6 and on the minimum requirement. WTRF of buttering using ERNiCr-3 is 0.65, and above the minimum requirement of the tube, boiler operated at 600 °C application.

Keywords: DMW · T91 · 347H · Buttering layer · Inconel

1 Introduction

Boiler power plants operate at high working temperatures and pressures, and use ferritic/martensitic steel, austenitic stainless steel, and nickel-based superalloy [1]. The application of different materials is due to the differences in the temperature and working pressure on the boiler tubes. The boiler area that operates at high temperatures uses mainly austenitic stainless-steel material for better creep strength. In contrast, the boiler area with lower working temperature and pressure uses ferritic/martensitic alloy steel. Materials with a ferritic/martensitic microstructure are preferred due to the good physical properties such as good thermal conductivity and higher heat shock resistance than austenitic stainless steels [2].

I. Tolj et al. (eds.), *Recent Advances in Mechanical Engineering*, Lecture Notes in Mechanical Engineering, https://doi.org/10.1007/978-981-19-0867-5_4

Since there are differences in the chemical and physical properties of the base materials, the welding process of the DMW is challenging. The differences in mechanical properties such as creep rupture strength, high-temperature corrosion, oxidation resistance, and microstructure stability must be considered [3]. Differences in the thermal expansion and carbon migration by differences of the chromium content have the potential to reduce the service life of DMW [4, 5]. Several studies have been done to reduce the failure rate of DMW. The selection of nickel-base FM is more recommended than austenitic-base filler metals (FM) for welding DMW materials used at high temperatures [6]. This reduces the risk of failure due to differences in the coefficient of thermal expansion (CTE), and decreased creep resistance in the weld joint area. The process of buttering the ferritic/martensitic side of the material using three layers of nickel-base FM interlayer has better creep-rupture resistance than welding one layer of buttering [7].

In the present work, the effect of the filler metals (Inconel ERNiCr-3 and ERNiCrMo-3) and buttering process on the creep properties of DMW between ferritic steel T91 and austenitic stainless steel 347H characteristics is presented.

2 Materials and Experiments

2.1 Materials and Welding Process

A boiler tube of type ferritic-martensitic steel SA213 T91 was used to fabricate the DMW joint with austenitic stainless steel SA213 347H. The tube of length 150 mm, outer diameter 60 mm, and 8 mm thickness was used to prepare the weld joint. The DMW joint in this study was fabricated with Ni-base consumables Inconel 82 (ERNiCr-3 of 1.6 mm dia.) and Inconel 625 (ERNiCrMo-3 of 1.6 mm diameter). The chemical composition of base metals (BM) and filler metals (FM) in weight percentage is listed in Table 1. Both of the tube pieces were machined for the required joint geometry, as shown in Fig. 1. A portable gas burner is used as preheating source. Preheat maintained at temperature 250 °C monitored with *Fluke* handheld digital thermometer for 15 min before welding is conducted on the T91 side. Two different welding methods were used in the study: direct welding (DW) and buttering using Inconel 82 and Inconel 625 FM, as listed in Table 2. All buttering is conducted on the T91 side using three layers of Ni-base FM. The Illustration of welding methods used in this research is shown in Fig. 2.

Table 1. Chemical composition of base metals and filler metals (wt %)

Material	Al	C	Cr	Cu	Fe	Mn	Mo	Nb + Ta	Ni	Si	Ti
T91 (BM)	–	0.1	9.0	–	Bal	0.5	1.0	–	–	0.4	–
347H (BM)	–	0.1	17.31	0.13	Bal	1.61	–	0.7	10.85	0.9	–
ERNiCr–3 (FM)	–	0.1	20.0	0.4	3.0	3.0	–	2.5	68.0	0.8	0.8
ERNiCrMo–3 (FM)	0.3	0.1	21.0	0.5	4.6	0.4	9.0	3.5	60.0	0.4	0.4

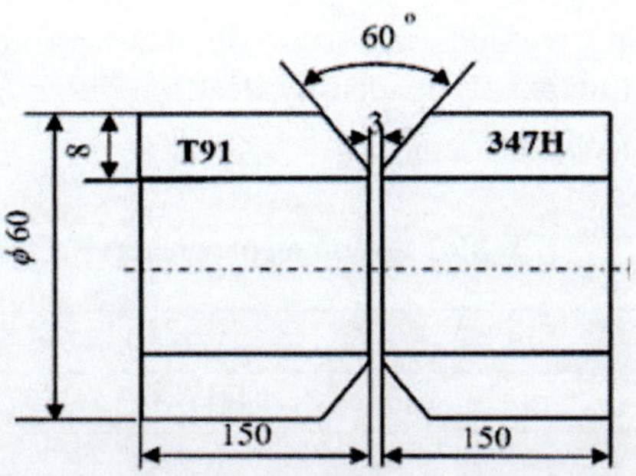

Fig. 1. Weld joint geometry

Table 2. Treatment code used in the research

Filler Metal	Welding methods	
	direct welding	Buttering
ERNiCr-3	A1	B1
ERNiCrMo-3	A2	B2

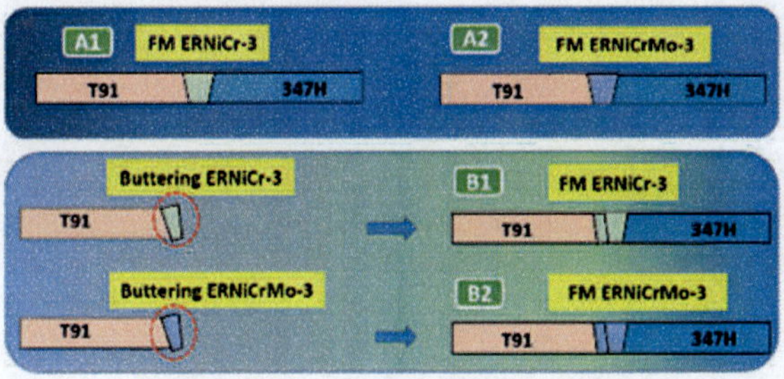

Fig. 2. Schematics of welding joints

The present work is focused on the comparison between ERNiCr-3 (Inconel 82) and ERNiCrMo-3 (Inconel 625) Ni-base filler metal (FM) in the DMW between ferritic T91 steel tubes and 347H austenitic stainless steel boiler tubes. The present work focused on the creep properties. All welding in this study is conducted with the manual Gas Tungsten Arc Welding (GTAW) process using portable welding machine *ESAB Buddy TIG 160*. Welding parameters used in the study are shown in Table 3. Tube pieces are placed in the welding positioner to maintain consistency during the welding processes.

2.2 Creep Testing

A uniaxial isothermal cross-weld creep rupture testing has been conducted using Electromechanical Creep Machine Testing Kappa 50 DS at 600 °C and stress levels of

200 MPa for all welding methods in this study. The specimen with gauge section diameters of 2.5 mm is uniaxially loaded by deadweight and heated by a three-zone furnace.

Table 3. Welding parameters

Parameters	Value
Welding process	GTAW
Position	1G/2G
Polarity	DCEN
Shielding gas	Argon UHP 99.99%
Backing shielding gas	Argon UHP 99.99%
Gas flow	12 cubic feet per minute
Tungsten electrode	Thoriated tungsten ϕ 2.4 mm
Voltage	13 V
Amperage	90 A
Travel speed	200 mm/minute

3 Result and Discussion

3.1 Creep Test

The creep test carried out in this study was a short-term creep test using a uniaxial isothermal cross weld. The results of creep testing with a stress of 200 MPa at a test temperature of 600 °C in the form of creep strain and creep rate as a function of time for all DMWs samples are shown in Fig. 3. The test results show that all treatments have almost similar characters in the form of creep deformation in minor strain initial loading, slight transient creep strain, secondary creep where the strain is in steady state, and tertiary creep area. The creep strain accumulation in tertiary creep is quite large compared to the transient and secondary regions, with a rapid accumulation of creep strain before failure (Fig. 3).

The results showed that there were differences in each treatment. The failure time of direct welding is relatively faster when compared to buttering. For example, the failure time of direct welding with FM ERNiCr-3 is 9.4 h, whereas for buttering, the failure time increases about 2 times to 18.6 h. The same thing also happened with the use of FM ERNiCrMo-3, where the failure time for direct welding was 9.7 h, whereas when buttering was done, the failure time increased by about 2 times to 17.4 h. This shows that there is an increase in creep resistance with buttering treatment for both types of FM. If a comparison is made between the two types of FM, then there are slight differences. Creep behavior in direct welding treatment showed that FM type ERNICrMo-3 was slightly better in creep resistance, namely 9.7 h when compared to ERNiCr-3 which was

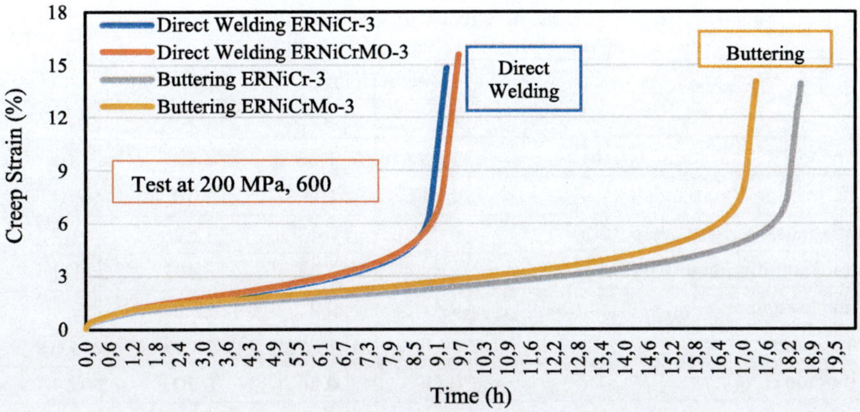

Fig. 3. Creep strain versus time curve of all material

9.4 h, but the primary and secondary creep behavior of both types of FM was relatively identical. The creep buttering behavior showed that ERNiCr-3 type FM was superior at 18.6 h compared to ERNiCrMo-3 at 17.4 h.

The creep life of the research shows that when compared with previous studies, for the loading of 200 MPa and the test temperature of 600 °C, the buttering treatment for both types of FM is relatively close to the T91 material creep test result [8]; this indicates an improvement in the properties of the welded material. If the material only undergoes direct welding, the resistance of the welded material to creep loads will decrease sharply due to the influence of heat input from welding which affects the resistance of the material. These results are also consistent with previous studies, where buttering 3 layers before welding increases the creep resistance of the welded material [9].

The creep deformation behavior in this study was carried out by determining the minimum creep rate ($_{min}$) from the creep curve to determine the time of initiation of tertiary creep [10] as shown in Fig. 5. This initiation time is evaluated as the intersection of the creep curve and the linear line with the minimum creep rate slope [10]. There is difficulty in determining the exact time of tertiary creep initiation, so a 0.2% offset approach is used [11]. The uncertainty of the timing of the occurrence of the minimum creep rate may affect the evaluation of the initiation of tertiary creep due to the difficulty in accurately taking the 0.2% anchor point. The results of the analysis of creep deformation behavior are tabulated in Table 4.

Based on the analysis of creep deformation behavior, further information was obtained regarding the material characteristics of each treatment. Time to reach deformation 1% ($t_{1\%}$) buttering treatment FM ERNiCR-3 was the highest compared to other treatments. These results indicate that the treatment has relatively high stability compared to other treatments. Further analysis was carried out by reviewing the life ratio, namely the comparison between the initiation time of tertiary creep and the total time required to reach rupture (t_i/t_R). Buttering treatment with FM ERNiCr-3 had the highest life ratio of 70%, followed by buttering treatment with FM ERNiCrMo-3 at 69%. This

Table 4. Creep behavior analysis

	Direct welding		*Buttering*	
	A1	A2	B1	B2
Time to 1% total strain, $t_{1\%}$(h)	1,05	1,06	1,32	1,06
Life fraction 1% strain (%)	11,17	10,93	7,10	6,09
Initiation of tertiary creep, t_i (h)	6,42	6,43	12,96	11,97
Strain at initiation tertiary creep (%)	2,89	3,03	3,132	3,37
Time to rupture, t_R (h)	9,4	9,7	18,6	17,4
Total strain	14,86	15,56	13,93	14,03
Life ratio (t_i/t_R)	0,68	0,66	0,70	0,69
Fracture location	T91	T91	T91	T91

shows that the material stability level of buttering treatment is better than direct welding. The life ratio also indicates a longer secondary creep stage for this treatment.

3.2 Weld Strength Reduction Factor

A relatively suitable method for testing the creep resistance characteristics of welded materials is the Weld Strength Reduction Factor (WSRF). The factor that underlies the use of this method is the degradation of the weld joint, especially the type IV failure in the HAZ area [12]. Welding heat input will reduce the strength of the material, especially in the heat-affected (HAZ) area. The thermal welding cycle causes a change in the material phase according to the temperature and the rate of heat transfer according to the thermal expansion of the material. Holmström proposed an equation using the Weld Time Reduction Factor (WTRF) to analyze the weld material's creep behavior [13]. This method is suitable for this study, where all treatments were tested at the same loading parameters (creep stress 200 MPa, test temperature 600 °C).

WTRF uses a comparison between the testing time of the welding material and the parent material for the same loading conditions. This is done to see the effect of welding on the strength of the material. Equation 1 is used to determine the welding reduction factor, where WTRF is a reduction factor in the failure time of the weld material; tu(W)/σ/T is the test time under creep stress conditions (σ) at temperature (T) compared to the collapse time of BM with the loading conditions were the same tu(BM)/σ/T. The results of the WTRF calculation for each treatment were then tabulated in Table 5.

$$WTRF = \frac{t_{uW}/\sigma/T}{t_{uBM}/\sigma/T} \tag{1}$$

The database of grade 91 material at the same loading (creep stress 200 MPa, temperature 600 oC) was used as the dividing factor (tu(BM)/s/T) in Eq. 4 [8]. The data used as a reference for WTRF in this study is the initiation time of tertiary creep (Ti), considering that the creep initiation time is the limit between the constant creep rate (creep stage 2) to the transient creep stage (creep stage 3) where the material still has resistance before collapsing due to creep load (creep-rupture) and is a factor in determining the material's resistance to creep loads.

There is some literature regarding the weld strength reduction factor analysis on the creep resistance of T91 material. The results of Abe's research (2014) show that the WTRF of the T91 welding material for loading 600 °C is 0.79 (plate), 0.67 (pipe) and 0.68 (forging) [14]. Weld reduction factor of 0.8 for 9Cr-1Mo (Grade 91) submitted by Becht (2006) using FM standard grade 91 referring to ASME Section B31.3 Year 2004. Weld reduction factor in the range of 0.8–0.6 for loading at a temperature of 550 °C–600 °C submitted by Swindeman (2007) referring to ASME Section III, Subsection NH for grade 91 steel. Based on references from previous studies, the value of 0.6 is used as the minimum limit for the reduction of weld strength compared to the parent material.

Table 5. Weld time reduction factor

	Initiation of tertiary creep—T*i* (h)	WTRF—T*i*
A1	6,42	0,32
A2	6,43	0,32
B1	12,96	0,65
B2	11,97	0,60
BM References T91 [8] $T_i = 20$ h [8]		

Based on the WTRF calculation results shown in Table 5, an overview of the creep resistance characteristics of the research material can be concluded. The Welding Strength Reduction Factor (WTRF) for the Direct Welding treatment, both using FM ERNiCr-3 and ERNiCrMo-3, was 0.32 and did not meet the minimum WTRF limit. This shows that the direct welding treatment is not suitable for use at a temperature of 600 °C. WTRF requirements were met by buttering treatment with WTRF values of 0.65 for FM ERNiCr-3 and 0.60 for FM ERNiCrMo-3. However, considering that the WTRF value of FM ERNiCrMo-3 is at the lower limit of the required range, the use of FM ERNiCrMo-3 buttering is not highly recommended for application at a temperature of 600 °C. Buttering treatment continued the value was at the upper limit of the requirements, so the use of this treatment met the application requirements at a temperature of 600 °C.

4 Conclusion

Dissimilar welding ferritic T91 and stainless steel 347H has been conducted using different filler materials and a buttering layer on the T91. Creep rupture time is increased almost two folds by the buttering layer on T91 compared with the direct welding. Weld Strength Reduction Factor is used to estimate weld joints' capability: Direct welding is not recommended for 600 °C application. Buttering with ERNiCrMo-3 is on the lowest limit of safe application (0.60). The present experiment and analysis show that buttering with ERNiCr-3 is recommended for boiler tube operating at 600 °C.

Acknowledgements. The author gratefully acknowledges the financial and technical support provided for this work by PT Perusahaan Listrik Negara (PT PLN) in cooperation with Departemen Teknik Mesin Institut Teknologi Sepuluh Nopember (ITS), Indonesia. The author also appreciates the useful technical discussions with Muhammad Nasruddin and Leonardo Bayu from PT Pembangkitan Jawa Bali, Indonesia. Support of material, welding process, and mechanical testing from PT Pembangkitan Jawa Bali are also gratefully acknowledged.

References

1. Ennis PJ (2003) Recent advances in creep-resistant steels for power plant applications. Journal 28:709–730
2. Cerjak H, Hofer P, Schaffernak B (1999) Influence of microstructural aspects on the service behaviour of advanced power plant steels. ISIJ 39:874–888
3. Huysmans S, Vekeman J, Hautfenne C (2017) Dissimilar metal welds between 9Cr creep strength enhanced ferritic steel and advanced stainless steels—creep rupture test results and microstructural investigations. Weld. World 61:341–350
4. Lundin, C. D.: Dissimilar Metal Welds-Transition Joints Literature Review Emphasis is on carbon migration, the stress/strain state of welds, and transition joint failure mechanisms. Weld. Res. Suppl., pp. 58s-63s (1982)
5. Al Hajri M, Malik AU, Meroufel A, Al-Muaili F (2015) Premature failure of dissimilar metal weld joint at intermediate temperature superheater tube. Case Stud. Eng. Fail. Anal. 3:96–103
6. DuPont JN (2013) Microstructural evolution and high temperature failure of ferritic to austenitic dissimilar welds. Int Mater Rev 57:208–234
7. Akram, J., Kalvala, P. R., Misra, M., Charit, I.: Creep behavior of dissimilar metal weld joints between P91 and AISI 304. Mater. Sci. Eng., A, vol. 688, pp. 396–406 (2017)
8. Kimura K, Kushima H, Sawada K (2009) Long-term creep deformation property of modified 9Cr-1Mo steel. Mater Sci Eng A 510–511:58–63
9. Akram J, Kalvala PR, Chalavadi P, Misra M (2018) Dissimilar Metal Weld Joints of P91/Ni Alloy: Microstructural Characterization of HAZ of P91 and Stress Analysis at the Weld Interfaces. J Mater Eng Perform 27:4115–4128
10. Chen YX, Yan W, Wang W, Shan YY, Yang K (2012) Constitutive equations of the minimum creep rate for 9% Cr heat resistant steels. Mater Sci Eng A 534:649–653
11. Kimura, K., Sawada, K.: Time to initiation of tertiary creep property of grade 91 steels. Proc. ASME 2016 Press. Vessel. Pip. Conf., vol. PVP2016, no. 63356, pp. 1–7 (2017).
12. Abson DJ, Rothwell JS (2013) Review of type IV cracking of weldments in 9–12%Cr creep strength enhanced ferritic steels. Int Mater Rev 58:437–473
13. Holmström S, Auerkari P (2006) Predicting weld creep strength reduction for 9% Cr steels. Int J Press Vessel Pip 83:803–808
14. Abe, F.: Grade 91 heat-resistant martensitic steel. vol. 91. Woodhead Publishing Limited (2014).

Analysis of Linkage Length Effect on the Dynamics of Series Active Variable Geometry Suspension

Rizal Bagus Prayogo[1] and Unggul Wasiwitono[2](✉)

[1] Undergraduate Program of Mechanical Engineering, Institut Teknologi Sepuluh Nopember (ITS), Surabaya, Indonesia

[2] Department of Mechanical Engineering, Institut Teknologi Sepuluh Nopember (ITS), Surabaya, Indonesia

unggul@me.its.ac.id

Abstract. The vehicle suspension system must isolate or reduce the vibrations that occur in the vehicle body due to road disturbance. Series Active Variable Geometry Suspension (SAVGS) is a suspension system that offers much better performance than passive suspension and can overcome the disadvantages of active suspension. This study will discuss the effect of single link length on SAVGS, especially on comfort and road-holding performance. The suspension system is modeled as a quarter car, where the effect of a single link length is analyzed in the form of equivalent stiffness and damping. The linear quadratic regulator is considered for the control design. The control system's performance was tested on a non-linear model with road disturbances consisting of bumpy roads and highway roads. The simulation results show that the longer the single link, the greater the vehicle's comfort and road holding on the vehicle increase. However, the longer the single link, the more the control input is needed.

Keywords: Variable geometry suspension · Linear quadratic regulator · Active suspension

1 Introduction

The comfortability and safety of the driver is the most crucial factor in addition to the reliability of the car engine itself. The suspension system is an essential component in a vehicle that can provide comfortability and stability (road holding) while driving. In meeting the comfort criteria of a vehicle, the vehicle suspension must be designed in such a way as to be able to deal with road changes, speed, and load mass.

Much research has been done on vehicle suspensions to resolve the conflicting performance between passenger comfort and road holding [1–3]. The development of the suspension system continues to be carried out, one of which is the discovery of an active suspension where there is an additional actuator that can provide direct force according to changes in the road conditions. However, due to weight, dimensions, and power

I. Tolj et al. (eds.), *Recent Advances in Mechanical Engineering*, Lecture Notes in Mechanical Engineering, https://doi.org/10.1007/978-981-19-0867-5_5

requirements, this suspension type is not the primary choice, even though the performance provided is excellent. Series Active Variable Geometry Suspension (SAVGS) is introduced in [4–7], where this suspension system has a much better performance than passive suspension and can overcome the disadvantages of active suspension.

Research to improve the performance of the SAVGS, starting with the quarter-car model approach, is proven to provide better suspension performance than passive suspension [8]. A previous study shows that the linkage geometry affects the dynamics of the active suspension [9, 10]. However, the previous study only analyzes the effect of the lower arm length on the suspension performance, while other parameters can affect the performance of the SAVGS. Therefore, in this study, a simulation will be carried out by varying the length of the SAVGS single link to analyze comfortability and stability while driving.

This paper is composed of several sections. Section 2 describes the quarter car's modeling and linearization, and Sect. 3 describes the analysis of the suspension. Then Sect. 4 presents the method for obtaining a synthesis controller used in system simulation. The simulation results will be discussed in Sect. 5. Finally, conclusions and suggestions are in Sect. 6.

2 Modeling and Linearization of a Quarter Car

The vehicle will be modeled as a quarter car using a double-wishbone configuration. The SAVGS concept is a development of the suspension system by adding a single link (*SL*) to the passive suspension, as shown in Fig. 1. This link connects the chassis (point G) and the upper end of the suspension (point F) and can rotate on the rotary axis or longitudinal axis (point G) with the help of an actuator which will produce a torque (T_{SL}).

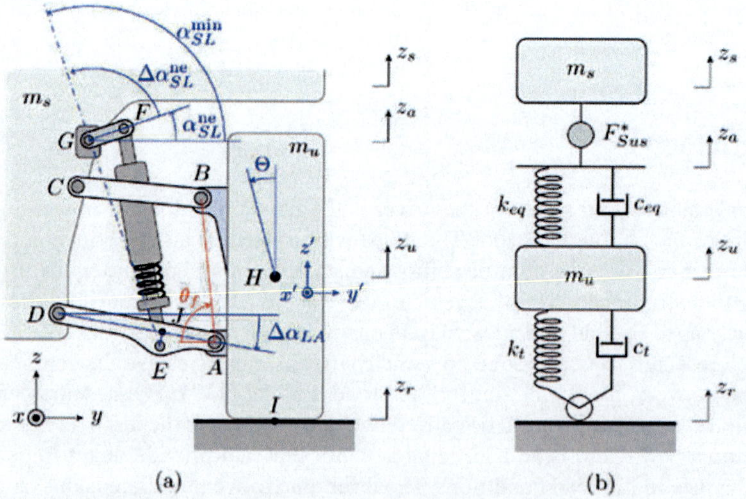

Fig. 1. **a** Non-linear quarter-car model **b** Linear quarter-car model

The vehicle suspension is a non-linear system. For the controller design, linear modeling of the suspension system is needed. In the linearization process, several assumptions were used based on reference [6]. The first assumption is geometric equivalence, where the vehicle's suspension deflection and tire deflection have the same reference point between the linear model and the non-linear model, as shown in Fig. 1 (a) and (b).

$$l_{SD} = z_a - z_u \tag{1}$$

$$l_t = z_u - z_r \tag{2}$$

where l_{SD} is spring damper deflection or suspension length (vertical distance between single link and unsprung mass) and l_t is tire deflection or wheel deflection (vertical distance between the road surface and unsprung mass).

The second assumption is the power balance on the actuator (power balance), where the power requirement of the actuator in the non-linear model (a) must be equal to the actuator power in the linear model (b). The energy balance equation can be written as follows:

$$T_{SL}\dot{\alpha}_{SL} = F^*_{sus}u \tag{3}$$

$$u = \dot{z}_s - \dot{z}_a \tag{4}$$

where F^*_{sus} is the suspension force and u is the linear velocity of the actuator linear model. Meanwhile, $\dot{\alpha}_{SL}$ is the angular velocity of the single link and T_{SL} is the torque applied to the single link (model a). The T_{SL} equation can be written as follows

$$T_{SL} = F_{sus}\frac{dl_{sus}}{d\alpha_{SL}} \tag{5}$$

The suspension force F_{sus} in model (a) and F^*_{sus} in linear model (b) are related to each other, according to the following equation

$$F^*_{sus} = R_{sus}F_{sus} \tag{6}$$

The change in position of point F in model (a) can be transformed into z_a in linear model (b). The parameter describes the non-linear geometry of the single link and is a function of the single link under non-linear and linear conditions.

$$\Gamma = \frac{u}{\dot{\alpha}_{SL}} \tag{7}$$

By substituting Eqs. (5) and (7) into Eq. (3), we obtain

$$u = \frac{1}{R_{sus}}\frac{dl_{sus}}{d\alpha_{SL}}\dot{\alpha}_{SL}, \Gamma = \frac{1}{R_{sus}}\frac{dl_{sus}}{d\alpha_{SL}} \tag{8}$$

The third assumption is that the rate of energy changes stored due to energy dissipation in the non-linear model must be the same as in linear model (b). In other words, the equivalent spring equation is by Hooke's law, while the equivalent damper has the same energy dissipation as the original.

$$k_{eq} = k_{sus}R_{sus}^2 - F_{sus}\frac{dR_{sus}}{dl_z}, c_{eq} = c_{sus}R_{sus}^2 \tag{9}$$

where k_{eq} is the equivalent spring stiffness and c_{eq} is the equivalent damping coefficient. The values of k_{eq} and c_{eq} for each configuration have been defined in the nominal state, so the equation of motion of the linear equivalent model for the quarter car using SAVGS is

$$m_u\ddot{z}_u = k_{eq}((z_s - z_u) - u) + c_{eq}((\dot{z}_s - \dot{z}_u) - \dot{u}) - k_t(z_u - z_r) - c_t(\dot{z}_u - \dot{z}_r) \tag{10}$$

$$m_s\ddot{z}_s = -k_{eq}((z_s - z_u) - u) - C_{eq}((\dot{z}_s - \dot{z}_u) - \dot{u}) \tag{11}$$

3 Suspension Analysis

The parameters of the vehicles used are listed in Table 1. In the SAVGS, the placement of the suspension has different angles and lengths, which will affect the suspension's geometry. Due to the difference in the length of the single link, the suspension angle for each configuration is different, as shown in Fig. 2. Configuration 1 has the largest angle with 9.54°, while configuration 4 has an angle of 0°.

Table 1. Quarter-car parameters

Parameter (unit)	Value
Sprung mass, m_s (kg)	0.460
Unsprung mass,m_u (kg)	0.190
Suspension stiffness coefficient, k_{sus} (N/m)	782.940
Suspension damping coefficient, c_{sus} (N.s/m)	42.260
Tire vertical stiffness, k_t (N/m)	2912.500
Tire damping coefficient, c_t (N.s/m)	5.100
Suspension unloaded stroke, l_{SD0} (mm)	18.095
Single link length, l_{SL} (mm)	5, 10, 15, 20
Low pass filter cutoff frequency, ω_C (rad/s)	2.600
Upper arm length, $\overline{BC}$ (mm)	91.430
Lower arm length, $\overline{AD}$ (mm)	111.260

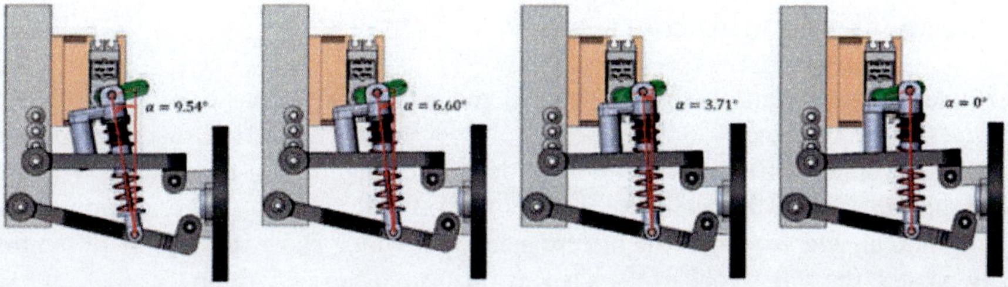

(a) (b) (c) (d)

Fig. 2. Suspension mounting angle **a** configuration 1, **b** configuration 2, **c** configuration 3, and **d** configuration 4

The values of k_{eq} and c_{eq} of each configuration have different values due to the difference in length and angle of installation of the suspension. Figure 3 are estimated values of k_{eq} and c_{eq} of all configurations, respectively. It can be seen from the figures that k_{eq} and c_{eq} are significantly influenced by the single link length. For the controller design, it is required to choose the nominal value of these parameters. The nominal value for each configuration is selected at the asterisk point shown in the figures. Table 2 shows the estimated nominal value of k_{eq} and c_{eq}.

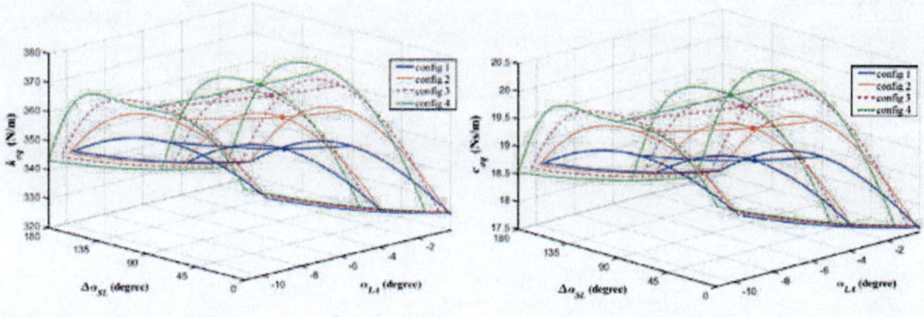

(a) (b)

Fig. 3. Variation for different configurations of **a** equivalence stiffness coefficient **b** equivalence damping coefficient

Table 2. Value of k_{eq} and c_{eq} of the suspension system

Configuration	k_{eq}	c_{eq}
1	339.88 N/m	18.35 N.s/m
2	352.02 N/m	19.00 N.s/m
3	361.81 N/m	19.53 N.s/m
4	368.53 N/m	19.89 N.s/m

4 Controller Synthesis

The state-feedback control scheme is considered in this study. Considering the following state variables $x_1 = z_a - z_u$ and $x_2 = z_u - z_r$ are the deflection of the suspension (suspension deflection) and wheel deflection (tire deflection) on the vehicle,$x_3 = \dot{z}_s$ is the sprung mass velocity, while $x_4 = \dot{z}_u$ is the vertical velocity of the unsprung mass on the vehicle, and $w = \dot{z}_r$ is the disturbance input which shows the profile of the road being passed, the state space of the suspension system can be written as

$$\dot{x}(t) = Ax(t) + B_u u(t) + B_w w(t) \tag{12}$$

where u is a control input to the system, provided the speed and the position of *SL*. The input feedback (linear equivalent speed) commanded by the controller is given as

$$u^* = -Kx \tag{13}$$

where K is feedback gain. The block diagram of the control system is shown in Fig. 4. u^* defined in Eq. (13) is not incorporated the bandwidth limitation. u considered on bandwidth limitation is generated after u^* surpassed the low-pass filter *SAVGS non-linear system* in Fig. 4. The dynamics of the filters incorporate the bandwidth of the rotary actuator.

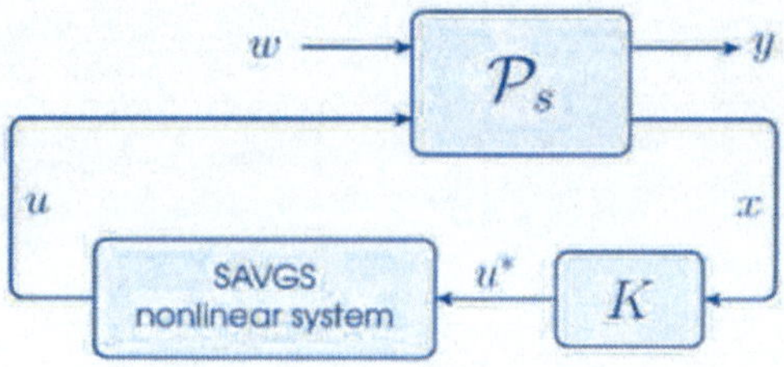

Fig. 4. State-feedback control scheme

5 Simulation Results

Figure 5 shows the simulation result on bump road for sprung mass acceleration compared to passive case. In configuration 1, the RMS response in sprung mass acceleration is $1.698x10^{-1}m/s^2$, while in configuration 4, the RMS is $1.528x10^{-1}m/s^2$ where there is a difference of 20% between configuration 1 and configuration 4. Meanwhile, the RMS passive suspension configuration 1 is 4.8%, and configuration 4 is 11.6% smaller, respectively. So, it is found that the longer the single link, the road comfort of the vehicle will increase.

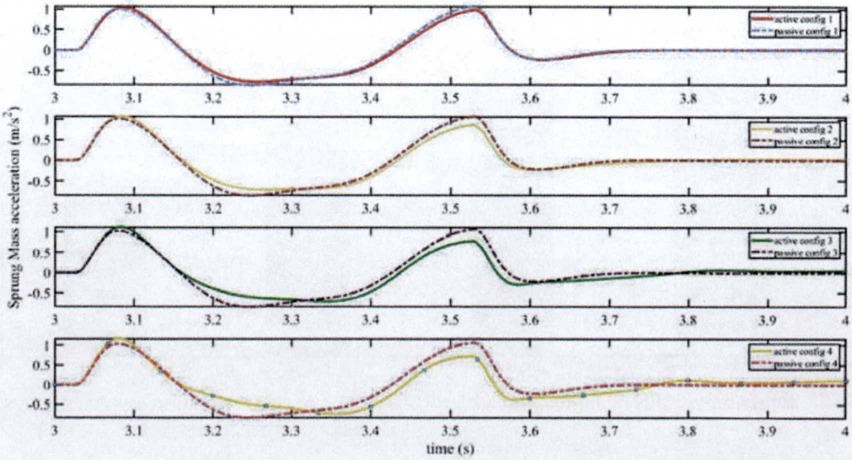

Fig. 5. Simulation results on bumpy road for sprung mass acceleration compared to passive

Figure 6 shows the comparison of the response for each configuration. Overall, configuration 4 provides a 10.1% performance increase compared to configuration 1 based on the RMS value. However, the longer the single link, the power required to drive the single link is higher.

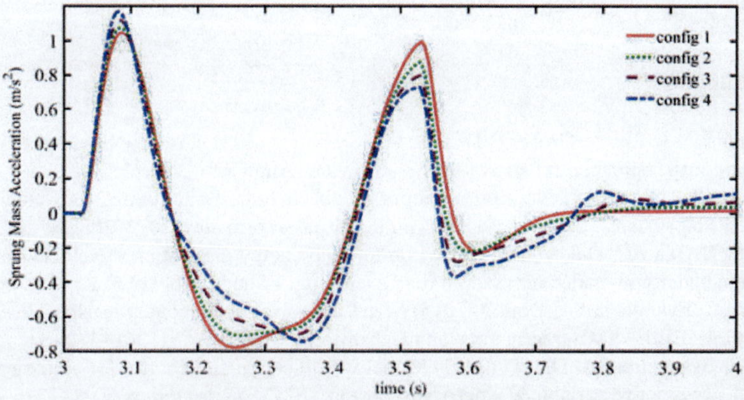

Fig. 6. Simulation results on bumpy road for sprung mass acceleration in each configuration

Then a road disturbance is given, representing the actual road conditions approximated by power spectral density (PSD) as a medium quality road. The result is shown in Fig. 7. It can be seen that the longer the single link, the smaller the RMS of sprung mass acceleration.

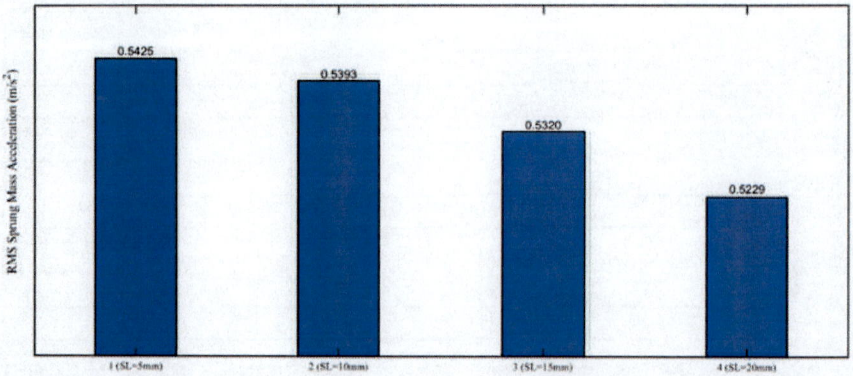

Fig. 7. Simulation results on highway road for RMS sprung mass acceleration each configuration

6 Conclusion

The simulation results show that the greater the performance improvement from the sprung mass acceleration when a longer single link is used, which shows that the comfort while driving will increase. The increase in performance compared to passive suspension ranges 4.8%–12.4% for sprung mass acceleration and 3.5%–12.4% for tire deflection. Meanwhile, when compared to configuration 1 with other arrangements, the performance increases 1.8%–10.1%. However, this is inversely proportional to the power required, where the longer the linkage or single link, the greater the power required.

References

1. Theunissen J, Tota A, Gruber P, Dhaens M, Sorniotti A (2021) Preview-based techniques for vehicle suspension control a state-of-the-art review. Annu Rev Control: 206–235
2. Pan H, Sun W (2018) Non-linear output feedback finite-time control for vehicle active suspension systems, IEEE transactions on industrial informatics: 2073–2082
3. Zhang B, Tan AC, Dai T (2021) Ride comfort and energy dissipation of vehicle suspension system under non-stationary random road excitation. J Sound Vib, vol 511
4. Arana C, Evangelou SA, Dini D (2015) Series active variable geometry suspension for road vehicles. IEEE/ASME transactions on mechatronics, Vol 20, No. 1, 361–372
5. Yu M, Evangelou SA, Dini D (2017) Model identification and control for a quarter car test rig of series active variable geometry suspension. IFAC Papers Online: 50–1, 3376–3381
6. Arana C, Evangelou SA, Dini D (2017) Series Active variable geometry suspension application to comfort enhancement. Control Eng Pract 59: 111–126
7. Cheng C, Evangelou SA, Arana C, Dini D (2015) Active variable geometry suspension robust control for improved. 2015 American Control Conference (ACC), pp 3440–3446
8. Yu M, Arana C, Evangelou SA, Dini D (2019) Quarter-car experimental study for series active variable geometry suspension. IEEE transactions on control systems technology: 743–759
9. Samnang C (2020) Modeling and simulation of series active variable geometry suspension on a quarter car model. Master thesis, Institut Teknologi Sepuluh Nopember
10. Wasiwitono U, Pramono AS, Sutantra I N (2018) Study on influences of linkage geometry on actively controlled double wishbone suspension.In: AIP Conference Proceedings, Vol 1983

Design Optimization of Braces for Three-Legged Minimum Jacket Offshore Structure

I. Putu Dipa Dhaneswara, Rudi Walujo Prastianto(✉), and Daniel Muhammad Rosyid

Ocean Engineering Department, Institut Teknologi Sepuluh Nopember, Surabaya, Indonesia
rudiwp@oe.its.ac.id

Abstract. The innovative design of offshore structures has developed rapidly in line with the increasing demand for energy from oil and gas in the world. One of the innovations is the effort to use a modified or minimum jacket structure. The minimum jacket structure is a concept that is very suitable to use in shallow water marginal fields, because the structural design is more economical in terms of cost, reusable, and easier to move than the conventional jacket. This study presents design optimization of the existing jacket platform in terms of brace pattern (configuration) and dimension selection with the objective function being to minimize the weight of the jacket platform. The optimization process was carried out with 2 steps of optimization scenarios which are selecting the most optimal jacket brace configuration among various variations and determining the most optimal jacket brace dimensions through a static in-place analysis of the jacket structure. The brace configuration variations considered are V-brace, N-brace, and K-brace patterns. The second optimization step is the determination of the brace dimensions including outside diameter (OD) and wall thickness (WT) by considering 12 model variations. The results showed that the optimization result on the existing minimum jacket has complied with the whole criteria and yielded a lighter structure weight of 3,542.11 kips. This optimized minimum jacket structure became 15.9% lighter than the weight of the existing initial jacket structure, which is 4,211.96 kips.

Keywords: Topology optimization · Minimum weight · Brace pattern · Minimum jacket offshore structure

1 Introduction

The development of oil and gas technology is always accompanied by innovations, especially in the jacket structure design. Design of minimum facility platforms (MFPs) is expected to produce a minimum jacket structure in terms of the weight of the structure and more economical cost, so that it is suitable for marginal field exploration.

Previously, there have been many studies discussing the modification of the minimum jacket structure. One of them was using a modular design approach for minimum low-cost facilities. The study has conducted modifications to obtain a minimum jacket structure which used conductors as the main support for the topside in which it is called Conductor Supported Minimum Offshore Structures (CoSMOs) [1]. Another research on minimum

I. Tolj et al. (eds.), *Recent Advances in Mechanical Engineering*, Lecture Notes in Mechanical Engineering, https://doi.org/10.1007/978-981-19-0867-5_6

jacket platforms has also been carried out by designing a minimum jacket for marginal areas in western India in which the research compared the form of a minimum jacket structure with leg variations [2].

Optimization is a very important process at the initial stage of designing the jacket structure because the design optimization process aims to obtain an optimal jacket structure both in terms of shape and strength. Previous research with the structural optimization method has been carried out by optimizing the tie-brace design of the blasting dolphin structure at the oil/condensate terminal of the Idol strait due to the increasing load on the export activity of the tanker [3]. Moreover, a study about the optimum arrangement of braces on jacket platforms based on strength and ductility has been done. By using a simple logical method for investigating the strength and ductility of the jacket structure, it is shown that the global geometry and configuration of the braces are very important and effective in both strength and ductility parameters [4]. In addition, research at Bohai Gulf proposed an acceleration-oriented design optimization of ice-resistant jacket platforms. This approach focused on the dynamic performance of the jacket platforms in terms of the deck acceleration and worked on the structural optimization technique to achieve economical and rational design [5].

Another study is about Optimization Approach in Offshore Wind Energy Supporting Structure Design in which optimization is used in the conceptual design stage of the wind turbine. The optimization process is carried out by two methods, which are artificial intelligence and optimization selection methods [6]. Moreover, there is a study about topology optimization design for offshore platform jacket structure in which the optimization results are compared to the original platform for static performance, dynamic performance, and Ultimate Carrying Capacity (UCC) [7]. In addition, another study on the design optimization process of a fixed jacket offshore platform under environmental loads was done. This study utilized an objective function of the amount of steel material used by the structural members with variations in their diameter and thickness [8].

The present study discusses design optimization of the brace configuration and dimension of an existing three-legged jacket structure in order to obtain the minimal weight of the jacket. The optimization process was carried out in 2 steps of optimization scenarios. The first step is to make a selection from a variety of brace configurations and then followed by sizing the outside diameter (OD) and wall thickness (WT) of the members for the selected brace pattern.

2 Method

The flowchart of the design optimization of braces for the minimum jacket is shown in Fig. 1. In this study, optimization was carried out to design a three-legged minimum jacket structure with a minimum weight in which the weight of the jacket must be lighter than the existing conventional jacket. Based on the flowchart, the first step in this analysis is modeling the jacket structure using SACS 12 software and validating with the weight control data from the company. After modeling the minimum jacket, input the loads on the structure such as all equipment loads, operational loads, live loads, and environmental loads.

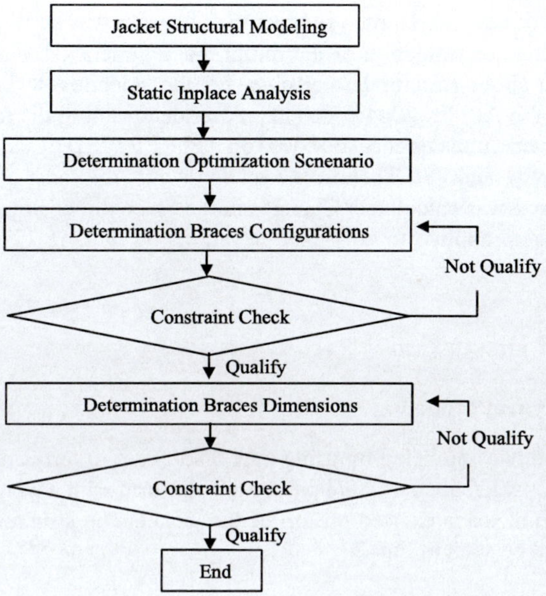

Fig. 1. Flowchart of the design optimization of brace system

In-place analysis is a static analysis used to ascertain whether the structure can withstand the load when it is being operated, either gravity load (dead load, live load, and equipment load) or environment load. In this analysis, a ratio between member stress and allowable stress, called UC (unity check), is used to assess the strength adequacy of all structural members of the jacket structure where the UC value must be less than 1.

In this study, optimization was carried out to design a three-legged minimum jacket structure with minimum weight. Determination of the optimization scenario on a minimum jacket structure refers to API RP2 A WSD code [9]. The following equation was used to calculate the weight of the jacket structure.

$$wjacket = \sum_{i=1}^{n} Yi.Li.Ai \tag{1}$$

where *Yi* is the density of the material, and *Li* and *Ai* are the length and cross-sectional area of the structural members. The optimization scenario of the minimum jacket structure is by selecting the variation of the brace configuration and the variation of the brace dimension, either outside diameter (OD) or wall thickness (WT).

The shape of braces to be considered consists of several types, such as K-braces, V-braces, N-braces, and X-braces. Each brace has its advantages and disadvantages. The selection of the brace type should be based on the environment, redundancy, and structural characteristics of the jacket [10]. Changes to the jacket configuration will be carried out at an elevation of -110 ft to -224 ft with the existing configuration being the X-braces.

Determination of the jacket structure bracing dimension was carried out in 2 stages. The first stage is the determination of the initial brace dimension variation. Based on API RP 2A WSD about structural steel pipe, for the selection of braces dimension variation refers to the API 5L 2004 code [11]. After determining the initial dimension, the tubular joint criteria must be checked based on API RP 2 WSD by using the geometric parameters of β, τ, γ, and D/t. Then in the second stage, the most optimum value of the braces dimension was calculated. Changes in the brace dimensions (horizontal and diagonal braces) were applied to all jacket elevations by making 12 variations of the braces dimension.

3 Results and Discussion

3.1 Jacket Structural Modeling

The jacket structure was modeled by using SACS software in terms of the jacket substructure only. A topside/superstructure part of the platform with a weight of 1,050 kips was modeled as joint loads exerted on top of the jacket. The structural model of the jacket structure can be seen in Fig. 2.

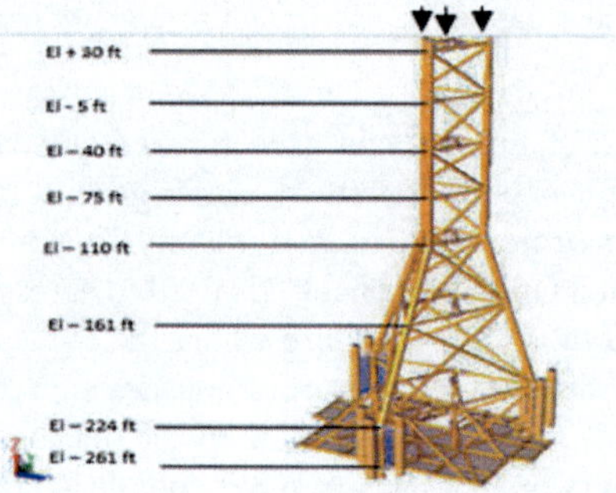

Fig. 2. Model of the minimum jacket platform for the analysis

3.2 Static In-Place Analysis

Based on the in-place analysis carried out on the minimum jacket structure, it is obtained that all members have UC < 1, in which the member with the largest UC is at member 0022–0025 with a UC of 0.261. This largest UC is still relatively small, therefore an optimization can be carried out to obtain the optimum jacket structure in terms of the weight of the structure.

3.3 The Optimization Scenario

The jacket optimization was carried out by 2 optimization scenarios, which are by selecting the variation of the brace's configuration at an elevation of –110 ft to -224 ft and the second scenario is the determination of the brace dimension (horizontal and diagonal braces) at elevations (+) 30 ft to (−) 224 ft of the jacket.

The objective function of the jacket optimization is the weight of the jacket structure. Meanwhile, the constraints for the optimization with the brace configuration selection are stress (UC < 1) and slenderness ratio (kL/r < 90) of the member, while the brace dimension optimization constraints are the joint deflection check (deflection < 0.875), joint punching shear stress check (UC < 1), and member stress unity check (UC < 1).

3.4 Determination of the Braces Configuration or Pattern

Braces configuration optimization (first optimization scenario) was carried out for the elevation of (+) 110 ft to (−) 224 ft. Figure3 shows results of the modeling and in-place analysis on the SACS software with K-brace, N-brace, and V-brace configurations.

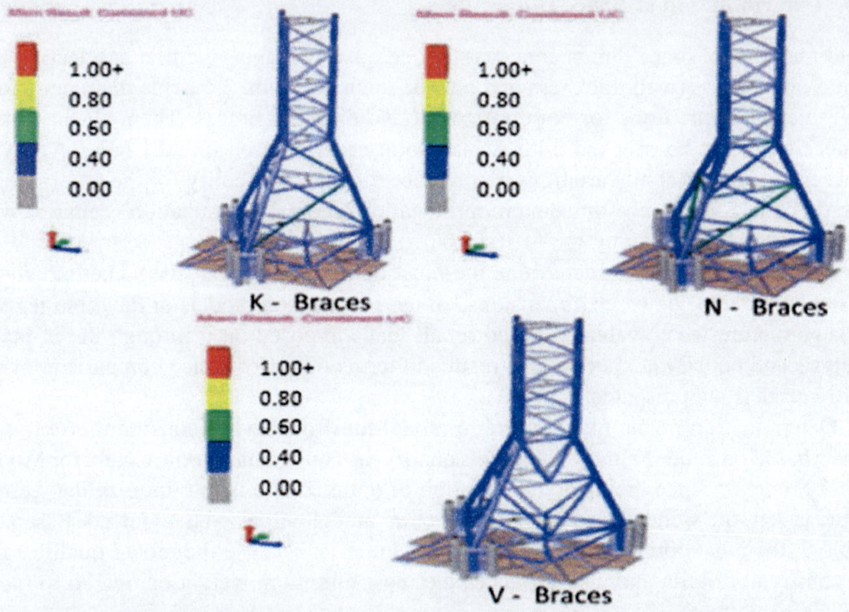

Fig. 3. In-place analysis result with K-brace, N-braces, and V-braces

Determination of the most optimum brace configuration should refer to the constraint, but it is also necessary to consider the weight of the jacket structure because the objective function of this optimization is to minimize weight. Table 1 shows a comparison of the performance of each brace configuration considered based on the maximum constraint criteria.

Based on Table 1, the maximum UC for each configuration of K-braces, N-braces, and V-braces qualifies the constraint because of all of them resulting in UC < 1, while in the slenderness ratio constraint, only the V-brace configuration qualifies the constraint

Table 1. Comparison of each brace configuration performance

Brace configuration	Max UC	Max slenderness ratio	Weight (kips)
K-brace	0.316	110.459	3,961.70
N-brace	0.386	110.459	3,968.87
V-brace	0.326	75.909	4,084.81

because of the slenderness ratio (kL/r) < 90. Then the V-brace configuration is considered as the most optimum configuration at the elevation of (−) 161 ft to (−) 224 ft because it qualifies all constraint criteria and also the resulting weight is lighter than the initial weight of the jacket structure.

3.5 Determination of Brace Dimensions

Determination of brace dimensions was carried out in 2 stages; the first is selecting the initial dimensions (wall thickness and outside diameter) with 3 outside diameter and 4 wall thickness variations for both horizontal and diagonal braces. Then, all the initial dimensions must be checked with tubular joint criteria based on API RP 2 A WSD. The result shows that all variations in the brace dimensions qualify for the tubular joint criteria check. The brace dimension optimization (second optimization scenario) was carried out for the elevation of (+) 30 ft to (−) 224 ft.

The second stage is to determine the most optimum model of the jacket structure. Based on the selected brace dimensions, all the 12 variation models of the three-legged jacket structure were evaluated based on all loads imposed on it through the in-place analysis and constraint checks. The results in terms of performance comparison of all the 12 models are presented in Table 2.

Determination of the most optimum model must qualify all constraint criteria, in which based on Table 2 almost all models qualify the constraint criteria, except for Model 4 and Model 8. Since the objective function of optimization is to minimize the weight of the jacket, the weight of each jacket structure model will also be reviewed. Based on Table 2, the most optimum option finally is Model 12 because the model qualifies for all constraint criteria and also produces the most minimum weight compared to other models.

Figure4 presents the final form of the minimum jacket structure before and after the optimization, while Table 3 outlines a comparison between the minimum jacket structure before and after the optimization process for all parameters considered.

Table 2. Comparison of all the 12 models based on all constraints considered

Model name	Vertical brace (OD × WT) inch	Horizontal brace (OD × WT) inch	Weight (kips)	Member stress UC	Criteria	Joint punching UC	Criteria	Deflection (ft)	Criteria
Model 1	28 × 1	22 × 0.75	3,872.7	0.345	Ok	0.482	Ok	0.447	Ok
Model 2	28 × 0.875	22 × 0.625	3,791.0	0.392	Ok	0.525	Ok	0.479	Ok
Model 3	28 × 0.688	22 × 0.5	3,681.4	0.461	Ok	0.757	Ok	0.525	Ok
Model 4	26 × 0.5	22 × 0.375	3,569.9	0.557	Ok	1.247	Not ok	0.618	Ok
Model 5	26 × 1	20 × 0.75	3,837.6	0.377	Ok	0.433	Ok	0.468	Ok
Model 6	26 × 0.875	20 × 0.625	3,748.0	0.429	Ok	0.516	Ok	0.492	Ok
Model 7	26 × 0.688	20 × 0.5	3,647.8	0.504	Ok	0.746	Ok	0.548	Ok
Model 8	26 × 0.5	20 × 0.375	3,545.8	0.63	Ok	1.22	Not ok	0.648	Ok
Model 9	24 × 1	18 × 0.75	3,795.7	0.419	Ok	0.429	Ok	0.49	Ok
Model 10	24 × 0.875	18 × 0.625	3,728.9	0.477	Ok	0.503	Ok	0.513	Ok
Model 11	24 × 0.688	18 × 0.5	3,635.7	0.56	Ok	0.799	Ok	0.574	Ok
Model 12	24 × 0.5	18 × 0.375	3,542.1	0.698	Ok	0.92	Ok	0.683	Ok

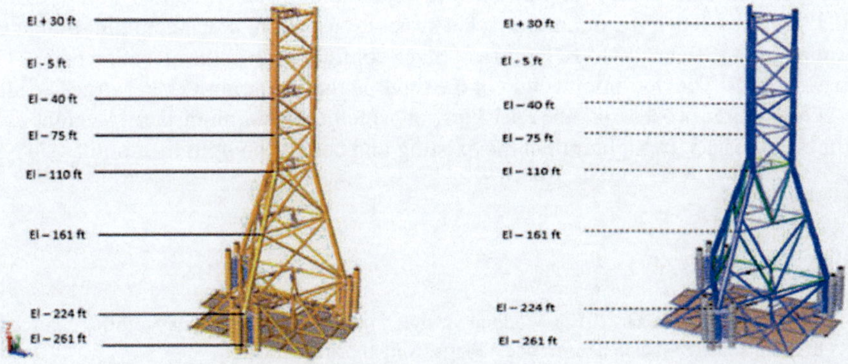

Fig. 4. Comparison of the three-legged minimum jacket structure model: before the optimization (left figure), and after the optimization (right figure)

Table 3. Comparison of a three-legged minimum jacket structure before and after the optimization in terms of weight, strength, dimensions, and brace configuration

Description	Minimum jacket before optimization	Minimum jacket after optimization
Self-weight (Kips)	4,211.96	3,542.13
Maximum Joint Deflection (ft)	0.354 ft	0.683
Maximum Member Stress (UC)	0.287	0.698
Maximum Joint Punching Shear (UC)	0.227	0.92
Outside Diameter Diagonal Braces (inch)	28–30	24
Wall thickness Diagonal Braces (inch)	1.25–1.375	0.5
Outside Diameter Horizontal Braces (inch)	22–26	18
Wall Thickness Horizontal Braces (inch)	0.75–1.00	0.375
Brace configuration at elevation (+) 110 ft to (−) 224 ft	X-braces	V-braces

4 Conclusion

In this study, the optimization aims to produce a lighter minimum jacket structure weight based on the optimum brace configuration system and dimensions. The optimal dimensions of the braces for the optimum three-legged minimum jacket structure are for diagonal braces with an outside diameter of 24 inches and a wall thickness of 0.5 inches, while the horizontal braces have an outside diameter of 18 inches and a wall thickness of 0.375 inches. Then the optimum jacket structure used a K-brace configuration at the elevation of (+) 30 ft to (−) 110 ft and a V-brace configuration at the elevation of (−) 110 ft to (−) 224 ft. The optimum weight of the three-legged minimum jacket structure after the optimization process is 3,542.13 kips, in which the minimum jacket weight after optimization is 15.9% lighter than the existing one before the optimization of 4,211.96 kips.

References

1. Helle Y, Nicholson G (2012) Modular design for low cost minimum facilities platforms. Offshore south east asia conference. Ghana, Singapore November
2. Nallayarasu S (2013) Structural concepts for minimum facility platforms for marginal field development in western offshore. Ocean engineering indian institute of technology Madras-36, India

3. Puspitorini DA (2017) Optimization of Tie-Brace design for breasting dolphin structure at the berhala strait oil/condensate terminal, Final project of the department of ocean engineering FTK ITS, Surabaya
4. Tabeshpour MR, Fatemi F (2020) Optimum arrangement of braces in jacket platform based on strength and ductility. J Mar Struct 71 102734
5. Liu X, Li G, Yue Q, Oberlies R (2009) Acceleration – oriented design optimization of ice-resistant jacket platforms in the bohai gulf, J Ocean Eng 36:1295–1302
6. Pramadhika Y, Murdjito, Rosyid DM (2009) Optimization approach in offshore wind energy supporting structure design. final project of the department of ocean engineering FTK ITS, Surabaya
7. Tian X, Wang Q, Liu G, Liu Y, Xie Y, DengW (2019) Topology optimization design for offshore platform jacket structure. Elsevier Appl Ocean Reasearch 84:35–50
8. Nasseri T, Shabakhty N, Afshar MH (2014) Study of fixed jacket offshore platform in the optimization design process under environmental loads. Int J MaritE Technol 75–84
9. API (2010). Recommended practice for planning, Designing and constructing fixed offshore platform - working stress design. 22nd Ed. American petroleum institute. Washington DC
10. Chakrabarti SK (2005) Handbook of Offshore Engineering. Elsevier, Amsterdam
11. API (2004). Specification for line pipe. 4 th edition. American Petroleum Institute Washington DC

Plant Maintenance Budgeting Prioritization Based on Reliability Prediction of Repairable System

Fandi Setia[1(✉)], Agus Wibawa[2], and Muhammad Nur Yuniarto[3]

[1] Graduate Program of Mechanical Engineering, Institut Teknologi Sepuluh Nopember, Surabaya, Indonesia
fandisetia@indonesiapower.co.id
[2] PT Pembangkitan Jawa Bali, Jl. Ketintang Baru No. 11, Surabaya 60231, Indonesia
[3] Department of Mechanical Engineering, Institut Teknologi Sepuluh Nopember, Surabaya 60111, Indonesia

Abstract. A reliability prediction study has been carried out using failure data from the gas turbine system at a combined cycle power plant in Indonesia. From this study, the prediction value of the equipment reliability of the gas turbine subsystem was obtained along with the value of *Failure Cost* (FC), *Component Risk* (CR), and *Maintenance Cost* (MC) which is then processed further to produce a *Benefit to Cost Ratio* (BCR) which is useful for preparing priority rankings for the allocation of very limited maintenance budgets. This research also leads to a conclusion that the more the failures experienced by the equipment in one observation period, the lower the predictive value of its reliability. In addition, it is also discovered the fact that subsystems with the same number of failures but with different times of failure will produce different reliability values. And lastly, the subsystem risk rating obtained from *Component Risk* (CR) does not necessarily become a budget priority rating; this rating can change when the risk rating is combined with *Maintenance Cost* (MC) to produce a more objective budget priority rating with broader considerations. The methodology that combines the prediction of equipment reliability with budget prioritization in this research can also be used in other power generation plants that predominantly manage physical assets in order to obtain optimal asset life cycle management without compromising the performance of the equipment in the power plant due to limited maintenance budgets.

Keywords: Power generation plant · Combined cycle power plant · Reliability prediction · Reliability block diagram · Budget prioritization · Life cycle management

I. Tolj et al. (eds.), *Recent Advances in Mechanical Engineering*, Lecture Notes in Mechanical Engineering, https://doi.org/10.1007/978-981-19-0867-5_7

1 Introduction

A power plant consists of many systems and pieces of equipment that are interconnected with each other to form a final function, i.e. to generate electric power. To ensure that plant equipment do their function as intended, the owner must do systematic and comprehensive maintenance activities consisting of a set of tasks which can be categorized as Preventive Maintenance (PM), Predictive Maintenance (PdM), and Corrective Maintenance (CM) tasks.

Each task needs resources in the form of manpower, time, materials, etc. which can be converted to financial quantification and need to be predetermined long before those maintenance activities are undergone, for example, next year's plant maintenance budget needs to be determined this year; this leads to ineffective budgeting plan due to bad assumptions conducted while compiling the budget proposal. This situation often gets worse while the budget is limited. The recent situation shows little scientific work related to this subject.

In a combined cycle power plant, the gas turbine system and its subsystems are often considered as a repairable system in general. In order to predict its future reliability level, it cannot be treated using traditional life data analysis methods due to the nature of the failure events which are not independent and identically distributed. The gas turbine system, as a repairable system, is subject to recurring data analysis hence it has to be treated using a relevant method for recurring data analysis. Crow [1] did some analysis to predict the reliability of repairable systems using the Weibull-Poisson process or better known as the Non-homogeneous Poisson Process (NHPP) while Nelson [2] used a different approach, the so-called Mean Cumulative Function (MCF) and Mettas et al. [3] explored the same subject through the General Renewal Process (GRP) method.

The main objective of this research paper is that one can do budgeting prioritization for the next maintenance period based on reliability prediction which is analyzed through NHPP analysis. The result will be expected as a systematic approach that can be practically used in the field for compiling a sound plant maintenance budgeting proposal.

2 Methodology

2.1 Simplification and Division of Gas Turbine System

In terms of repairable systems, the failed components are repaired instead of replaced by new ones, hence it will be subjected to a recurring data analysis method.

The complex gas turbine system consists of several subsystems as a lower hierarchy as shown in Table 1 and then be simplified to a schematic repairable system consisting of six main functioning subsystems, i.e. air intake system, compressor, combustion chamber, gas turbine, electric generator, and main transformer as shown in Fig. 1.

The schematic then is simplified again as a Reliability Block Diagram (RBD) in which each component has its own historical data, both failure data and maintenance cost data. The RBD is shown in Fig. 2, and each component in the RBD has its own reliability parameter which is interacting in a serial connection from the first component all the way through the last component.

Table 1. The hierarchy of the gas turbine system consisting of six subsystems as a higher hierarchy and subsystems as a lower hierarchy

Code	Higher hierarchy (subsystem)	Lower hierarchy
1	Air Intake System (AI)	Inlet air filter
		Field instrumentation
		Variable guide vane
2	Compressor (CP)	Blades and vanes
		Cooling system
		Bleed valves
		Lube oil system
3	Combustion Chamber (CC)	Fuel gas supply
		Safety system
		Combuster basket
		Nozzle
		Safety system
4	Gas Turbine (GT)	Lube oil system
		Electrical aux
		Gt control system
		Safety system
		Cooling system
		Starting motor
5	Electric Generator (GN)	Lube oil system
		Excitation system
		H2 Cooling system
		Winding generator
6	Main Transformer (MT)	Cooling system
		Tap changer
		grounding
		Safety system

2.2 Failure Data Acquisition

The failure data was obtained from PLTGU Cilegon's Computerized Maintenance Management System (CMMS) from 2018 to 2020. The types of failure data collected are not limited to forced outage type only, but also all failures that require resources for recovery to return to function. Figure 3 shows the failure data recapitulation of the gas

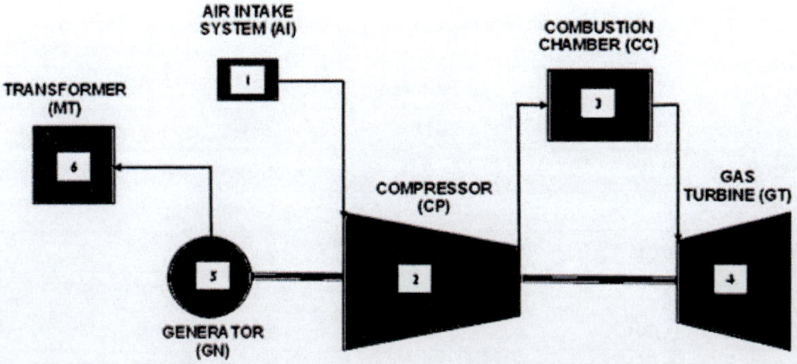

Fig. 1. Simplified schematic gas turbine system

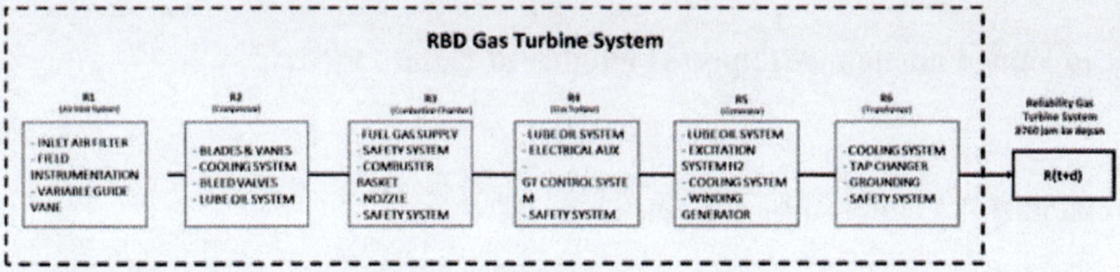

Fig. 2. Reliability block diagram of a simplified gas turbine system as shown in Fig. 1

turbine system's subsystem, while Table 2. Detailed failure data of the gas turbine system's subsystem shows the detailed and tabulated failure data of the gas turbine system's subsystem.

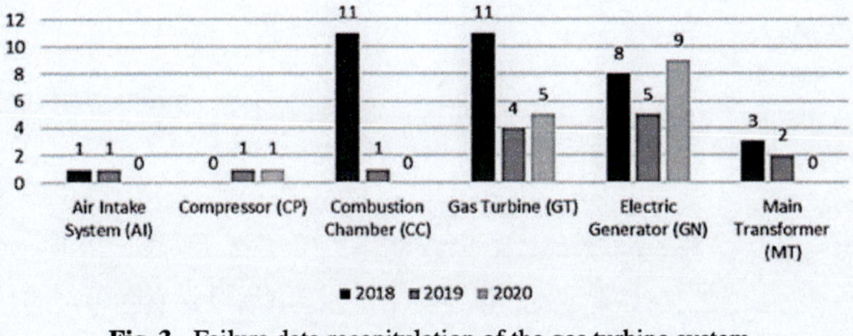

Fig. 3. Failure data recapitulation of the gas turbine system

2.3 Non-homogeneous Poisson Process (NHPP) [4–7]

Intensity function u(t)

$$u(t) = \lambda \beta t^{\beta - 1} \mathrm{t} > 0 \tag{1}$$

Table 2. Detailed failure data of the gas turbine system's subsystem

Subsystem	Failure number		
	2018	2019	2020
Air Intake System (AI)	1	1	0
Compressor (CP)	0	1	1
Combustion Chamber (CC)	11	2	0
Gas Turbine (GT)	11	4	5
Electric Generator (GN)	8	5	9
Main Transformer (MT)	3	2	0
Total	34	15	15

Mean Value Function, or Expected number of Failure E[N(t)]

$$E[N(t)] = \lambda t^{\beta} \tag{2}$$

Probability N Failures over time (0, T), Prob[N(t)=n]

$$Prob[N(t) = n] = \frac{\left(\lambda t^{\beta}\right)^{n} e^{-\lambda t^{\beta}}}{n!}, \text{ (n = 0, 1, 2, ...)} \tag{3}$$

Mission Reliability of age t with length d, R(t)

$$R(t) = e^{-[\lambda(t+d)^{\beta} - \lambda t^{\beta}}} \tag{4}$$

Maximum Likelihood Estimates of λ and β

General Form

$$\widehat{\lambda} = \frac{\sum_{q=1}^{K} {N_q}^{k}}{\sum_{q=1}^{K}\left(T_q^{\widehat{\beta}} - S_q^{\widehat{\beta}}\right)}, \text{ (n = 0, 1, 2,)} \tag{5}$$

$$\widehat{\beta} = \frac{\sum_{q=1}^{K} {N_q}^{k}}{\widehat{\lambda}\sum_{q=1}^{K}\left(T_q^{\widehat{\beta}} \ln T_q - S_q^{\widehat{\beta}} \ln S_q\right) - \sum_{q=1}^{K}(..q)\sum_{i=1}^{N_q}(\ln X_{iq})} \tag{6}$$

Time Truncated Form with S_q=0 dan T_q=T (where q=1, k)

$$\widehat{\lambda} = \frac{\sum_{q=1}^{K} {N_q}^{k}}{kT^{\beta}} \tag{7}$$

$$\widehat{\beta} = \frac{\sum_{q=1}^{K} {N_q}^{k}}{\sum_{q=1}^{K}(..q)\sum_{i=1}^{N_q} ln\left(\frac{T}{X_{iq}}\right)} \tag{8}$$

Failure Truncated Form (with k = 1, S= 0, $X_{N,1}$= T_1):

$$\hat{\lambda} = \frac{N_1}{T_1^{\beta}} \tag{9}$$

$$\hat{\beta} = \frac{N_1}{\sum_{i=1}^{N_1} ln\left(\frac{T_1}{X_{i1}}\right)} \tag{10}$$

ML Estimate of the intensity function:

$$\hat{u}(t) = \hat{\lambda}\hat{\beta}t^{\hat{\beta}-1}, t > 0 \tag{11}$$

2.4 Benefit to Cost Ratio (BCR)

Benefit to Cost Ratio (BCR) combines *Component Risk* (CR) and *Maintenance Cost* (MC), while the former is a product of *Expected number of failure,* E[N(t)] and *Failure Cost* (FC) and the latter is the cost to maintain the equipment in gas turbine system excluding capital expenditure.

Component Risk (CR)

$$CR = \text{Expected number of failure} \times \text{Failure Cost(USD or IDR)} \tag{12}$$

Failure Costs are incurred as a result of failure events, including direct costs to restore equipment function and consequential costs as a result of the failure of the equipment.

Maintenance Cost (MC)

$$\text{MC} = \text{PMCost} + \text{PdMCost} + \text{CMCost, (inUSDorIDR)} \tag{13}$$

With PM, PdM dan CM Cost are Preventive Maintenance, Predictive Maintenance, dan Corrective maintenance Cost, respectively.

Benefit to Cost Ratio (BCR)

$$BCR = \frac{\text{Component Risk(CR)}}{\text{Maintenance Cost(MC)}} \tag{14}$$

The BCR is a number without units, where the higher the score, the higher the priority.

3 Results and Discussion

Figure 4 shows the predicted mission reliability for each subsystem with R(t) at 26,280 h, i.e. after operating for 3 years consecutively from 2018 to 2020. As the graphic suggests, it can be seen that the more the failures experienced by a subsystem, the lower the predicted reliability value for the next 8760 h of mission time even though the overall system had been enduring a mature age of 26,280 h. It is also observed in Fig. 5 that the expected number of failures E[N(t)] was proportionally related to the reliability of each subsystem.

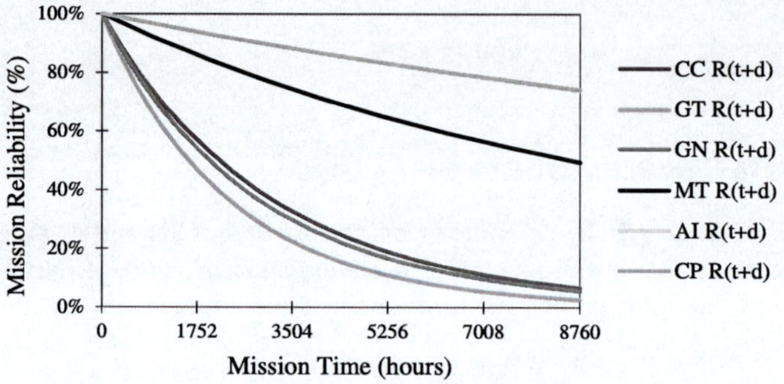

Fig. 4. Graphical results of R(t+d) for all subsystems, respectively

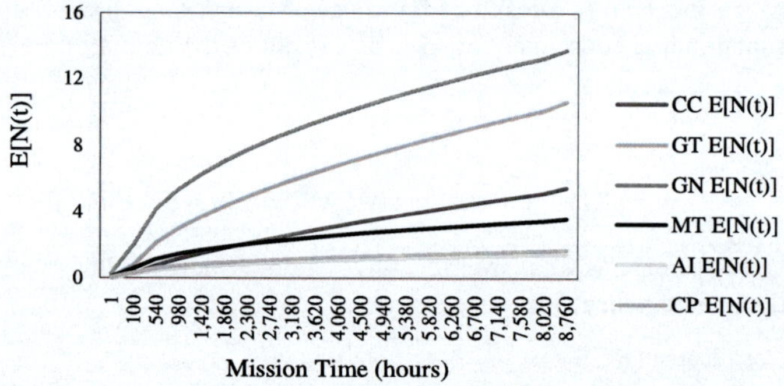

Fig. 5. Graphical results of expected number of failures, E[N(t)]

Table 3 Component Risk (CR) ranking for all subsytem lists the risk ranking generated from lambda and beta combined with the expected number of failure and failure costs, and then Table 4. Budget priority ranking based on BCR lists the priority ranking based on BCR. One can suggest how maintenance cost affects the priority ranking which in some cases the maintenance cost will be able to modify the risk ranking, in other words, the priority ranking will be different from the risk ranking.

Table 3. Component Risk (CR) ranking for all subsytems

Subsystem	$\widehat{\lambda}$	$\widehat{\beta}$	Expected no. of failure(*)	Failure cost (FC)	Component risk (CR)	Risk ranking
Air Intake System (AI)	0.032	0.423	0.739	6,654,128	9,782,671	6
Compressor (CP)	0.054	0.380	0.741	86,134,683	147,003,917	4
Combustion Chamber (CC)	0.009	0.712	0.065	11,578,701,389	63,530,657,892	1
Gas Turbine (GT)	0.059	0.573	0.028	37,262,108	397,029,090	3
Electric Generator (GN)	0.280	0.429	0.056	79,278,873	1,090,049,950	2
Main Transformer (MT)	0.086	0.411	0.493	6,999,902	25,116,528	5

(*) t = 26,280 h is system age after 3 years (2018–2020); d = 8760 h is projected hours for 1 year forward

Table 4. Budget priority ranking based on BCR

Subsystem	Component Risk (CR)	Maintenance Cost (MC)	Benefit to Cost Ratio (BCR)	Priority Ranking
Air Intake System (AI)	9,782,671	5,495,074,586	0.002	6
Compressor (CP)	147,003,917	8,372,113,568	0.018	4
Combustion Chamber (CC)	63,530,657,892	9,163,638,092	6.933	1
Gas Turbine (GT)	397,029,090	22,121,530,859	0.018	3
Electric Generator (GN)	1,090,049,950	15,443,845,993	0.071	2
Main Transformer (MT)	25,116,528	5,981,966,571	0.004	5

4 Conclusions

From this study, a new approach has been developed which combines the equipment reliability prediction of a repairable system with its respective failure and maintenance cost to determine the budget prioritization ranking for the subjected set of equipment. This approach can be convenient to tackle the challenge due to budget cutting for maintaining equipment in a power plant.

In addition, it is also concluded that the more the failures experienced by an equipment in one observation period, the lower the predicted reliability value, but on the other hand the same number of failures but with different *Time to Failure* (TTF) will result in different reliability value predictions.

References

1. Crow LH (1990) Evaluating the reliability of repairable systems. In: Annual proceedings on reliability and maintainability symposium, pp 275–279
2. Wibawa A, Ichsani D, Yuniarto MN (2021) A novel approach of a power plant life cycle management (LCM) combining financial and engineering analysis to determine optimum O&M strategies for a power plant during its lifetime. Int J Technol
3. Nelson W (1988) Graphical analysis of system repair data. J Qual Technol 20:1 24–35
4. Mettas A, Zhao W (2005) Modeling and analysis of repairable systems with general repair. In: Proceedings of the annual reliability and maintainability symposium 176–182
5. Crow LH, An extended reliability growth model for managing and assessing corrective actions. In: proceeding annual symposium reliability and maintainability (2004 – RAMS) 73–80
6. Loll V (1998) From reliability-prediction to a reliability-budgeted. In: Proceeding International symposium on product quality and integrity 421–427
7. Guo R, Asher H, Love E (2014) Generalized models of repairable systems: A survey via stochastic processes formalism. ORiON 16

Analysis of the Reduction of Vibration and Chatter Effect in Boring Process Due to the Addition of *Spring Radial Vibration Damper* (SRVD) on the Workpiece

Wiwiek Hendrowati[1](✉) and Mumtaza Rizky Iswanda[2]

[1] Laboratory of Vibration Engineering and Automotive System, Department of Mechanical Engineering, Institut Teknologi Sepuluh Nopember, Surabaya, Indonesia
wiwiek@me.its.ac.id

[2] Graduate Program of Mechanical Engineering, Institut Teknologi Sepuluh Nopember, Surabaya, Indonesia

Abstract. Modern machining processes have been rapidly evolving into much more sophisticated forms. However, even with such sophistication in hand, the effect of chatter remains to be a significant problem. To date, engineers keep referring to the traditional chatter stability lobe to address the problem of limiting themselves in creativity to achieve high efficiency. The research aims to observe the vibration reduction along the boring process achieved by adding one type of DVA called Spring Radial Vibration Damper (SRVD) onto the workpiece. The workpiece is a cylindrical rod with the ratio of overhang length to a diameter at 6:1. The experiment conducted in different depths of cut (DoC) varies at 0.25 mm, 0.2 mm, and 0.15 mm. The experiment results show a comparison between the main system and without the SRVD in a graphical representation of the dynamic response, percentage of RMS reduction in each parameter, and surface finish of each parameter. This paper concludes that SRVD can be beneficial for the cutting process within the unstable area of the chatter stability lobe. It will worsen the cutting process if the parameters still lie within the stable area.

Keywords: Boring · Chatter · Reduction of vibration · Spring radial vibration damper · Dynamic vibration absorber

1 Introduction

Modern machining processes have been rapidly evolving into their sophisticated forms as we know them today. However, even with such sophistication in hand, the effect of chatter remains significant to this day. The excessive vibration from the chattering effect reduces the effectiveness of the machining process and the durability of the machine [1, 2].

Engineers keep on referring to the traditional chatter stability lobe to address the problem. A combination of particular spindle speed and specific cutting depth is needed to avoid the chatter generation. Several attempts had been made in the past, such as

I. Tolj et al. (eds.), *Recent Advances in Mechanical Engineering*, Lecture Notes in Mechanical Engineering, https://doi.org/10.1007/978-981-19-0867-5_8

using Dynamic Vibration Absorber (DVA) in the boring bar [3–5]. By taking the DVA concept further, one way to reduce the excessive vibration in lathe machines would be using Spring Radial Vibration Damper (SRVD). This study aims to examine the reduction of chatter given by the SRVD through experimental methods. The experiment was conducted in different depths of cuts (DoCs) with variation of 0.25 mm, 0.2 mm, and 0.15 mm. The purpose of observing the experiment result would be to compare the surface finish with the naked eye.

The scope of this study includes: the workpiece used is cylindrical rod ST-41 steel; the workpiece's overhang length to diameter ratio is 6:1; SRVD's weight bending effect can be neglected; the workpiece is assumed to be homogenous; no deflection at the boring bar; lathe machine and the tool insert are assumed to be in good condition; friction between the mass and its pin is neglected; chuck of the lathe machine is assumed to be the system's boundary. The boring bar must be sufficiently stiff to minimize tool deflection and thus maintain dimensional accuracy and avoid vibration and chatter. For this reason, a material with a high elastic modulus (such as tungsten carbide) is desirable [6, 7].

Uncontrolled vibration and chatter of the cutting tools and the machining components have adverse effects on product quality, such as poor surface finish, loss of dimensional accuracy of the workpiece, premature wear, chipping, and failure of the cutting tool [8, 9].

In 2017, Ufuk Yigit, Ender Cigeroglu, and Erhan Budak researched the effect of piezoelectric shunt damping (PSD) on chatter vibrations in a boring process. This study stated that regenerative chatter is considered the criteria for determining cutting stability as it is less stable than mode coupling. The phase difference between previously machined and the new cutting surfaces creates variation in chip thickness [10–12]. A sample stability lobe diagram is given in Fig. 1, which illustrates the unstable (chatter) and stable (chatter-free) regions.

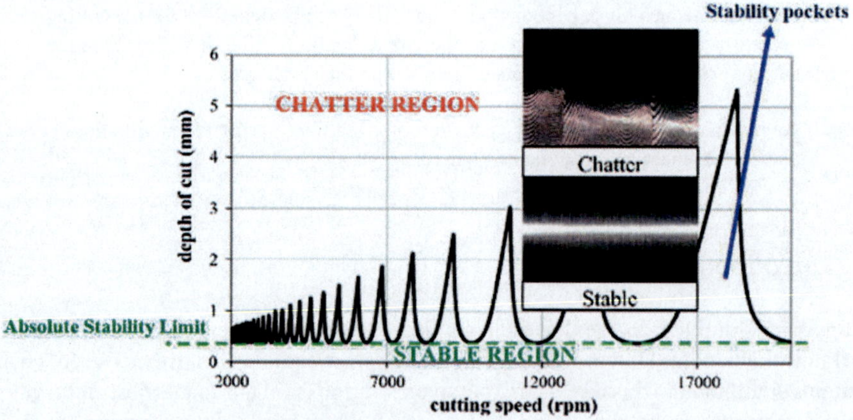

Fig. 1. Stability lobe diagram for chatter problem [10]

2 Methods

2.1 Developing the SRVD Design

The vibration dampers will be designed to overcome the vibrations that arise in the boring process. The input of this research is the design of SRVD and variations in the working frequency, mass of the SRVD, and stiffness constant from the spring of SRVD attached to the workpiece. The output of the experiment is the main system vibration response before and after SRVD installation.

The modelling of the dynamic system in this research consists of two steps: first the modelling of the main system without the SRVD and the other is the modelling of the main system with the SRVD. The physical model embodies the experiment model with cutting force as the input to the main system, as demonstrated in Fig. 2. As for the dynamic models of the system without and with SRVD are shown in Fig. 3.

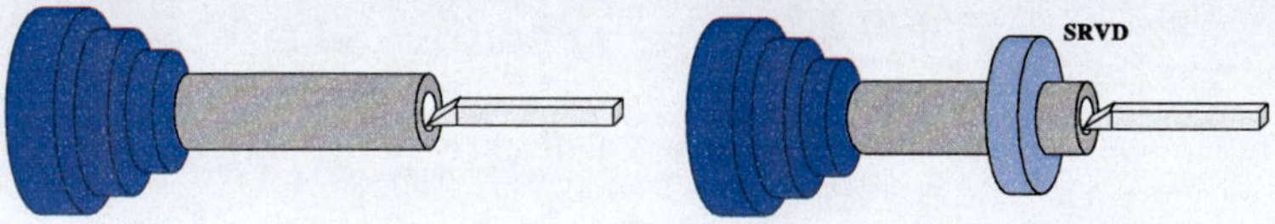

Fig. 2. Physical model of the workpiece with and without the SRVD

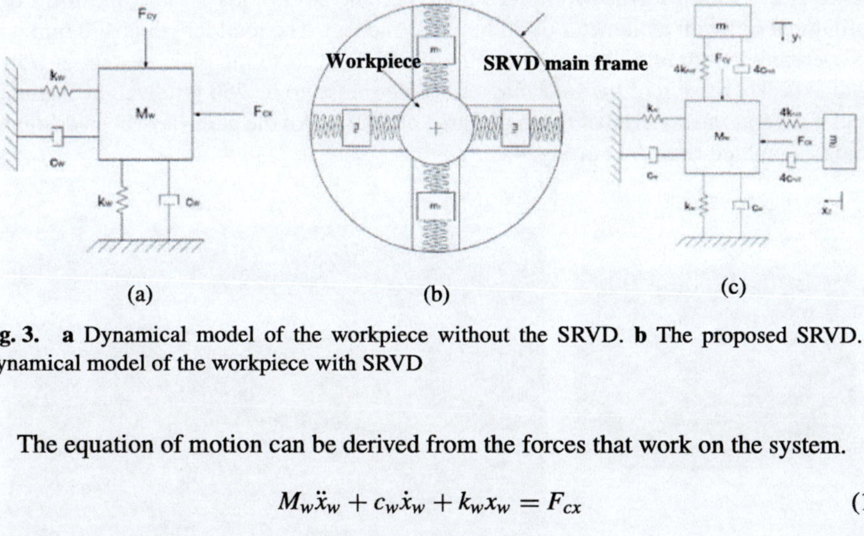

Fig. 3. **a** Dynamical model of the workpiece without the SRVD. **b** The proposed SRVD. **c** Dynamical model of the workpiece with SRVD

The equation of motion can be derived from the forces that work on the system.

$$M_w\ddot{x}_w + c_w\dot{x}_w + k_wx_w = F_{cx} \tag{1}$$

$$M_w\ddot{y}_w + c_w\dot{y}_w + k_wy_w = F_{cy} \tag{2}$$

The vertical and horizontal axis of motion can be determined by the forces that work on the system. Equation 3 and 4 are the equation of motion for the SRVD mass on both vertical and horizontal axis. Meanwhile, the value of all parameters is shown in Table 1.

Table 1. Parameters

Parameter	Value	Units	Notes
M_w	1.7	kg	Mass of workpiece
m_1	0.231	kg	Vertical axis SRVD mass
m_2	0.231	kg	Horizontal axis SRVD mass
k_w	4.55×10^6	N/m	Workpiece stiffness coefficient
k_{rvd}	650.25	N/m	SRVD stiffness coefficient
c_w	0	Ns/m	Workpiece damping coefficient
c_{rvd}	0	Ns/m	SRVD damping coefficient

$$m_1\ddot{y}_1 + 4k_{rvd}(y_1 - y_w) + 4c_{rvd}(\dot{y}_1 - \dot{y}_w) = 0 \tag{3}$$

$$m_2\ddot{x}_2 + 4k_{rvd}(x_1 - x_w) + 4c_{rvd}(\dot{x}_1 - \dot{x}_w) = 0 \tag{4}$$

2.2 Manufacturing of the Workpiece and SRVD

ST-41 steel is used as the workpiece material, and SRVD mass takes the form of a cylindrical rod with a diameter of 38 mm (1.5 inches). The total length is 400 mm with an overhang length of 240 mm. The ratio of SRVD mass to the main system is 1/5 for each axis. Then, each of the four masses is determined to be 250 g in weight. Stainless steel is used as the material of the mainframe of SRVD. All the parts have been designed, being assembled as SRVD in Fig. 4.

(a) (b)

Fig. 4. **a** Manufactured SRVD. **b** Equipment installation during the experiment (1) Spindle and chuck; (2) Workpiece; (3) SRVD; (4) Accelerometer probe; (5) Turret; (6) Boring bar Bode diagram of the main system at L/D ratio

2.3 Data Sampling Instrument and Installation

The experiment is conducted using three different depths of cut, namely 0.25 mm, 0.2 mm, and 0.15 mm. The spindle rotational speed is 720 rpm, and the feed of the tool is 0.1 mm/rev. On each variation, sampling is done using 1 cm of the cutting process.

3 Results and Analysis

3.1 Analysis of Inferential Statistics

The dynamic responses of the main system with and without the SRVD are plotted in an overlaying manner within the same time frame. Judging from the behaviour of the data sample that oscillates around zero in nature, the author decided to use RMS as the comprehensive yet simple statistical tool to analyze the data. It represents the averaged deviation of each amplitude for a given sampling period, which is why it is suitable to determine each parameter's level of "vibration stability". The dynamic response of the main system with and without SRVD are plotted in an overlaying manner within the same time frame as shown in Fig. 5.

The dynamic response of the main system without the SRVD at DoC of 0.25 mm experienced a beating phenomenon that is shown by the large amplitude that builds up and diminishes in a regular pattern. Table 2 represents the vertical RMS value of the workpiece with and without the SRVD and the percentage of reduction of those values after the addition of SRVD. Negative percentage values (highlighted cells) show that the workpiece's dynamic response with the SRVD is greater than that without the SRVD.

All the RMS points of the dynamic response in the vertical and horizontal axis are presented in Table 2 and then plotted to graphs to see whether a trend exists as shown in Fig. 6. This finding shows that, in both vertical and horizontal axes, the addition of SRVD will only effectively reduce vibration that happens at high DoC. In contrast, the addition of SRVD in low DoC will only worsen the vibration of the workpiece (Fig. 6).

3.2 Surface Finish of the Workpiece

This section aims to report the surface finish of the boring process for both with and without the SRVD. Based on the acquired results, the vibration reduction can be seen by comparing the smoothness of the surface finish (Table 3). The smooth surface finish indicates that low vibration had occurred during the cutting process, and the coarse surface finish indicates the high intensity of vibrations or chatter.

4 Conclusion

The data results validate the author's hypothesis that the addition of SRVD does reduce chatter in the boring process. In a detailed manner, this report concludes the following:

1. The addition of SRVD affects the surface finish of the boring process. At DoC 0.2 mm and 0.25 mm, the surface finish from the SRVD-added process exhibit a smoother surface than the surface finish without the SRVD.

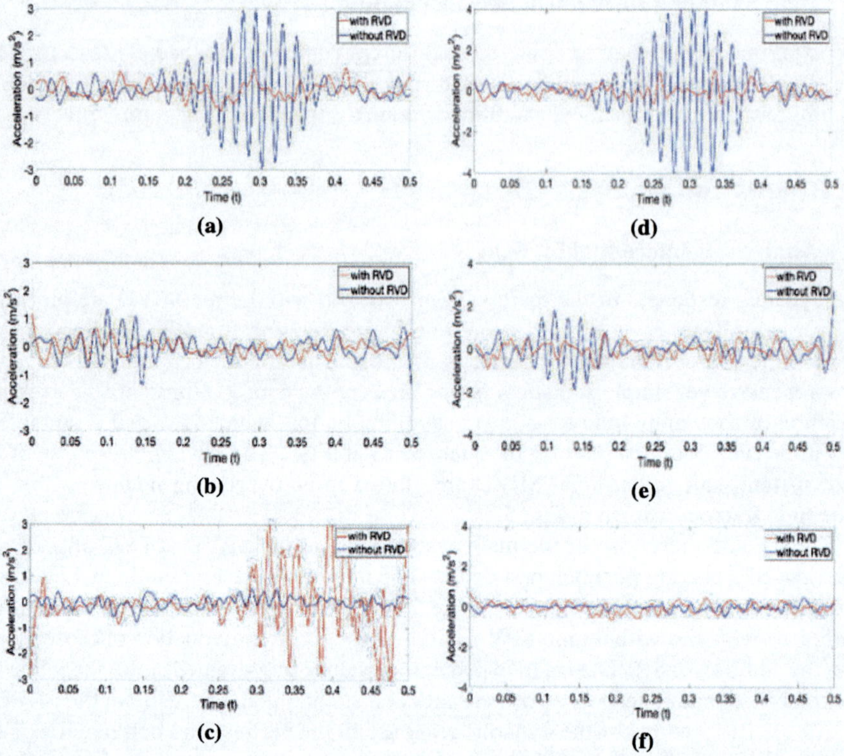

Fig. 5. Dynamic response of the main system with and without the SRVD at the vertical axis **a** DoC of 0.25 mm. **b** DoC of 0.2 mm. **c** DoC of 0.15 mm. Dynamic response of the main system with and without the SRVD at the horizontal axis. **d** DoC of 0.25 mm. **e** DoC of 0.2 mm. **f** DoC of 0.15 mm

Table 2. RMS of the dynamic response

DoC (mm)	RMS of Acceleration (m/s^2)		
	Without SRVD	With SRVD	Percentage of Reduction (%)
		Vertical Axis	
0.25	0.9225	0.287	68.89
0.2	0.3942	0.2455	37.72
0.15	0.2212	1.0633	-380.7
		Horizontal Axis	
0.25	1.3869	0.2667	80.77
0.2	0.5737	0.24	41.73
0.15	0.138	0.2346	-70

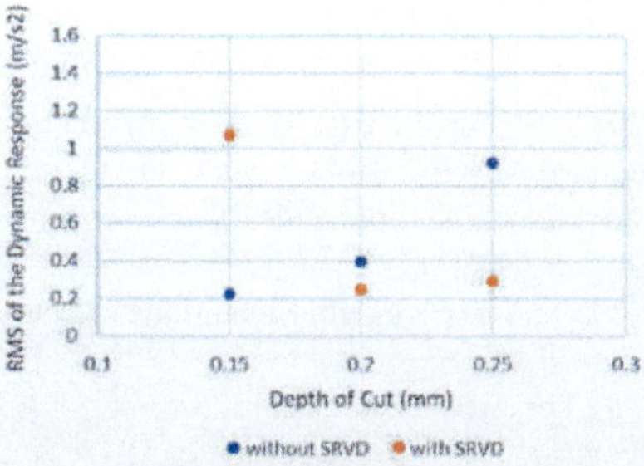

(a)

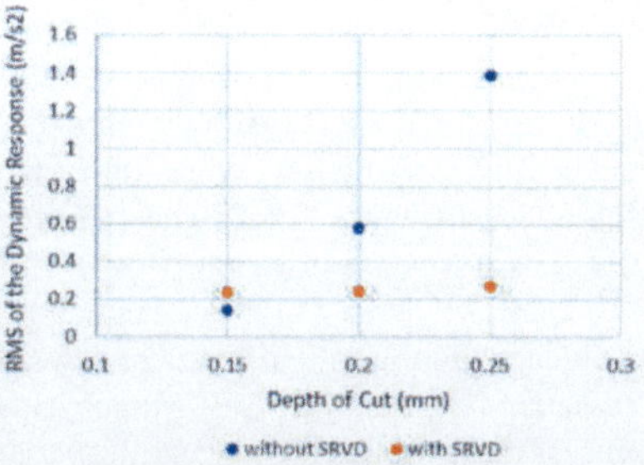

(b)

Fig. 6. RMS of the dynamic response **a** vertical axis, **b** horizontal axis

Table 3. The surface finish of the workpiece at a particular DoC

Without SRVD	With SRVD
DoC of 0.25 mm	
	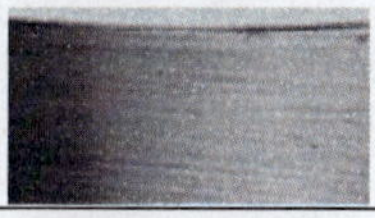
DoC of 0.2 mm	

DoC of 0.1 mm	
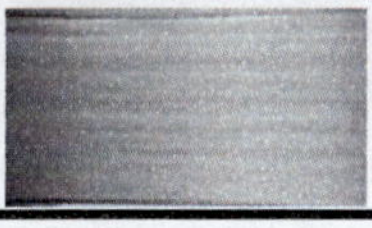	

2. The addition of SRVD reduces the dynamic response of the boring process.
3. The addition of SRVD is observed to reduce vibration only at the unstable area of the chatter stability lobe. The addition of SRVD appears to worsen the vibration of the boring process when it is operated at a stable area of the chatter stability lobe. Further research needs to be done to validate this finding.
4. The results indicate a positive correlation between the depth of cut to the dynamic response at the boring process without the SRVD. The results indicate a negative correlation between the depth of cut to the dynamic response at the boring process with the SRVD.

References

1. Kashyzadeh KR, Ostad-Ahmad-Ghorabi MJ (2012) Study of chatter analysis in turning tool and control methods—a review. Int J Emerg Technol Adv Eng II(4)
2. Tobias SA, Fishwick (1958) Theory of regenerative machine tool chatter. Engineer, Lond, pp 199–238
3. Urbikain G, Olvera D (2019) Prediction methods and experimental techniques for chatter avoidance in turning systems: a review. MDPI
4. Yussefian NZ, Imani BM, El-Mounayri H (2008) The prediction of cutting force for boring process. Elsevier 48:1387–1394
5. Sadílek M, Dubsky J, Sadílková Z, Poruba Z (2015) Cutting forces during turning with variable depth of cut. Elsevier
6. Kalpakjian S, Schmid SR (2009) Manufacturing engineering and technology sixth edition. Pearson Education, Upper Saddle River
7. Isakov E (2019) Boring: calculating deflection
8. Tlusty J, Polacek M (1963) The stability of machine tools against self excited vibrations in machining. Int Res Prod Eng 465–474
9. Badosa GQI (2009) Stability lobes diagram identification and surface roughness monitoring in milling process. Girona
10. Yigit U, Cigeroglu E, Budak E (2017) Chatter reduction in boring process by using piezoelectric shunt damping with experimental verification. Elsevier, pp 312–321
11. Hendrowati W, Guntur HL, Daman AAA, Lestari H (2019) Analytical and experimental study of translational vibration response's reduction on aluminum (Al) drilling process using translational mass vibration absorber (TMVA) system. AIP Conf Proc Surabaya
12. Shankar R, Brahadeeshwaran V (2017) Design optimization of dynamic vibration absorber for boring process. Int J Sci Technol Eng IV(5)

Analysis of Vibration Characteristics Due to the Effect of Adding Rubber Radial Vibration Damper (RRVD) to the Boring Process

Wiwiek Hendrowati[1](✉), Aida Annisa Amin Daman[1], Harus Laksana Guntur[1], and Fuad Adi Darmawan[2]

[1] Laboratory of Vibration Engineering and Automotive System, Department of Mechanical Engineering, Institut Teknologi Sepuluh Nopember, Surabaya 60111, Indonesia
wiwiek@me.its.ac.id

[2] Undergraduate Program of Mechanical Engineering, Institut Teknologi Sepuluh Nopember, Surabaya 60111, Indonesia

Abstract. There will be vibrations caused by friction between the rotating workpiece and the insert on the boring bar during the boring process. Vibrations that occur in the workpiece can be suppressed with a Dynamic Vibration Absorber (DVA). Therefore, the boring process produces a rough surface and damages the components of the lathe. In this research, DVA is placed on a workpiece by utilizing force damping in the radial direction and it is called as Rubber Radial Vibration Damper (RRVD) with rubber as a damping element. An analysis of the effect of RRVD addition on the vibration reduction response of the workpiece was carried out. The variations in this study are mass of RRVD (AISI 4340 and Copper Alloy), cutting parameters, and the ratio of length per diameter (L/D) of the workpiece overhang 7, 8, and 9. The simulation results showed that the main system's highest natural frequency and amplitude occurred at the L/D 7 without RRVD or with RRVD. In addition, the RRVD Copper Alloy mass variation has a lower natural frequency and amplitude than the AISI 4340. Based on this research, it was found that the most stable and well-reduced system was in the main system with L/D 8 and using RRVD Copper Alloy mass.

Keywords: Boring process · Chatter · Vibration reduction · Rubber radial vibration damper · Dynamic vibration absorber

1 Introduction

In the boring process, a cutting tool (insert) is held by a boring bar. During the process, there will be vibrations caused by friction between the rotating workpiece and the insert on the boring bar [1]. If the vibration amplitude that occurs is very high, the system will experience chatter. Furthermore, the resulting product will have a rough surface and damage the lathe components [2, 3]. Therefore, a vibration damper is needed, such as a Dynamic Vibration Absorber (DVA).

I. Tolj et al. (eds.), *Recent Advances in Mechanical Engineering*, Lecture Notes in Mechanical Engineering, https://doi.org/10.1007/978-981-19-0867-5_9

The development of an efficient Dynamic Vibration Absorber (DVA) is still being developed to this day. Research conducted by Ayatullah Faiz Aufa [4, 5] was Vibration reduction analysis using Radial Vibration Damper (RVD). The main system was in the form of a long thin shaft with a diameter of 1 cm and a length of 106 cm. Afterward, the RVD was applied to the main system, and the position of RVD against the main system was varied. Lie [6] analyzed the vibration characteristics in boring bars and various rigidity in rubber bush damping. This study used boring bars with a diameter of 40 mm and a length of 310 mm. The rigidity value of the damping varied by turning the knob that provides axial compression on the rubber bush as the stiffness of the damping.

Therefore, this research analyzed the vibration reduction using Rubber Radial Vibration Damper (RRVD) in the boring process. This analysis is conducted by simulating the related system using ANSYS Harmonic Response. The ratio of the workpiece's overhang length to its diameter (L/D) was varied. Moreover, the variations in this study were the mass of RRVD and cutting parameters such as feed rate, depth of cut, and spindle rotation speed. The expected results in this study are to determine the design and mechanism of RRVD, the effect of RRVD mass variations, and the ratio of overhang length per diameter (L/D) of the workpiece to the reduction of the workpiece vibration response.

2 Methods

This research has modeled and simulated the harmonic response of the main system by adding the RRVD to the workpiece. The position of RVD was based on the overhang length. The detail of the workpiece, the RRVD model, and the arrangement is described below.

2.1 Workpiece Material

Figure 1 shows the modeling of the workpiece (main system) without and with RRVD. The RRVD is placed at the free end of the workpiece. Meanwhile, the workpiece's overhang ratio (L/D) variations are 7, 8, and 9. Modeling and simulation of the main system without and with RRVD was carried out using the Ansys software [7].

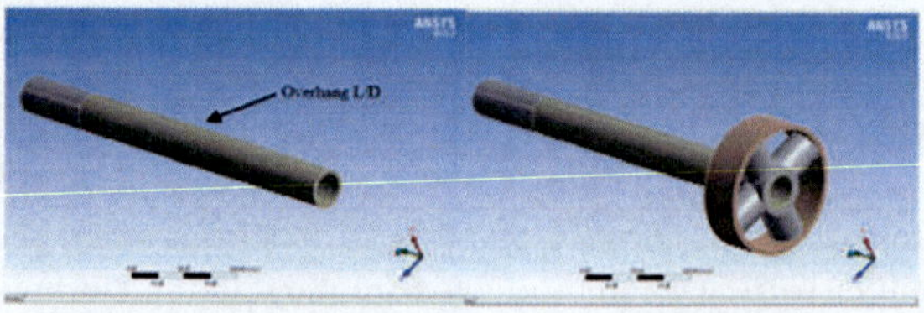

(a) (b)

Fig. 1. Main system at L/D variation of 7 **a** Without RRVD dan **b** With RRVD

Variations in the overhang ratio will cause different workpiece stiffness. The deflection is obtained from static analysis by applying a load of 200 N at the free end of the workpiece. Furthermore, the stiffness of each overhang variation is shown in Fig. 2 and obtained by Eq. (1). From the deflection value (Δx), the workpiece stiffness value (K_w) is obtained as follows:

$$K_w = F \big/ \Delta x \tag{1}$$

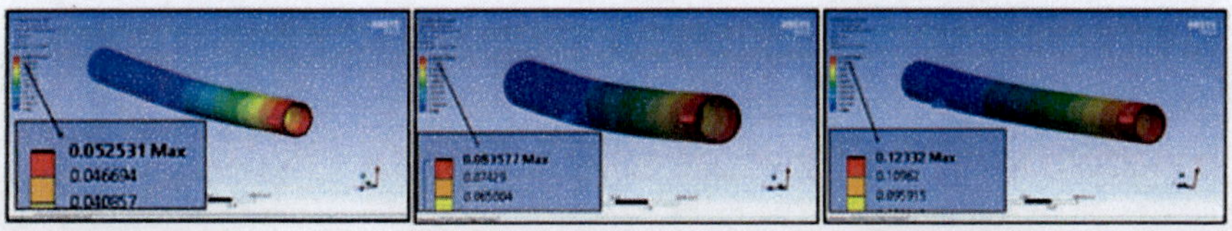

(a) (b) (c)

Fig. 2. Deflection in the workpiece with L/D variations **a** 7, **b** 8, and **c** 9 in the X-axis direction

As the stiffness value of one workpiece differs from other workpieces, the natural frequency of each variation in the overhang ratio is also different. The increase in the overhang ratio (L/D) provides a decrease in its natural frequency. Furthermore, the equivalent mass value of the workpiece is obtained by the following equation:

$$M_w = \frac{K_w}{\omega_n^2} \tag{2}$$

In this study, the damping ratio is obtained from the damping ratio ($\zeta = 0.0045$) of the workpiece in the form of AISI 4340 material. Then the workpiece damping constant value is obtained with the following equation:

$$C_w = 2\zeta\sqrt{M_w K_w} \tag{3}$$

2.2 Design RRVD Mechanism

The RRVD mass materials used in this research are AISI 4340 and Copper Alloy with specific dimensions. The mass of the RRVD used is 1/20 of the mass of the workpiece. The mass of the workpiece is 3.1024 kg, while the mass of the RRVD is 0.155 kg consisting of 2 small blocks.

The type of rubber used is silicon rubber which has a damping ratio of 0.058. The value of rubber stiffness (K_r) can be determined by the following equation:

$$K_r = \frac{M.{\omega_n}^2}{2} \tag{4}$$

The RVD damping constant is obtained from the multiplication of the rubber damping ratio (ζ_r) with the root of the mass multiplied by stiffness constants that have been previously obtained. The equation of the RRVD damping constant value is written as

$$C_{RVD} = 2\zeta_r\sqrt{M.K_r} \tag{5}$$

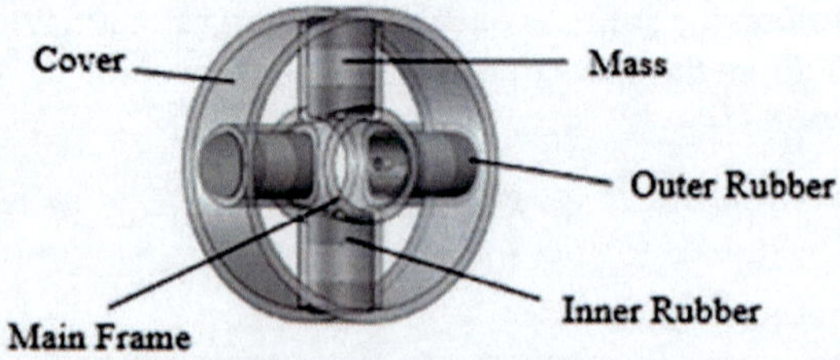

Fig. 3. RRVD mechanism

The RRVD design consists of inner rubber, outer rubber, RRVD mass, mainframe, and cover, as shown in Fig. 3. The mainframe is made of aluminum surrounded by four cylinders. Meanwhile, the outer shell (cover) is made of aluminum with a diameter of 120 mm and a thickness of 5 mm. The inner and outer rubber has a thickness of 11 mm and 13 mm, respectively. Both rubbers have a diameter of 27 mm and are made of silicone. The RRVD mass is cylindrical with 27 mm in diameter and 16 mm in thickness. In contrast, the mass variation comes from different materials in that one is made of AISI 4340 and the other variation is made of Copper Alloy.

2.3 Cutting Parameter

Cutting parameters used in this study are spindle speed (N), feed rate (f), and depth of cut. The variations of spindle rotation (N) are 920 rpm, 1250 rpm, and 2000 rpm. Then, it is used to obtain cutting speed (V). The V value affects the cutting time. Cutting speed (V) variations based on the function of spindle rotation are 95.3304, 129.525, and 207.24 m/min.

The frequency of chatter/operational frequency is obtained by substituting the parameters, i.e., cutting speed (V), feed rate (f), and depth of cut (d) into Eq. (6), as in the following equations [8, 9]:

$$\omega_c = 17.0 - 0.566V + 3971f + 155d \tag{6}$$

The excitation force given to the workpiece is a force in the machining process (F_m), which consists of cutting force (F_c), radial force (F_r) and thrust force (F_t), as shown in Fig. 4. The cutting force (F_c) derives from Eq. (7).

$$F_c = K_c.d.f \tag{7}$$

Cutting force (F_c) is provided from the multiplication between specific cutting force (K_c) with feed rate (f) and depth of cut (d), which is the value of $K_c(f)$ obtained from technical data ISO 13399 standard that is at $f = 0.05$ mm/rev, $K_c = 4275$ MPa. Meanwhile, the cutting force $K_c = 4050$ MPa is provided at feed rate $f = 0.1$ mm/rev and $K_c = 3825$ MPa is at $f = 0.15$ mm/rev. Radial force (F_r) and thrust force (F_t) are obtained from a journal as written on Eqs. (8) and (9), respectively. The magnitude of each cutting parameters is presented in Table 1.

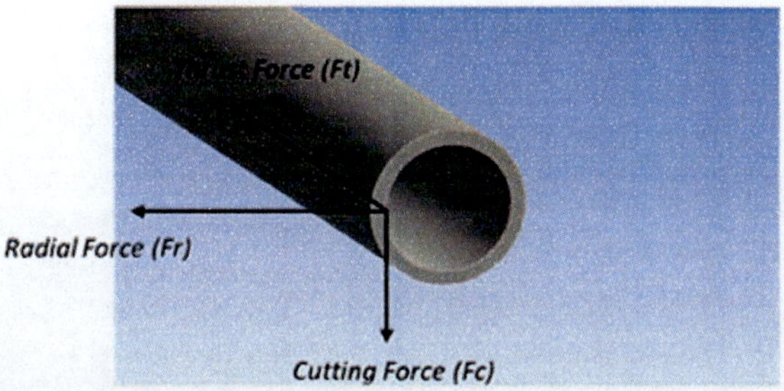

Fig. 4. Forces applied to the boring process

$$F_r = 0.69F_c \tag{8}$$

$$F_t = 0.38F_c \tag{9}$$

After building the workpiece and RRVD model, the next step is to work on each part's material properties data, as shown in Table 2. After that, the meshing is chosen according to the workpiece's dimension, and then the system is simulated using static structural analysis, as presented in Fig. 5. The output is a graph of the main system's total deformation and harmonic response without RRVD and with RRVD.

After that, the system's boundary is set as well as the nodal force. Furthermore, the harmonic response is simulated based on parameter variations, namely the L/D overhang ratio and the RRVD mass.

3 Results and Analysis

The simulation is carried out using the ANSYS Workbench model and Harmonic Response software. The output of the simulation was the dynamic deformation response of the workpiece in the direction of the *y-axis* (Radial) and *z-axis* (Tangential). The deformation data were taken at the frequency of 1–1000 Hz on each variation.

3.1 System Vibration Response on X-Axis and Z-Axis

The main system motion is analyzed with L/D variation of overhang 7, 8, and 9 on the x-axis and z-axis at frequencies 0–650 Hz. The mass of RRVD is made of AISI 4340 or copper alloy. Figure 6 shows the effect of adding RRVD to the workpiece according to L/D ratio variations on the x-axis and the z-axis. The response results for each variation can be seen in Table 3.

Figure 6 and Table 3 show that the addition of RRVD can reduce vibration in the main system. In general, there is no significant difference when using the different materials of DVA mass on the x and z-axis. For workpieces with an overhang ratio of L/D 7, it is shown that the main system has the same vibration response when RRVD mass material

Table 1. Cutting parameter

Variation	N (rpm)	V (m/min)	f (mm/rev)	d (mm)	Kc (MPa)	ω_C (rad/s)	Fc (N)	Fr (N)	Ft (N)
A1	920	92.489	0.05	0.1	4275	178.702	21.38	14.75	8.12
A2	920	92.489	0.05	0.2	4275	194.202	42.75	29.50	16.25
A3	920	92.489	0.05	0.3	4275	209.702	64.13	44.25	24.37
A4	1250	125.664	0.05	0.1	4275	159.924	21.38	14.75	8.12
A5	1250	125.664	0.05	0.2	4275	175.424	42.75	29.50	16.25
A6	1250	125.664	0.05	0.3	4275	190.924	64.13	44.25	24.37
A7	2000	201.062	0.05	0.1	4275	117.249	21.38	14.75	8.12
A8	2000	201.062	0.05	0.2	4275	132.749	42.75	29.50	16.25
A9	2000	201.062	0.05	0.3	4275	148.249	64.13	44.25	24.37
B1	920	92.489	0.1	0.1	4050	377.252	40.50	27.95	15.39
B2	920	92.489	0.1	0.2	4050	392.752	81.00	55.89	30.78
B3	920	92.489	0.1	0.3	4050	408.252	121.50	83.84	46.17
B4	1250	125.664	0.1	0.1	4050	358.474	40.50	27.95	15.39
B5	1250	125.664	0.1	0.2	4050	373.974	81.00	55.89	30.78
B6	1250	125.664	0.1	0.3	4050	389.474	121.50	83.84	46.17
B7	2000	201.062	0.1	0.1	4050	315.799	40.50	27.95	15.39
B8	2000	201.062	0.1	0.2	4050	331.299	81.00	55.89	30.78
B9	2000	201.062	0.1	0.3	4050	346.799	121.50	83.84	46.17
C1	920	92.489	0.15	0.1	3825	575.802	57.375	39.589	21.803
C2	920	92.489	0.15	0.2	3825	591.302	114.750	79.178	43.605
C3	920	92.489	0.15	0.3	3825	606,802	172.125	118.766	65.408
C4	1250	125.664	0.15	0.1	3825	557,024	57.375	39.589	21.803
C5	1250	125.664	0.15	0.2	3825	572,524	114.750	79.178	43.605
C6	1250	125.664	0.15	0.3	3825	588,024	172.125	118.766	65.408
C7	2000	201.062	0.15	0.1	3825	514,349	57.375	39.589	21.803
C8	2000	201.062	0.15	0.2	3825	529,849	114.750	79.178	43.605
C9	2000	201.062	0.15	0.3	3825	545,349	172.125	118.766	65.408

Table 2. Engineering data

Mechanical properties	Workpiece	RVD body	Rubber	RVD mass	
Material	AISI 4340	Aluminum	Silicon rubber	Copper alloy	AISI 4340
Density (g/cm^3)	7.83	2.69	1.14	8.3	7.83
Young modulus (GPa)	200	68.3	0.05	110	200
Poissons ratio	0,3	0.34	0.49	0.34	0.3
Bulk modulus (GPa)	167	71.1	0.83	114.5	167
Shear modulus (Gpa)	76.9	25.5	0.17	41.04	76.9

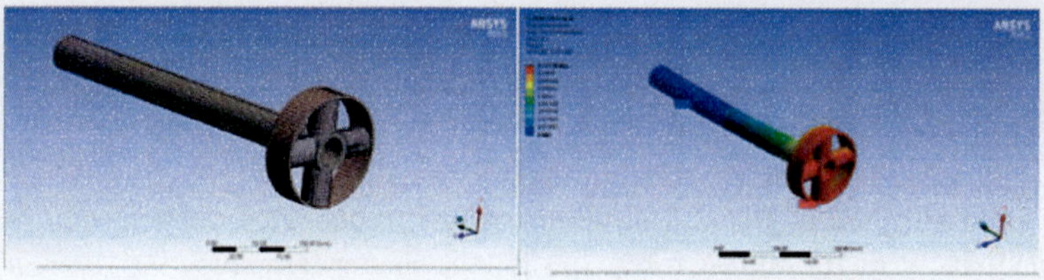

(a) (b)

Fig. 5. Model of the workpiece with RRVD **a** Meshing model **b** Deflection of the workpiece

Table 3. Vibration response in the main system without RRVD and with RRVD

Variation L/D	Axis	Without RRVD		AISI 4340		Copper alloy	
		Frequency (Hz)	Amplitude ratio	Frequency (Hz)	Amplitude ratio	Frequency (Hz)	Amplitude ratio
7	x	463	107,49	271	100,10	269	99,76
	z		110,57		103,14		103,23
8	x	343	105,21	213	90,54	215	101,87
	z		106,51		94,80		100,24
9	x	270	108,89	213	63,74	215	71,25
	z		105,86		65,40		68,99

is made of AISI 4340 or copper alloy. However, in workpieces with an L/D overhang ratio of 8 and 9, the addition of RRVD with a mass made of AISI 4340 material is able to reduce main system vibration more than the copper alloys do.

3.2 The Effect of Adding Variation of Cutting Parameters on Stability

In the boring process, there are various kinds of cutting parameters that can affect the quality of the boring results, as described previously. This parameter will affect the

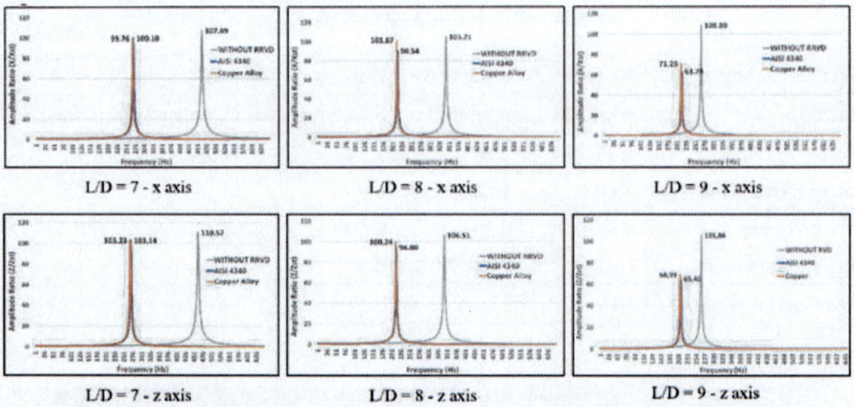

Fig. 6. Bode diagram of the main system at x-axis **a** L/D 7, **b** L/D 8, and **c** L/D 9 and at y-axis **d** L/D 7, **e** L/D 8, and **f** L/D 9

amount of cutting force (F_c) and different chatter frequencies (ω_c). In this study, the relationship between the cutting force (F_c) and the chatter frequency (ω_c) were analyzed along with the results of motion response and stability.

There are 27 cutting parameters of the boring process in each workpiece overhang ratio (L/D), where these parameters influence the cutting force and chatter frequency, as shown in Table 1. Stability is the benchmark for a system to work stably so that the system is not only reduced, but also safe to use. A limit is needed to set a stable system or not by using a stability limit. In Boring Bar with an overhang ratio (L/D) of 4, the main system is steady when the material removal rate (MRR) is 25 cm^2/min. Therefore, the simulation was carried out on the Boring Bar so that the x-axis response was 0.0087961 mm and the z-axis response was 0.58348 mm. This value is used in this study as the limit for the system stability.

There are only 13 among 27 parameter variations that have responses below the stability limit given above. The stable variations and the system responses are written in Table 4. The deflection of the main system with RRVD in 13 parameter variations is lower than the deflection of the workpiece without RRVD. It can be seen in Table 4 that the reduction in the x-axis and y-axis is above 0. The reduction value is varied in each variation.

Furthermore, the lowest reduction, 19.79%, is found at the operational frequency of 132.75 Hz with L/D 9 and RRVD mass made of AISI 4340. Meanwhile, the most significant reduction value occurs at the working frequency of 346,798 Hz with L/D of 8 and RRVD mass made of Copper Alloy, where this variation has a reduction value of 98.77%. Based on the stability analysis results, it can be seen that the main system with overhang ratio (L/D) variations of 7 and 8 tends to experience a significant decrease. In contrast, the deflection of the main system with an overhang ratio of L/D 9 reduces slightly. The workpiece with L/D 9 has a smaller stiffness constant than other overhang ratio variations. Consequently, the workpiece deflection is large and the RRVD addition could not optimally reduce the deflection. In addition, Table 4 shows that the lowest

Table 4. The cutting parameters to produce a stable system response

Variation name	Response								x-axis reduction (%)	y-axis reduction (%)
	Stable Main System					Lowest Response				
	Overhang ratio (L/D)			RRVD Mass Material		(Overhang, Mass Material)	x-axis (mm)	z-axis (mm)		
	7	8	9	AISI 4340	Copper Alloy					
A7	-	-	✓	✓	✓	L/D 9, AISI 4340	0.016	0.007	24.54	25.26
A8	-	-	✓	✓	✓	L/D 9, AISI 4340	0.018	0.008	19.79	20.30
B1	-	✓	-	✓	✓	L/D 8, Copper Alloy	0.014	0.007	91.62	91.26
B2	✓	✓	-	✓	✓	L/D 7, Copper Alloy	0.016	0.008	78.58	76.98
B3	✓	✓	-	✓	✓	L/D 8, Copper Alloy	0.011	0.005	86.76	86.07
B4	-	✓	-	✓	✓	L/D 8, Copper Alloy	0.017	0.008	95.80	95.65
B5	-	✓	-	✓	✓	L/D 8, Copper Alloy	0.014	0.007	91.40	91.05
B6	✓	✓	-	✓	✓	L/D 7, Copper Alloy	0.017	0.008	76.53	74.82
B9	-	✓	-	-	✓	L/D 8, Copper Alloy	0.018	0.008	98.77	98.74
C4	✓	-	-	✓	✓	L/D 7, Copper Alloy	0.005	0.003	89.64	87.29
C7	✓	-	-	✓	✓	L/D 7, Copper Alloy	0.006	0.003	93.01	91.85
C8	✓	-	-	✓	✓	L/D 7, Copper Alloy	0.006	0.003	91.55	89.97
C9	✓	-	-	✓	✓	L/D 7, Copper Alloy	0.005	0.003	90.41	88.42

response is mostly found in the mass variation of RRVD Copper Alloy, which means that a mass with a higher density and damping ratio will absorb vibrations better.

4 Conclusion

Based on the research that has been conducted, it can be concluded that:

1. The variation of RRVD density and overhang ratio (L/D) significantly affects the boring process. In each L/D variation, the magnitude of the natural frequency and the vibration amplitude of the main system without RRVD is greater than the main system with RRVD.
2. The addition of RRVD can reduce the magnitude of the vibration amplitude, especially with the RRVD mass made of AISI 4340 can optimally reduce the magnitude of the vibration amplitude in the system.
3. The reduction percentage is varied in each variation. However, the lowest reduction, 19.79%, occurred in the addition of RRVD, in which the mass is made of AISI 4430. Meanwhile, the RRVD mass made of Copper Alloy provides the highest reduction percentage, accounting for 98.77% on the x-axis and 98.74% on the z-axis.

References

1. Kalpakjian S, Schmid SR (2009) Manufacturing engineering and technology, 6th edn. Pearson Education, Upper Saddle River
2. Kashyzadeh KR, Ostad-Ahmad-Ghorabi MJ (2012) Study of chatter analysis in turning tool and control methods—a review. Int J Emerg Technol Adv Eng II(4)
3. Urbikain G, Olvera D, López de Lacalle LN, Beranoagirre A, Elías-Zuñiga A (2019) Prediction methods and experimental techniques for chatter avoidance in turning systems: A Review. Appl Sci 9(21):4718. https://doi.org/10.3390/app9214718
4. Aufa AF (2019) Design and analysis of the vibration reduction responses to the radial vibration damper (RVD) position variation on a long thin shaft
5. Hendrowati W, Guntur HL, Daman AAA, Lestari H (2019) Analytical and experimental study of translational vibration response's reduction on aluminum (Al) drilling process using translational mass vibration absorber (TMVA) system. In: AIP conference proceeding. Surabaya
6. Lie L, Sun B, Hua H (2019) Analysis of the vibration characteristics of a boring bar with a variable stiffness dynamic vibration absorber
7. Chaphalkar SP, Khetre SN, Meshram AM (2015) Modal analysis of cantilever beam structure using finite element analysis and experimental analysis
8. Ugochukwu O (2017) Evaluation of chatter vibration frequency in CNC turning of 4340 alloy steel material
9. Yussefian NZ, Imani BM, El-Mounayri H (2008) The Prediction of cutting force for boring process, vol 48

Ball Screw Regenerative Shock Absorber Design and Road Test for Small Class Multi Purposes Vehicle Suspension

Harus Laksana Guntur(✉) and Dhion Khairul Nugraha

Laboratory of Vibration Engineering and Automotive System, Department of Mechanical Engineering, Institut Teknologi Sepuluh Nopember, Surabaya 60111, Indonesia
haruslg@me.its.ac.id

1 Introduction

The use of energy losses in vehicle has been investigated and designed by the researcher over the decades. The goal of the research is to improve fuel efficiency. There are some mechanisms to harvest the energy losses from the vehicle such as recoveries of heat waste energy [1], regenerative brake systems using the kinetic energy recovery method [2], and vibration energy recovery system [3]. The vibration energy recovery system mainly recovers the energy from vehicle suspension and the mechanism is named as Regenerative Shock Absorber (RESA). The potential energy harvesting from vehicle suspension has been investigated by Abdelkareem [4]. Up to 420 W of vibration losses, energy can be recovered with standard driving style. Zhang [5] predicted that the potential fuel improvement efficiency is 2–3% for typical vehicle passenger, 3–5% for heavy-duty vehicle, and more than 6% for military vehicle. The improvement in fuel efficiency is even higher for an electric vehicle, up to 7–10%.

A regenerative vehicle suspension using a combination of 16-slots and 16-poles is modeled by Kim [6] and the experiment works of a high-performance 16-Slots and 8-Poles magnetic shock absorber are reported [7]. It can provide average power of 83.1 W and improve power by 59.7%. The harvesting conversion efficiency is the highest among other RESA designs. However, the damping density for the linear magnet is low and it is not appropriate for the vehicle [8]. In order to increase the damping, Singh [9] added the fluid damper combined with a linear magnet.

Zhang [10] made a prototype RESA using rack and pinion transmission. The prototype gain average power of 4.302 W and average mechanical efficiency of 44.21% in a benched test. Liu [11] designed RESA using a ball screw and conducted the road test in a pickup truck. In order to gain more efficiency, Liu used transmission gear to rectify the generator rotation. The average output power was 13.3 W at 64.4 km/h on a paved road. Nugraha [12] made a simulation of RESA using ball screw for various types of vehicles. Salman [13] designed RESA using helical gear as a transmission mechanism to convert translational movement into rotational motion. The result estimated 270 W in a full-scale prototype. The RESA design has a hydraulic tube, gear transmission, hydraulic motor, and generator.

I. Tolj et al. (eds.), *Recent Advances in Mechanical Engineering*, Lecture Notes in Mechanical Engineering, https://doi.org/10.1007/978-981-19-0867-5_10

The performance of the regenerative absorber is simulated, analyzed, and reported by Guntur [14] by varying the hydraulic and gear parameter. The road test was also conducted in a pickup truck. The impact of the battery charge with the damping force and the characteristic of the hydraulic regenerative shock absorber is analyzed. Further, Guntur [15] designed hydraulic RESA for a specific heavy-duty vehicle. The efficiency gain for the design is around 30–40%. The hydraulic RESA has the closest characteristic and performance to a conventional shock absorber because of the viscous fluid damping. However, the hydraulic RESA has a higher loss in hydraulic system, oil leakage, and higher manufacturing cost.

Recently, Wang [16] designed a prototype of RESA for range extended electric vehicles. The design uses twin ball screws and gear transmission to increase generator speed. The twin ball screws are used to convert reciprocating vibration input into unidirectional rotation. The prototype tested with sinusoidal input generated an average power of 3.701 W in 1 Hz and 3 mm vibration input. The average efficiency reported in the bench test was 36.4%. However, Wang did not conduct the road test. The use of the ball screw to harvest energy has been widely used in many applications. The ball screw provides high mechanical efficiency and high positional accuracy [17].

We present a mechanical-based RESA design using ball screw and rotation motion rectifier. The road test was conducted in a pickup truck. The RESA was modeled and simulated to know the theoretical power obtained and compare the result with the road test. To know the effect of road type and vehicle speed in power harvesting, there are several road types with different speeds and traffic conditions. The acceleration of the pickup chassis was measured to know the RESA performance and compare it with a conventional shock absorber. The result and discussion will be presented in the paper.

2 Methods

2.1 Design and Working Principle

The RESA design consists of mechanical and electrical components. The mechanical components are the ball screw, shaft, bearing, and gear, while the electrical component is the generator, as seen in Fig. 1a. The ball screw is used to convert translational movement from vehicle suspension into rotational motion. A bevel gear is assembled as a rotational motion rectifier. Ball screw and bevel gear 1 will rotate bevel gear 2 and bevel gear 3. When the RESA move upward, bevel gear 2 rotate and bevel gear 3 loosing. When RESA move downward, bevel gear 3 rotate and bevel gear 2 loosing. When RESA move downward, bevel gear 3 will rotate the shaft and bevel gear 3 is loosing. The output shaft will be in the unidirectional rotation. There is a spur gear pair that is used to connect the unidirectional rotation output to the generator. The ratio of the spur gear is varied. The generator used for the RESA design is 3 phase generator.

2.2 RESA Mathematical Model

Figure 1a shows the free body diagram of the RESA. To get the damping force equation, each body of gear is derived from Newton's second law of motion. The ball screw torque

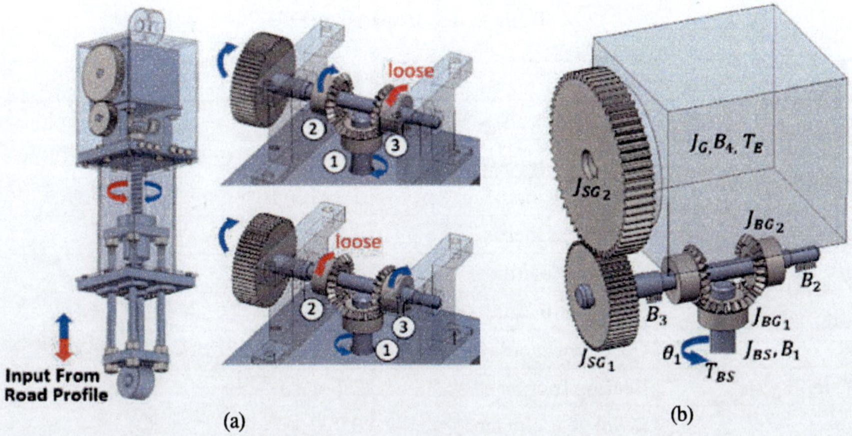

Fig. 1. **a** RESA Design Principle and **b** Free Body Diagram for RESA

is shown in Eq. (1)

$$T_{BS} = J_{BS}\ddot{\theta}_1 + J_{BG_1}\ddot{\theta}_1 + B_1\dot{\theta}_1 + N_1^2 J_{BG_2}\ddot{\theta}_1 + N_1^2 B_2\dot{\theta}_1 + N_1^2 J_{SG_1}\ddot{\theta}_1 + N_1^2 B_3\dot{\theta}_1 + J_{SG_2}N_1^2N_2^2\ddot{\theta}_1 + N_1^2N_2^2 B_4\dot{\theta}_1 + N_1^2N_2^2 J_g\ddot{\theta}_1 + N_1^2N_2^2 T_E \tag{1}$$

J is inertia for each component, B is the bearing friction, N_1 and N_2 are gear ratio for bevel gear and spur gear, respectively, and T_E is the electrical torque produced by the generator. θ_1 is the angular movement of the ball screw. The spur gear ratio is varied from 0.4; 0.6; and 1.5 to know the characteristic of the RESA and appropriate with a specific targeting vehicle. The electrical torque and the damping force of the RESA are shown in Eqs. (2) and (3), respectively.

$$T_E = \frac{1.5}{\eta_{BG}\eta_{SG}}\frac{k_t k_e}{R_{in} + R_{ex}}\dot{\theta}_g \tag{2}$$

$$F_D = \frac{4\pi^2}{\eta_{BS} l^2}\left(\left(J_{BS} + J_{BG_1} + N_1^2 J_{BG_2} + N_1^2 J_{SG_1} + J_{SG_2}N_1^2N_2^2 + N_1^2N_2^2 J_g\right)\ddot{x} + \left(B_1 + N_1^2 B_2 + N_1^2 B_3 + N_1^2N_2^2 B_4 + \frac{1.5}{\eta_{BG}\eta_{SG}} N_1^2N_2^2 \frac{k_t k_e}{R_{in} + R_{ex}}\right)\dot{x}\right) \tag{3}$$

where x is the linear movement of the RESA, l is the lead screw of the ball screw,F_a is thrust force of ball screw, η_{BS} is the efficiency of the ball screw,B_G is damping coefficient of generator, and $\dot{\theta}_G$ is generator angular velocity. The generator angular velocity is determined from Eq. (2). The damping coefficient is linear damping (B_L), while the generator damping is rotary damping (B_R). η_{BG} and η_{SG} are bevel gear and spur gear efficiency, respectively. k_t is the torque constant, k_e is voltage constant of the generator, R_{in} is internal resistance, and R_{ex} is external resistance. Table 1 shows the shock absorber parameters.

Table 1. RESA parameter

Notation	Parameter	Value	Unit
l	Ball screw lead	5	mm/rev
η_{BS}	Ball screw efficiency	0.92	–
J_{BS}	Ball screw inertia	1.325×10^{-6}	$kg.m^2$
J_{BG_1}, J_{BG_2}	Bevel gear inertia	7.221×10^{-7}	$kg.m^2$
J_{SG_1}	Spur gear inertia	1.599×10^{-5}	$kg.m^2$
J_{SG_2}	Spur gear inertia	9.734×10^{-5}	$kg.m^2$
J_g	Generator inertia	1.8×10^{-5}	$kg.m^2$
B_1, B_2, B_3, B_4	Bearing friction	2.1×10^{-4}	N.m.s/rad
η_{bg}	Bevel gear efficiency	0.9	–
η_{sg}	Spur gear efficiency	0.9	–
N_1	Bevel gear ratio	1	–
N_2	Spur gear ratio	0.4; 0.6; 1.5	–
k_t	Torque constant	0.1	N.m/A
k_e	Voltage constant	0.1	V.s/rad
R_{in}	Internal resistance	0.5	ohm
R_{ex}	External resistance	6	ohm

The generator used in this RESA is 3 phase generator and the electrical circuit for 3 phase generator is seen in Fig. 2. The voltage obtained from 3 phase generator is shown below

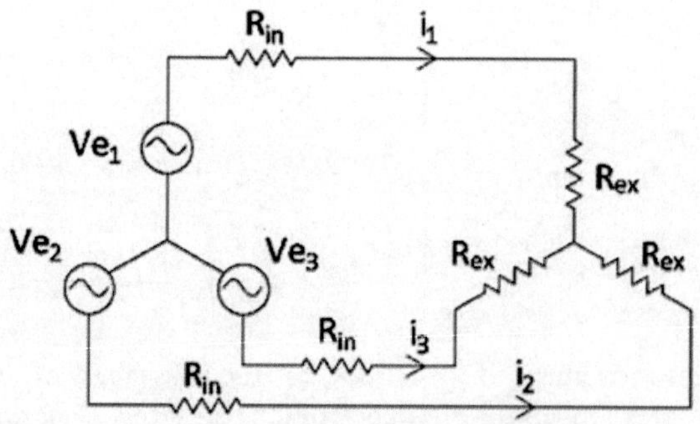

Fig. 2. Electric diagram for RESA

$$Ve_1 = E_m cos(\omega_e t - 120) \quad (4)$$

$$Ve_2 = E_m cos(\omega_e t) \tag{5}$$

$$Ve_3 = E_m cos(\omega_e t + 120) \tag{6}$$

where E_m is the induced voltage amplitude and ω_e is the angular frequency from the voltage by the generator. The total power acquired from these voltages is given in Eq. (7)

$$P = \frac{Ve_1^2}{R_{in} + R_{ex}} + \frac{Ve_2^2}{R_{in} + R_{ex}} + \frac{Ve_3^2}{R_{in} + R_{ex}} \tag{7}$$

Substitute Eqs. (4), (5), (6) into Eq. (7) yield the power generated by RESA, seen in Eq. (8).

$$P = 1.5\frac{E_m^2}{R_{in} + R_{ex}} \tag{8}$$

2.3 Quarter Car with RESA Mathematical Model

The quarter car model with RESA is simulated to investigate the potential generated electric power acquired by RESA when the vehicle passes the speed bump and the speed trap. The result is compared to the road test. The quarter car model is shown in Fig. 3. Table 2 shows the quarter car model parameters. The equation of motion of the quarter car model is shown in Eqs. (9) and (10)

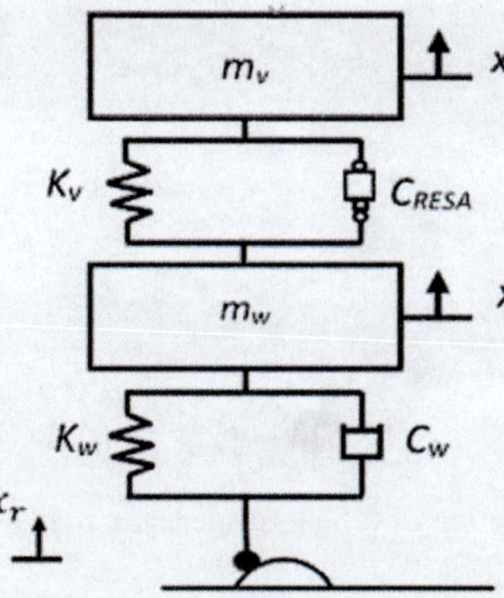

Fig. 3. Quarter car model of vehicle

$$m_v\ddot{x}_v + k_v(x_v - x_w) + F_D = 0 \tag{9}$$

$$m_w\ddot{x}_w + k_w(x_w - x_r) + C_w(\dot{x}_w - \dot{x}_r) - k_v(x_v - x_w) - F_D = 0 \tag{10}$$

Table 2. Quarter car model

Notation	Parameter	Value	Unit
m_v	¼ vehicle mass	505	kg
k_v	Vehicle spring constant	43,646	N/m
m_w	Wheel mass	31	Kg
k_w	Wheel stiffness	317,000	N/m
c_w	Damping wheel	3430	N.s/m

2.4 RESA Prototype and Road Test

The RESA prototype was installed in the rear suspension of a typical small class pickup truck. Figure 4 shows the installation of RESA, road bump, and speed trap. The pickup truck road test was carried out on a bumping road, speed trap, road class A, and road class C. The pickup truck was also tested at different vehicle velocities: 10, 15, and 20 km/h. The output voltage and current from RESA was measured. The vehicle body acceleration was also measured.

Fig. 4. RESA installation in vehicle suspension, road bump and speed trap

3 Results and Discussions

3.1 Damping Force Characteristics

Figure 5 shows the damping force characteristics of RESA (a) Force–Displacement and (b) Force–Velocity with a specific spur gear ratio of 0.4 and 1.5. The input of the simulation is sinusoidal input with various excitation, 1–3 Hz. The Force–Displacement indicates the potential harvested energy and dissipating energy of the RESA. The larger closed loop area in Fig. 6 shows larger potential harvested energy.

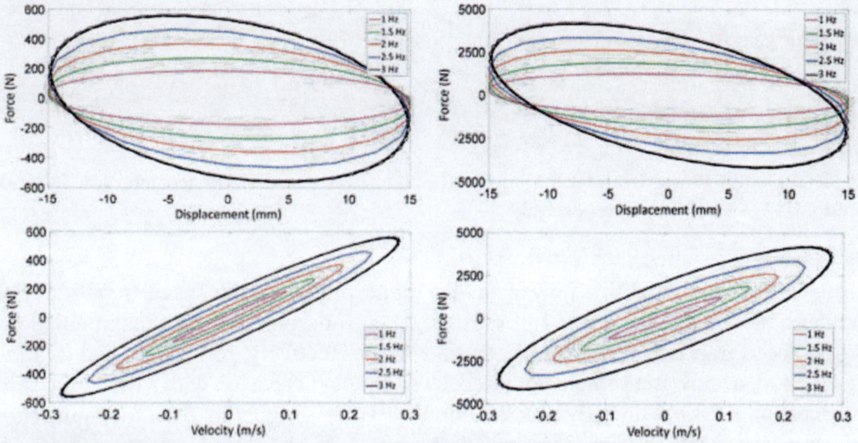

Fig. 5. Damping Force Characteristic **a** Force–Displacement and **b** Force–Velocity for spur gear ratio 0.4 and 1.5

As shown in Fig. 5, increasing frequency and gear ratio increases the potential harvested energy. Both frequency and spur gear ratio will lead to higher generator speed. As a result, a higher generator speed will increase the inertia effect. The inertia effect can be seen from the gradient of the closed-loop line. The Force–Velocity indicates the performance of the shock absorber. The damping coefficient value can be estimated by performing the gradient value of the closed-loop line. The results of the damping coefficient value are 1,901 and 12,770 N.s/m, respectively, for spur gear ratio 0.4 and 1.5. Those values are appropriate for city cars, pickup trucks, and heavy-duty trucks. The inertia effect from the Force–Velocity can be seen from the hysteresis of the loop line. With a higher spur gear ratio and frequency, the hysteresis is higher. Despite having different inertia effects, the damping coefficient value remains constant as indicated by the gradient loop line.

3.2 Road Test Results

Figure 6a shows the power generated from RESA passing the speed trap. The power scheme was quite different from speed bumps. There was repetitive power generated

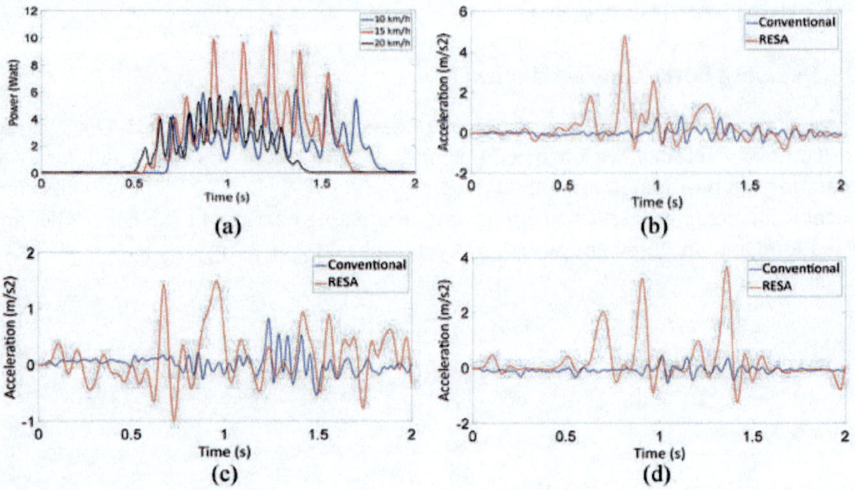

Fig. 6. **a** Output Power from RESA on speed bumb, **b**, **c**, and **d** Body accelerartion at various speed 10, 15, and 20 km/h, respectively

from RESA. This repetitive power has the same profile as the speed trap, where the number of the bump was seven. It is quite constant for around 1 s. The maximum power for the speed trap is 6.1 W, 10.5 W, and 5.8 W, respectively, for 10, 15, and 20 km/h. Figure 6b, c, d show the vehicle body acceleration for suspension with a hydraulic shock absorber and RESA. The body acceleration for RESA is higher than the hydraulic shock absorber. The RMS acceleration for RESA were 1.25, 1.57, and 1.45 m/s^2, while for hydraulic shock absorber, the RMS acceleration were 0.21, 0.32, and 0.31 m/s^2. The difference was high, around 5 times higher. For the hydraulic shock absorber, the result is relatively constant for different velocities. It can be seen that hydraulic shock absorbers perform much better to absorb vibration and are more stable.

Figure 7a shows the generated power from RESA when the vehicle passes the road class A and C at 10, 20, and 30 km/h. The output power from RESA for this road type is random, depending on the road surface. Since road class A is the finest road, the generated power from RSA is small. Maximum power is only 0.24 W for 30 km/h. While road class C is rougher than road class A, as a result, the power harvesting from RESA is higher. It can be seen from Fig. 7b. The maximum power is much higher than road class A. It can reach 13.8 W at 30 km/h and 60 times higher than road class A. The vehicle body acceleration at a vehicle speed of 30 km/h can be seen in Fig. 20. The maximum RMS acceleration for RESA was 1.19 m/s^2 and for hydraulic shock absorber was 0.76 m/s^2. Figure 8 shows the outpower from RESA at the first run (a) and the second run (b). The velocity range varied from 0 km/h to 40 km/h and depended on the road traffic. The maximum generated power is 17.2 W, which occurred at around 40 km/h on the bumpy road. The output power from RESA is random and depends on the road disturbance. The RMS power was 1.63 and 1.61 W, respectively, for the first run and the second run.

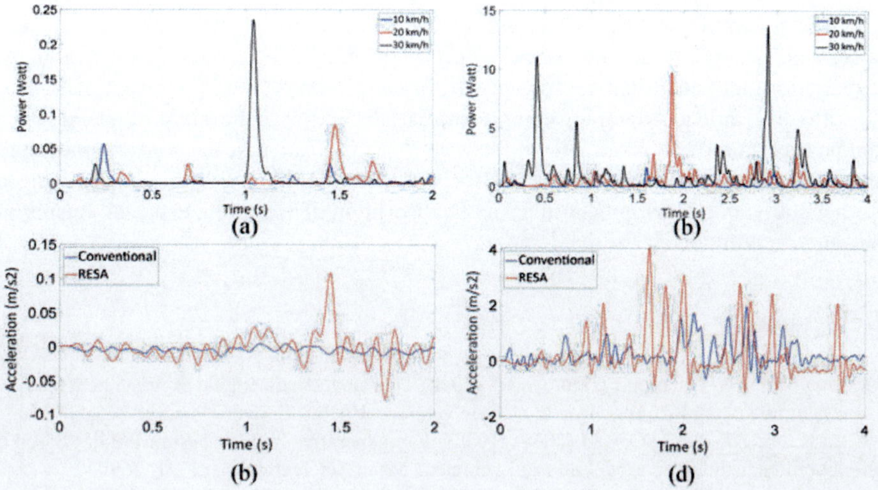

Fig. 7. Output Power from RESA on **a** Road class A and **b** Road class C for various vehicle speeds, and vehicle body acceleration on **a** Road class A and **b** Road class C, for vehicle speed of 30 km/h

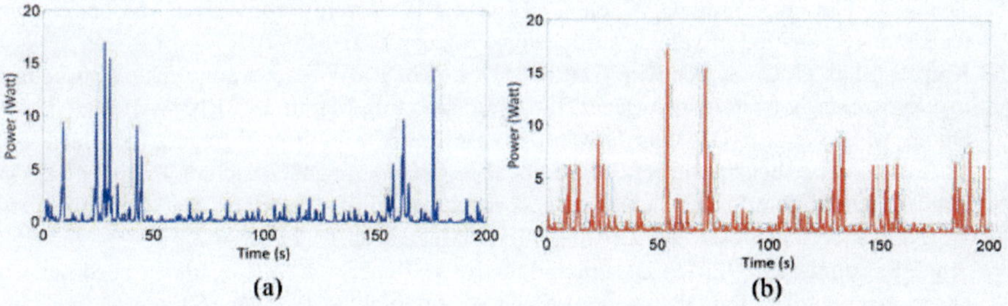

Fig. 8. Power Harvesting with Varied Road Class and Velocities **a** First Run and **b** Second Run

4 Conclusion

The RESA mechanical-based design and development has been simulated, tested, and analyzed. The proposed RESA design consists of a ball screw, gear transmission, and generator. The simulation result indicated that the variation of gear transmission ratio leads to the damping force characteristic and damping constant. The different damping constant values will influence different vehicle types. Increasing the gear ratio also increases the inertia effect and potential power generated. The RESA was tested in a pickup truck vehicle with various road types and different velocities. The peak power generated was 11.2 and 10.5 W, respectively, when the pickup truck passed the speed bump and the speed trap. When the pickup truck passed the road class A and C, the power was 0.24 and 13.8 W with vehicle speed at 30 km/h. Several test were conducted at different road type and different speed. The power generated was 17.2 W. From the road test, it can be seen that more road disturbance will generate higher power.

The RESA damping characteristic is quite different compared with conventional shock absorbers. The RESA damping is mainly caused by mechanical components (gear inertia and friction) and electrical components from the generator. While conventional shock absorbers use fluid as a damping component. The difference in the damping characteristic can be seen in the test result. All of the road types indicate that the conventional shock absorber performs better vibration absorption than RESA. With this result, it needs further study and development in terms of absorption of vibration, especially using the mechanical component for the damping.

References

1. Bou Nader W, Chamoun J, Dumand C (2020) Thermoacoustic engine as waste heat recovery system on extended range hybrid electric vehicles. Energy Convers Manage 215:11–29
2. Li N, Ning X, Wang Q, Li J (2017) Hydraulic regenerative braking system studies based on a nonlinear dynamic model of a full vehicle. J Mech Sci Technol 31:2691–2699
3. Seo J, Jhang K-Y, Lee H, Kim Y-C (2019) Vibration energy harvesting technology for smart tire monitoring. J Mech Sci Technol 33:3725–3723
4. Abdelkareem MAA, Xu L, Guo X, Ali MKA, Elagouz A, Hassan MA, Essa FA, Zou J (2018) Energy harvesting sensitivity analysis and assessment of the potential power and full car dynamics for different road modes. Mech Syst Signal Process 110:307–332
5. Zhang Y, Guo K, Wang D, Chen C, Li X (2017) Energy conversion mechanism and regenerative potential of vehicle suspensions. Energy 119:961–970
6. Kim JH, Shin YJ, Chun YD, Kim JH (2018) Design of 100W regenerative vehicle suspension to harvest energy from road surfaces. Int J Precision Eng Manuf 19(7):1089–1096
7. Duong M-T, Chun Y-D, Hong D-K (2018) Design of a high-performance 16-Slot 8-Pole electromagnetic shock absorber using a novel permanent magnet structure. Energies 11:3352
8. Tang X, Lin T, Zuo L (2014) Design and optimization of a tubular linear electromagnetic vibration energy harvester. IEEE/ASME Trans Mechatron 19(2):615–622
9. Singh S, Satpute NV (2015) Design and analysis of energy-havesting shock absorber with electromagnetic and fluid damp-ing. J Mech Sci Technol 29(4):1591–1605
10. Zhang Z, Zhang X, Chen W, Rasim Y, Salman W, Pan H, Yuan Y, Wang C (2016) A high-efficiency energy regenerative shock absorber using supercapacitors for renewable energy applications in range extended electric vehicle. Appl Energy 178:177–188
11. Liu Y, Xu L, Zuo L (2017) Design, modeling, lab, and field tests of a mechanical-motion-rectifier-based energy harvester using a ball-screw mechanism. IEEE/ASME Trans Mechatron 22(5):1933–1943
12. Nugraha DK, Guntur HL (2019) Design and analysis of regenerative shock absorber using ball screw mechanism for vehicle suspension. AIP Conf Proc 2187:050014
13. Salman W, Qi L, Zhu X, Pan H, Zhang X, Bano S, Zhang Z, Yuan Y (2018) A high-efficiency energy regenerative shock absorber using helical gears for powering low-wattage electrical device of electric vehicles. Energy 159:361–372
14. Guntur HL, Hendrowati W, Syuhri SNH (2020) Designing hydro-magneto-electric regenerative shock absorber for vehicle suspension considering conventional-viscous shock absorber performance. J Mech Sci Technol 34(1):55–67
15. Guntur HL, Hendrowati W, Daman AAA (2019) Hydro-regenerative shock absorber with two generators in series for vehicle suspension. AIP Conf Proc 2187

16. Wang Z, Zhang T, Zhang Z, Yuan Y, Liu Y (2020) A high-efficiency regenerative shock absorber considering twin ball screws transmissions for application in range-extended electric vehicles. Energy Built Environ 1:36–49
17. Abdelkareem MAA, Xu L, Ali MKA, Elagouz A, Mi J, Guo S, Liu Y, Zuo L (2018) Vibration energy harvesting in automotive suspension system: a detailed review. Appl Energy 229:672–699

Numerical Study on Dynamic Stabilities Prediction in a Small Unmanned Aerial Vehicle

Arif Wahjudi(✉) and Adi Susanto

Department of Mechanical Engineering, Institut Teknologi Sepuluh Nopember, Surabaya 60111, Indonesia

arif_w@me.its.ac.id

Abstract. In Unmanned Aerial Vehicle (UAV) developments, one of the demands is to design a UAV that has optimal dynamic stabilities. The implementation of a wing with appropriate geometries has a significant effect on the dynamic stabilities of UAVs. The aim of this study is to predict the dynamic stabilities of UAVs based on wing geometries. The wings geometries considered are dihedral and tip-twist angles, while the dynamic stabilities predicted are short periods and Dutch roll. Both dynamic stabilities were defined by the related eigenvalue which was obtained by XFLR5 numerical software. The relationships between wing geometries and dynamic stabilities were obtained using the Artificial Neural Network (ANN) method. Based on the ANN training, the eigenvalues of short period and Dutch rolls as the representation of the UAV dynamic stabilities can be predicted by using 2-9-3-1 and 2-6-5-1 of ANN structures, respectively, with the mean square error of 1.0712e–07 and 2.6170e–08, respectively.

Keywords: Dihedral angle · Tip-Twist angle · Short period · Dutch roll · Artificial neural network

1 Introduction

Today, Unmanned Aerial Vehicle (UAV) is widely used in human life, from hobby activities to state defense equipment. This UAV is applied for monitoring activities because a camera can be installed so that it can provide aerial views, either in real time or not. Therefore, a lot of research for the development of UAVs has been carried out to meet the needs of these uses. A UAV has been proposed for agricultural production management needs [1]. UAVs have also been developed for air quality measurement [2]. In fact, UAVs are also widely applied for military purposes [3].

To be flown, UAVs must have good aerodynamic performance and flight stability. Many studies have been conducted to get the best aerodynamic performance and stability of the UAV. Optimization for aerodynamic performance has been proposed to obtain a more efficient UAV wings geometry than the existing one [4]. Several factors such as wind and throttle settings have also been investigated for their effect on UAV flight stability [5]. However, the influence of the wings geometry factor has not been widely discussed in studies on the stability of the UAV, especially its dynamic stability.

I. Tolj et al. (eds.), *Recent Advances in Mechanical Engineering*, Lecture Notes in Mechanical Engineering, https://doi.org/10.1007/978-981-19-0867-5_11

In analyzing a UAV, it can be done by experimentation in a wind tunnel or numerically in a computer. A wind tunnel has been used to determine the leading-edge flow separation of the UAV by utilizing a flexible shear-stress sensor skin [6]. However, numerical software such as xlfr5 was also used to analyze the drag of a fixed-wing UAV [7]. Unfortunately, software such as xlfr5 has not been used to predict UAV flight stability.

Many ways are applied in various fields to predict the effect of some variables on other variables [8–10]. One of them that is currently being widely applied is the artificial neural network (ANN) method because of its advantage that it can predict complex functions with a small mean square error [11]. Therefore, the ANN method in this study was used to predict the dynamic stability of the UAV based on its wing's geometry. The wings geometry considered were the dihedral and the tip-twist angles, where these two angles were used to predict two dynamic stability, namely, short period and Dutch rolls.

2 Wings Geometries and Dynamic Stabilities of UAV

Some wings geometries such as root and tip chords, dihedral and anhedral angles, wingspan, and tip-twist might be set to obtain the desired performance and stability of a UAV. Dihedral and tip-twist angles were studied in this research to predict the UAV stabilities. The dihedral angle is the upward angle from the horizontal of the wings, while the tip-twist is the angle between the zero-lift angle of the tip and root chords.

There are several dynamic stabilities of a UAV such as short period and Dutch roll stabilities that need to be considered in this study. The short period stability is the stability of the pitching movement, while the Dutch roll stability is the stability of the combination of yawing and rolling movements. A system including UAV can be dynamically stable both for a short period and Dutch roll if each natural response is close to zero at the time approaching infinity.

The stability is indicated by the negative damping constant value of the eigenvalue [12]. An eigenvalue is a complex number that describes the frequency mode and its damping as shown in Eq. (1)

$$\lambda = \sigma_1 + \omega_N \tag{1}$$

where σ_1 is the damping constant and ω_N is the natural circular frequency. In this study, the damping constant of the eigenvalue was predicted based on the dihedral and tip-twist angles to determine the stability of the UAV.

3 Design of Simulation

To obtain an equation describing the relationship between the dihedral and tip-twist angles with respect to the damping constant of the short period and the Dutch roll eigenvalues, several trials have been carried out. Nine levels for each of the dihedral and tip-twist angles shown in Table 1 were combined according to a full factorial design, while some wing geometries such as wingspan, root chord, tip chord, UAV mass, and maximum stall speed were set constant. Then, the damping constant value was obtained using XLFR5 software. XFLR5 is a numerical vortex lattice method (VLM) software that can be used to simulate the flight test of a UAV. Based on previous studies, it was found that there was not enough significant evidence to reject the results obtained by XFLR5 being different from the experimental results [13, 14]. Figure 1 shows a UAV simulation model used in this study with a combination of dihedral and tip-twist angles of 5° and -2°, respectively.

Table 1. UAV simulation design

Parameters	Level
Wingspan (b)	800 mm
Root chord (C_r)	189 mm
Tip chord (C_t)	160 mm
Dihedral angle (Γ)	0, 1, 2, 3, 4, 5, 6, 7, 8°
Tip-twist (ϕ)	−2, −1.5, −1, −0.5, 0, 0.5, 1, 1.5, 2°
UAV mass (m)	1.04 kg
Max. stall speed (V_s)	12 m/s

After all the trials were run, the effect of dihedral angle and tip-twist as independent variables was examined using the ANOVA method on the value of damping constant as the dependent variable to determine whether the effect was significant or not. Determination of the equation that describes the relationship between the independent and dependent variables was carried out only if the effect is significant. The examination was conducted by testing two hypotheses as follows:

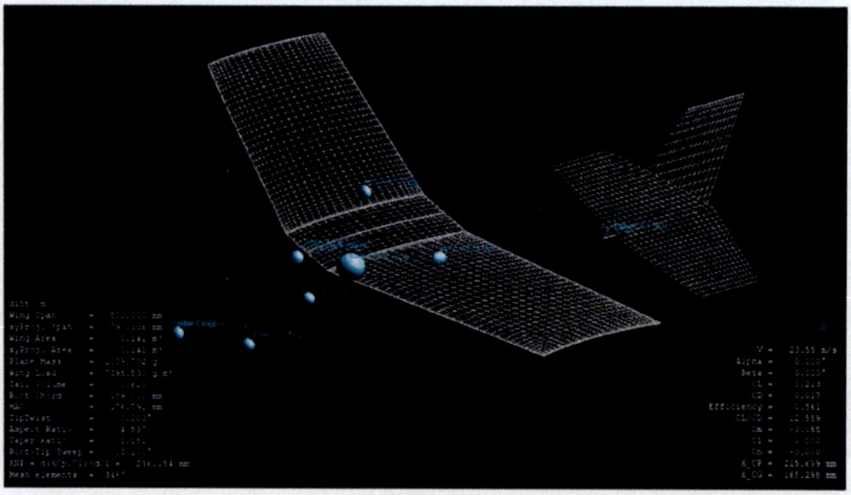

Fig. 1. UAV simulation model

H_0: the independent variable has no significant effect on the dependent variable.
H_1: the independent variable has a significant effect on the dependent variable.

ANOVA provides an F value where H_0 will be rejected if the F value is greater than the F_{table}. The results of the ANOVA calculation are shown in Table 2, where all F values are greater than F_{table}. Therefore, it can be concluded that both independent variables have significant effects on the two dependent variables so that they are feasible to determine the relationship equations.

Table 2. ANOVA result

Wing geometries	Damping constant of	F	F_{table}	Decision
Γ	Short period stability	82.67	2.087	Reject H_0
	Dutch roll stability	55.44	2.087	Reject H_0
ϕ	Short period stability	73,711.72	2.087	Reject H_0
	Dutch roll stability	1358.33	2.087	Reject H_0

4 Artificial Neural Network

Based on ANOVA, dihedral and tip-twist angles are known that they have a significant effect on both damping constant values of the short period and the Dutch roll stabilities. Hence, if the dihedral and tip-twist angles are changed, both stabilities will also change significantly. Then, a mathematical equation that describes the relationship between the dihedral and tip-twist angles and the value of the damping constant which corresponds to the dynamic stability of the UAV needs to be obtained. Estimation for suitable mathematical equations was carried out by applying the Artificial Neural Network method.

Artificial Neural Network is a method that tries to imitate the human nervous system, which consists of several circles called neurons and is connected to each other by a connector shown in Fig. 2. Every structure of ANN consists of three layers, i.e., input, hidden, and output layers. The input and output layers have a single layer, but the hidden layer may have a single or multi-layer(s). Every layer may have a single or multi-neuron (s). Moreover, every neuron in a certain layer is connected to the other neuron in the next layer by a connector that has a weight value. Hence, this ANN structure has the mathematical equations as shown in Eqs. (2–4).

$$a1 = \frac{2}{1 + \exp[-2(IW1\_1 * Xp1 + b1)]} - 1 \tag{2}$$

$$a2 = \frac{2}{1 + \exp[-2(LW2\_1 * a1 + b2)]} - 1 \tag{3}$$

$$a3 = LW3\_2 * a2 + b3 \tag{4}$$

where *Xp1* is the neuron in the input layer, *a1* and *a2* are neurons in the first and second layers, respectively, *a3* is the neuron in the output layer, $lW1\_1$ is the weight value of the connector between neuron in the input layer, and neuron in the first hidden layer, $LW2\_1$ is weight value of connector between neuron in the first and second hidden layer, $LW3\_2$ is weight value of connector between neuron in second hidden layer and neuron in the output layer, and $b1$, $b2$, and $b3$ are bias values of the first hidden layer, second hidden layer, and output layer, respectively.

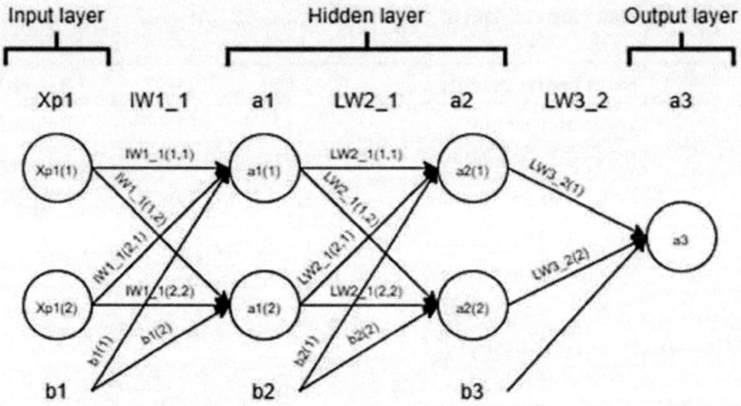

Fig. 2. Structure of artificial neural network

After being trained to obtain the network structure, weight, and bias values, both ANN have the mean square error of 1.0712e–07 and 2.6170e–08 for each of the short periods, and Dutch roll constant damping values using the network structure 2-9-3-1 and 2-6-5-1, respectively. The weight and bias values of the ANN structure which describe the relationship between dihedral and tip-twist angles and the damping constant of short period stability are

$$lW1_1 = \begin{bmatrix} 0.97\ -0.03\ -0.12 & 0.05\ 0.46\ -2.91 & 3.50\ -1.64\ -1.65 \\ -2.81\ -3.54\ -1.98 & -2.69\ 0.79\ 1.86 & -1.42\ 2.20\ -3.85 \end{bmatrix}^T \tag{5}$$

$$b1 = \begin{bmatrix} -4.30\ -3.18\ 1.42\ \ -0.63\ -1.37\ -0.33\ \ 2.36\ -3.28\ -3.72 \end{bmatrix}^T \tag{6}$$

$$LW2\_1 = \begin{bmatrix} -0.06\ 0.30\ 1.06 & 0.32\ -0.81\ 0.01 & 0.02\ -0.14\ 0.00 \\ 0.41\ 0.27\ -0.39 & -1.78\ 0.12\ -0.02 & -0.03\ 0.05\ 1.20 \\ -0.08\ -1.50\ 0.72 & -0.58\ -0.10\ 0.01 & -0.19\ -0.14\ 0.18 \end{bmatrix} \tag{7}$$

$$b2 = [-2.737\ -0.474\ -1.384]^T \tag{8}$$

$$LW2\_3 = \begin{bmatrix} -1.746\ 0.850\ 0.132 \end{bmatrix} \tag{9}$$

$$b3 = [-0.403]. \tag{10}$$

Moreover, they describe the relationship between dihedral and tip-twist angles and the damping constant of Dutch roll stability are

$$lW1\_1 = \begin{bmatrix} -1.524\ -0.084\ -1.016 & 0.095\ 0.440\ 1.251 \\ -1.055\ 0.403\ 0.089 & -1.987\ 1.478\ -2.487 \end{bmatrix}^T \tag{11}$$

$$b1 = \begin{bmatrix} 3.188\ -0.040\ 0.157\ \ -1.458\ 3.693\ 3.383 \end{bmatrix}^T \tag{12}$$

$$LW2\_1 = \begin{bmatrix} 0.727 & -1.028 & 0.002 & 1.589 & -0.697 & -0.622 \\ 1.452 & -1.566 & 1.463 & -0.377 & 0.396 & 0.461 \\ -0.430 & -1.757 & -0.013 & 0.158 & 1.006 & -0.228 \\ -0.451 & 0.938 & 0.029 & -0.125 & 0.290 & -0.043 \\ 1.139 & -0.594 & 1.052 & -0.323 & 0.453 & -0.909 \end{bmatrix} \tag{13}$$

$$b2 = \begin{bmatrix} -1.715\ -0.619\ -1.304\ -0.235\ 1.816 \end{bmatrix}^T \tag{14}$$

$$LW2\_3 = \begin{bmatrix} -1.169\ 0.220\ -1.138\ 0.747\ -0.955 \end{bmatrix} \tag{15}$$

$$b3 = [-0.585] \tag{16}$$

After predicting the damping constant values of both short period and Dutch roll that were based on the dihedral and tip-twist angles applied in the simulations, other values in the range of the analyzed dihedral and tip-twist angles have been conducted in this study. They are shown in Figs. 3 and 4. Based on these figures, it is known that all damping constant values in the studied dihedral and tip-twist angles have negative values. Hence, the UAV has good stability not only in the pitch movement, but also in the roll and yaw movements. In addition, if the change in the tip-twist and dihedral angles is compared, the change in the tip-twist angle will give a larger change in the value of the damping constant. Therefore, it is known that the tip-twist angle has a greater effect than the dihedral angle. Finally, based on these predictions, it is required to determine a combination of the value of tip-twist and dihedral angles which will provide the best stability of the designed small UAV.

5 Conclusions

The mathematical equations have been developed based on ANN to describe the relationships between dihedral and tip-twist angles as the wing geometries and the damping constant of the short period and Dutch roll as the UAV stabilities. Moreover, the UAV in the range of studied dihedral and tip-twist angles is stable based on the negative damping constant value.

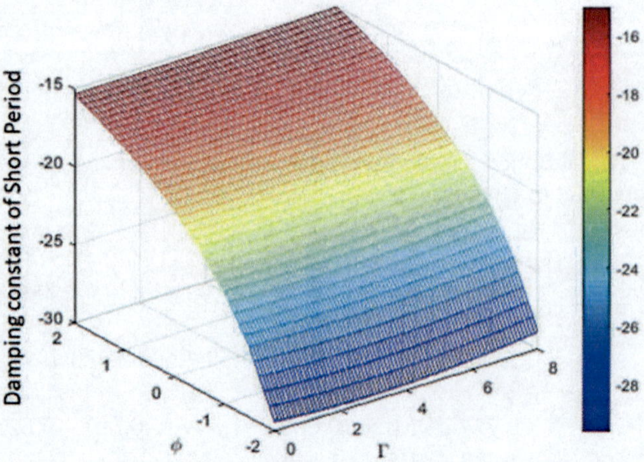

Fig. 3. The relationship between the dihedral and tip-twist angles on the damping constant of the short period stability

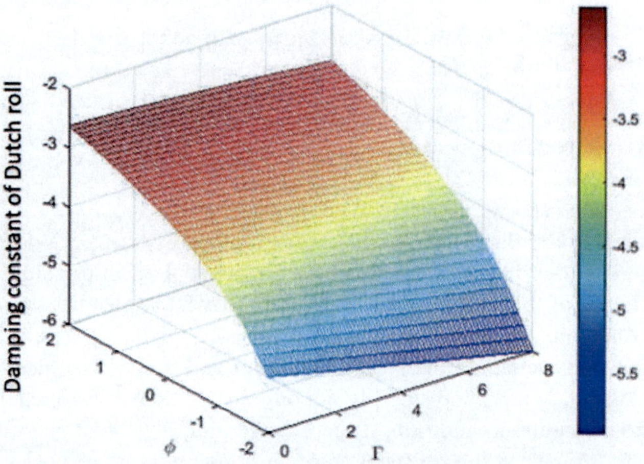

Fig. 4. The relationship between the dihedral and tip-twist angles on the damping constant of the Dutch roll stability

References

1. Huang Y, Thomson SJ, Hoffmann WC, Lan Y, Fritz BK (2013) Development and prospect of unmanned aerial vehicle technologies for agricultural production management. Int J Agricult Biol Eng 6(3):1–10
2. Haas PY, Balistreri C, Pontelandolfo P, Triscone G, Pekoz H, Pignatiello A (2014) Development of an unmanned aerial vehicle UAV for air quality measurement in urban areas. In: 32nd AIAA applied aerodynamics conference. Atlanta

3. Udeanu G, Dobrescu A, Oltean M (2016) Unmanned aerial vehicle in military operations. Scientific Research and Education in the Air Force-AFASES 2016 https://www.afahc.ro/ro/afases/2016/RP/UDEANU_DOBRESCU_OLTEAN.pdf
4. Eftekhari S, Mahdi Al-Obaidi AS (2019) Investigation of a cruising fixed wing mini unmanned aerial vehicle performance optimization. Indonesian J Sci Technol 4(2)
5. Hu X, Huang X (2017) Orthogonal design and optimization of flight stability test for the quadrotor unmanned aerial vehicle. In: IEEE International conference on unmanned systems (ICUS)
6. Xu Y, Jiang F, Newbern S, Huang A, Ho CM, Tai YC (2003) Flexible shear-stress sensor skin and its application to unmanned aerial vehicles. Sens Actuators A 105(3):321–329
7. Deulgaonkar V, Patil M, Patwardhan V (2020) Design and drag analysis of fixed wing unmanned aerial vehicle for high lift. Int J Vehicle Struct Syst 12(2):140–145
8. Harnany D, Batan IML, Wahjudi A, Ardiansyah MR (2016) The effect of biocomposite variation composition on thermal characteristic and cross-sectional structure as alternative material plastic product. In: AIP conference proceedings
9. Palmer PB, O'Connell DG (2009) Regression analysis for prediction: understanding the process. Cardiopulmonary Phys Therapy J 20(3):23–32
10. Yang A, Han Y, Pan Y, Xing H, Li J (2017) Optimum surface roughness prediction for titanium alloy by adopting response surface methodology. Results Phys 7:1046–1050
11. Salamoni TD, Wahjudi A (2018) Injection molding process modeling using back propagation neural network method. In: AIP conference proceedings
12. Nise NS (2011) Control systems engineering, danvers. Wiley, New York
13. Meschia F (2008) Model analysis with XFLR5. Radio Controlled Soaring Digest 25(2):27–51
14. El-Salamony M, Serokhvostov S (2016) Comparison between experimental, analytical, and numerical methods in natural frequencies of the longitudinal flight modes for small UAVs. In: International micro air vehicle conference. Beijing

Check for updates

Stent Geometry Optimization for In-Stent Restenosis Prevention Through Fluid–Structure Interaction Analysis

Varien Janitra Nuralif Susanto[1], Ryo Takeda[2](✉), Shinya Honda[2], and Katsuhiko Sasaki[2]

[1] Graduate School of Engineering, Hokkaido University, Sapporo 060-8268, Japan
[2] Faculty of Engineering, Hokkaido University, Sapporo 060-8268, Japan
r.takeda@eng.hokudai.ac.jp

Abstract. Heart attacks are the most common cause of death in the world in the last decade. Heart attacks are caused by a deficiency of nutrients in the heart muscle cells due to blood flow blockage. Blocked blood flow can be restored by the deployment of a medical device called a stent. However, many studies found that stent deployment may induce plaque formation, known as in-stent restenosis. Previously, there have been attempts to minimize in-restenosis occurrence on stents by redesigning the stent. Optimization algorithms and numerical simulations are employed to search for the best stent design according to preset objective functions and constraints. This paper studies the optimum stent configuration considering the extensively studied restenosis risk factors, such as intimal stress and wall shear stress. For the optimization, a genetic-based optimization algorithm (GA) was used. The risk factors occurrence is analyzed by coupling finite element method (FEM) and computational fluid dynamics (CFD) solvers. Several conflicting objective functions based on the analyzed risk factors and constraints are also applied to the optimization algorithm. As a result, optimum stent configurations are obtained, which have the best size and angles of struts.

Keywords: Fluid–structure interaction · Finite element method · Genetic algorithm · Ischemic heart diseases · In-stent restenosis · Stent

1 Introduction

Heart attacks are the most common cause of death in the world [1]. Heart attacks are caused by oxygen deficiency of the heart muscle cells due to blood flow blockage caused by plaque formation on the heart's arterial vessel wall. The plaque formation phenomenon is known as atherosclerosis. To restore the blood flow to its pre-blockage state, a tubular metallic device called stents are installed. However, during its installation, the stent might damage the vessel creating a new injury site for plaque growth. This new plaque formation is known as in-stent restenosis.

Occurrence of in-stent restenosis could be analyzed numerically by obtaining the vessel's contact von Mises stress using FEM solvers and the wall shear stress (WSS)

I. Tolj et al. (eds.), *Recent Advances in Mechanical Engineering*, Lecture Notes in Mechanical Engineering, https://doi.org/10.1007/978-981-19-0867-5_12

using CFD solvers. Based on the von mises stress and WSS values, a stent optimization study can be done by employing an optimization algorithm [2]., One of the suggested optimization algorithms for multiple conflicting objectives and a large number of design variables is genetic algorithms [3]. However, few studies had attempted to optimize stent designs to solve restenosis problems by implementing a multiple-objective genetic algorithm.

Therefore, this study aims to obtain the most optimum configurations by using a Palmaz-Schatz stent model as a reference model. The stent model is parametrized to be used as the design variable in the optimization algorithm. Stent optimization is conducted using a multiple-objective genetic algorithm (MOGA), which considers two stent performance metrics, vessel von Mises contact stress and WSS.

2 Methodology

This research employs a numerical simulation method using Ansys version 2021 R1 (Ansys, Inc.). This software package includes all the software required to conduct this research, such as the 3D CAD software, FEM, CFD, and optimization algorithm solvers. Figure 1 shows the workflow of this research.

In summary of the workflow, the optimization solver randomly generates an initial population of stent configuration individuals. An individual is defined as a set of design variables that define a certain stent configuration. Figure 2 shows the design variables set used in this study.

Sequentially, the CAD software generates a simulation geometry for the FEM and CFD solvers based on the generated design variables. The FEM and CFD solvers analyze the numerical problem and obtain the stent performance metrics values. The obtained values are used by the optimization solver to select the best stent configuration within the population according to the MOGA algorithm. The workflow iterates until the improvement rate over generations becomes insignificant.

The plague and vessel geometry is modeled as a simplified version of a middle section of the left anterior descending (LAD) region of a coronary artery. The model

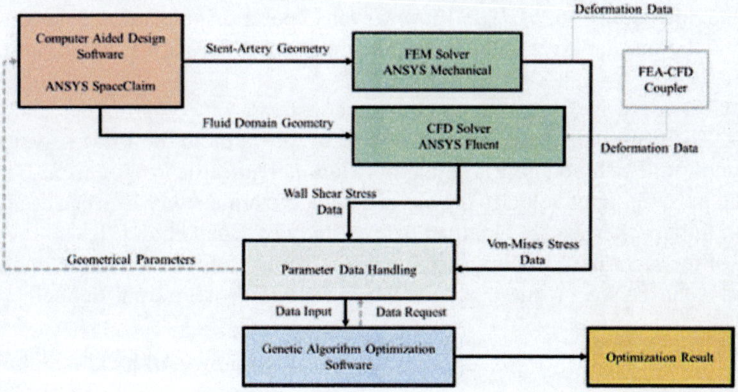

Fig. 1. Overall research workflow

sizing is done according to measurements obtained by a study [4]. This study uses a 180° longitudinal section model to reduce computational cost. The blood vessel model has an inner diameter of 3 mm with a 0.5 mm thick plaque. The plaque length is modeled to be the same as the stent length. The artery is modeled as a three-layer material, in which the nodes on the layer's boundary are shared between layers.

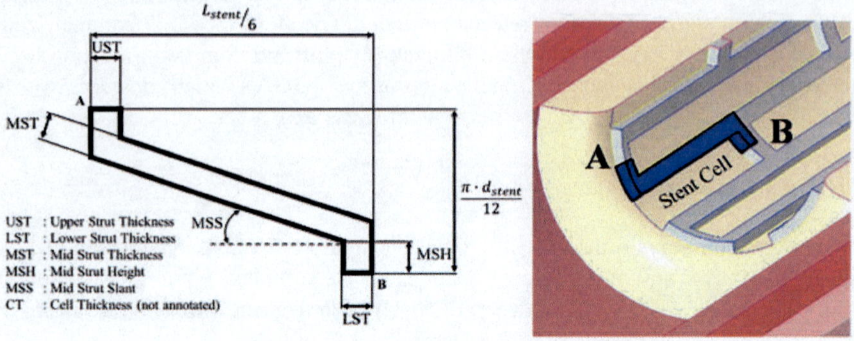

Fig. 2. Definition of optimization design variables

The vessel with plaque and stent part of the CAD-generated geometry is used for the FEM solver. The geometry was discretized into around 40,000 nodes and 8,000 elements. The mesh density is defined empirically considering the grid independence test and optimization workload time constraints. The time step size of 0.05 s is also chosen considering the optimization time constraints. The vessel and plaque are modeled as an Ogden hyperelastic material, whose parameters were obtained from a previous study [5]. Meanwhile, the stent is modeled as an AISI 316L stainless steel, whose stress–strain graph is obtained from another research [6]. The stent deployment is modeled as a pressure loading applied directly to the stent surface. The pressure increases linearly from 0 to 0.9 MPa in 1.0 s. The pressure is kept constant after 1.0 s. This study employs contact modeling between the stent and plaque, and the friction coefficient is obtained from a previous study [6, 7]. The FEM solver exports displacement data to the CFD solver, and the intimal layer and stent's maximum von Mises stress to the optimization solver.

The CFD mesh is an unstructured mesh consisting of 70,000 nodes and 360,000 elements. The time step size is set to be equal to the value in the FEM solver to allow solver coupling. The blood flow is defined as a laminar pulsatile flow of a non-Newtonian fluid. The pulsatile inlet velocity profile is obtained from a study by Torii et al. [8] and the blood fluid properties are obtained from a study by Chan et al. [9].

One of the stent performance metrics that is being considered for the optimization workload is the TAWSS (time-averaged WSS) ratio. TAWSS ratio is defined as

$$TAWSS\ Ratio = \left(\frac{Area_{WSS\leq 3.8Pa}}{Area_{total}}\right)_{Plaque} \times 100\% \tag{1}$$

where $Area_{WSS \leq 3.8Pa}$ is the area of the analyzed region where WSS is less than 3.8 Pa and $Area_{total}$ is the total area of that region. This study uses plaque as the region analyzed for restenosis occurrence. Pressure drop of the vessel is also considered as the stent performance metric. The performance metrics mentioned above are exported to the optimization solver.

The optimization solver searches for an optimum design according to the defined search space. The search space is defined to provide the largest number of feasible designs within the geometric constraints of the stent cell shapes. The generated design variables are constrained by a set of equations so that the struts do not collide with each other and/or have an excessive thickness.

This study considers the following objective functions as listed in Table 1.

Table 1. Optimization objectives and constraints

	Objective function/Constraint	Objective
a	Intimal maximum von mises stress	Minimize
b	Time-averaged wall shear stress (TAWSS) ratio	Minimize
c	Time-averaged blood flow pressure drop	Minimize
d	Stent maximum von mises stress	Constraint $\leq$ 600 MPa

a and b in Table 1 are the main stent metrics of interest that would be extensively discussed. Meanwhile, c and d are the set as the stent's functional constraint that constrained the allowable pressure drop and maximum stent stress during expansion.

The optimization solver generates an initial population of 60 individuals and 40 individuals in the subsequent generations. Genetic mutation and crossover probabilities are set to 5% and 98%, respectively, after trial and error. The optimization workload is assumed to be converged once the difference in the population's mean and deviance value of each stent performance metric between the last generation and that of the previous generation is less than 1%.

This study applies objective function weighting to the two main objectives based on the solver's built-in post-processing function. The weighting is done to study the objective function selection. The weighing criterions applied here are 'default importance' and 'higher importance'. Application of the weighting function does not affect the genetic algorithm mechanism.

3 Results

From all of the candidates obtained in the optimization, the best stent designs considering the intimal stress and TAWSS ratio are found when the TAWSS ratio is weighted as important objective function. All stent design candidates yield better stent metric performance compared to the original design. However, the improvements are traded off by an increase of pressure drop on every candidate. The stent performances data of each candidate are summarized in Fig. 3.

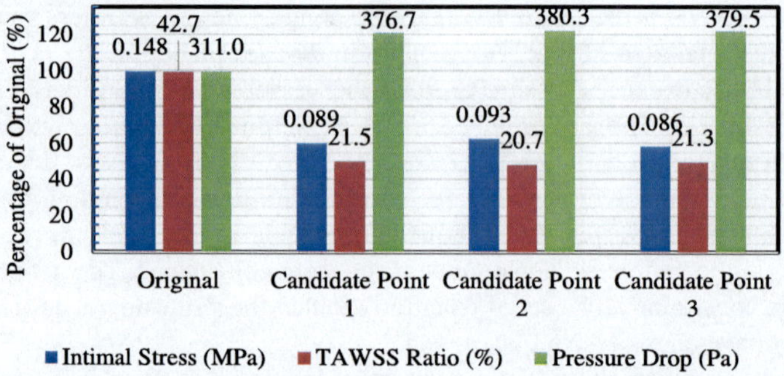

Fig. 3. Performance summary of the optimum candidates

In terms of intimal-stress reduction, candidate 3 reduces contact stress the best by 41.6% compared to the original design. The next best performing candidate is candidate 1 that yields a reduction of 39.7%. Candidate 2 yields the least reduction of 37.3%. Meanwhile, in terms of TAWSS reduction, candidate 2 reduces the TAWSS ratio by 51.5% compared to the original. The next best performing candidate is candidate 3 that yields a reduction of 50.2%. The least reduction is yielded by candidate 1, which reduces only 49.8% of the original design.

Compared to the candidates obtained by an unweighted and intimal-stress weighted objective function, the candidates obtained from the TAWSS ratio-weighted objective function have the best intimal stress and TAWSS ratio minimization performance. However, the pressure drop caused by the candidates from the TAWSS ratio-weighted objective function is worse than the other two.

Figure 4 shows that the candidates, labeled as DP, which have thinner and apparent slanted struts, compared to the original design, are labeled as ORI. The inter-strut spacing is also found to be wider. Table 2 summarizes the values of the design variables in comparison to the original design and each candidate.

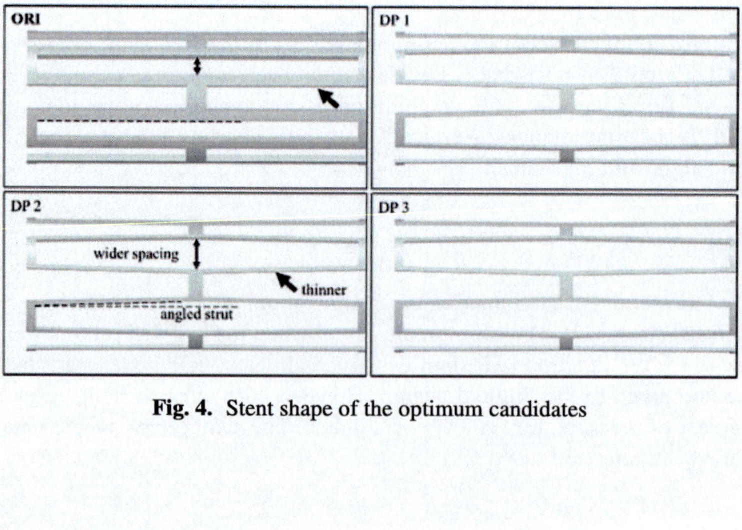

Fig. 4. Stent shape of the optimum candidates

Table 2. Summary of design variables values

	UST	LST	MST	MSH	MSS	CT
Original	0.15	0.15	0.1668	0.1668	90.1	0.1
DP 1	0.14	0.11	0.06	0.15	91.4	0.05
DP 2	0.14	0.11	0.06	0.17	91.4	0.05
DP 3	0.13	0.1	0.06	0.16	91.4	0.05

Figure 5 shows the intimal von mises contour produced by the candidates. A red color indicates larger stent-vessel contact stress. The stress distribution is lower on all optimum candidates compared to the original design. Also, some of the stress concentration spots shown by the circle disappeared on the contour plot of all candidates.

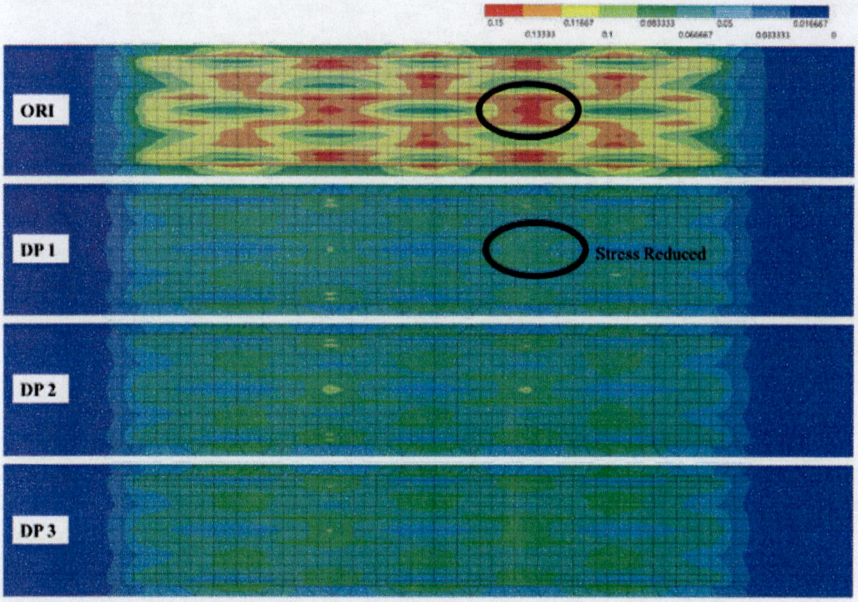

Fig. 5. Comparison of intimal von mises stress contour between original and optimum stents

Figure 6 shows the TAWSS ratio contour produced by the candidates. The yellow and red color in Fig. 6 shows a region in which the TAWSS value is below a restenosis threshold value specified by Korshunov et al. [10]. The green color shows a region with a TAWSS value above the threshold value. Therefore, fewer regions of yellow and red color are preferred. In Fig. 6, the arrows on the middle and right section show that the optimum candidates have fewer low TAWSS regions compared to the original design.

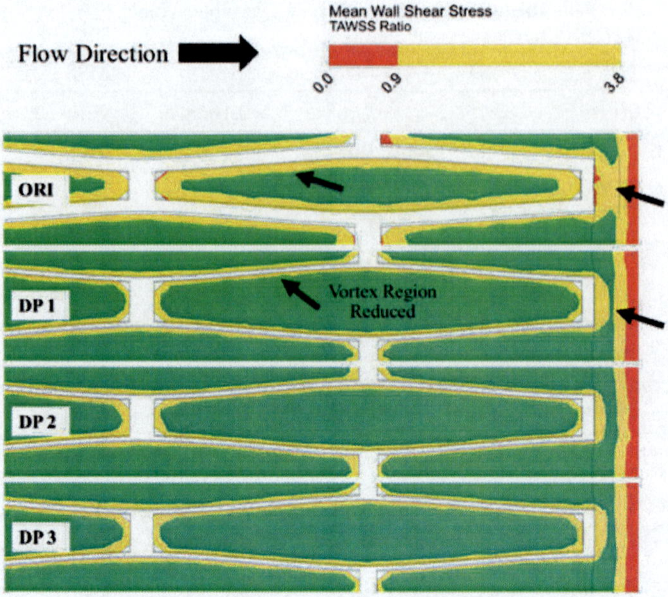

Fig. 6. Comparison of TAWSS ratio between original and optimum stents

Comparing the results to other studies, a clinical study by Morton et al. [11] reported that the occurrence of restenosis is lower in thinner stents than the thicker counterpart. Wassif et al. [12] explained that the lower stress yielded by thinner stents is due to lower vessel stretching exerted on the vessel. Morton et al. [11] also reported that stents with larger inter-strut spacing minimize restenosis occurrence.

Meanwhile, for the blood flow, another numerical study by Ladisa et al. [13] shows that low TAWSS regions are mainly concentrated around the stent struts. The study explained that low TAWSS regions are caused by vortex region formation due to stagnate flow and boundary layer separation on the upstream and downstream of the struts, respectively. It is suspected that thinner stents yield smaller vortex regions, as demonstrated by Hatipoglu et al. [14].

From the discussion above, it is found that the yielded design candidates here agree with the previously mentioned studies. To improve the stent design even further, a study on a greater number of design variables is suggested. This suggestion is based on the previous findings that modifying the stent design by changing a certain combination of design variables produces a different effect from changing a single design variable.

4 Conclusions

This research aims to obtain optimal stent configurations that minimize in-stent restenosis occurrence through numerical methods, which findings could be summarized into several points as follows:

1. Restenosis-minimizing stents have the characteristics of a thin stent profile and thin cells with slightly angled struts.
2. Restenosis-minimizing stents have an increased pressure drop as a tradeoff.
3. The best performing stent in terms of restenosis minimizations were as follows:

   (a) Intimal stress metric: Intimal stress = 0.086 Pa, TAWSS ratio = 21.3%
   (b) TAWSS ratio metric: Intimal stress = 0.093 Pa, TAWSS ratio = 20.7%

References

1. The top 10 causes of death, https://www.who.int/en/news-room/fact-sheets/detail/the-top-10-causes-of-death. Last accessed 23 Dec 2020
2. Bressloff N (2016) Design optimisation of coronary artery stent systems. Ann Biomed Eng 44(2):357–367
3. Blouza A (2008) Multiobjective optimization of a stent in a fluid-structure context. In: 10th annual conference companion on genetic and evolutionary computation. Association for Computing Machinery, New York, pp 2055–2060
4. Holzapfel G (2005) Determination of layer-specific mechanical properties of human coronary arteries with nonatherosclerotic intimal thickening and related constitutive modeling. Am J Physiol Heart Circulat Physiol 289(5):2048–2058
5. Schiavone A (2015) A study of balloon type, system constraint and artery constitutive model used in finite element simulation of stent deployment. Mech Adv Mater Modern Process 1(1):1–15
6. Syaifudin A (2018) Development of asymmetric stent for treatment of eccentric plaque. Bio-Med Mater Eng 29(3):299–317
7. Syaifudin A (2015) Effects of plaque lengths on stent surface. Bio-Med Mater Eng 25(2):189–202
8. Torii R (2009) Fluid-structure interaction analysis of a patient-specific right coronary artery with physiological velocity and pressure waveforms. Commun Numer Methods Eng 25(5):565–580
9. Chan W (2007) Modeling of non-newtonian blood flow through a stenosed artery incorporating fluid-structure interaction. Anziam J 47:507
10. Korshunov V (2007) Vascular remodeling: hemodynamic and biochemical mechanisms underlying Glagov's phenomenon. Arterioscler Thromb Vasc Biol 27(8):1722–1728
11. Morton A (2004) The influence of physical stent parameters upon restenosis. Pathol Biol (Paris) 52(4):196–205
12. Wassif H (2013) Restenosis. SCAI interventional cardiology board review, 2nd edn, vol 4(2), pp 10–15
13. Ladisa J (2003) Three-dimensional computational fluid dynamics modeling of alterations in coronary wall shear stress produced by stent implantation. Ann Biomed Eng 31(8):972–980
14. Hatipoglu F (2002) Flow around a partly buried cylinder in a steady current. Ocean Eng 30(2):239–249

Work Exchange Network Synthesis Based on Superstructure Approach

Yu-Ming Zhang and Vincentius Surya Kurnia Adi(✉)

Department of Chemical Engineering, National Chung Hsing University, Taichung 40227, Taiwan
vska@nchu.edu.tw

Abstract. There has been relatively little research on the work exchange network synthesis (WENS). It is challenging to implement the outcomes in the industrial process as the assumptions are not mature enough. Although the WENS is analogous to the heat exchanger network synthesis (HENS), the exchange equipment is quite different. HENS considers heat exchangers, heaters, and coolers, while pressure manipulating equipment classified into direct and indirect devices plays a more critical role in WENS. Each of them has its advantages and disadvantages. Therefore, the choice of appropriate equipment is very significant to the WENS. This study will address these issues by improving the design approaches and constraints using the WENS superstructure model. The phase change issues are avoided, and rigorous thermodynamics is implemented to enhance the current WENS. Besides, the efficiency of pressure manipulating equipment will be incorporated. The minimum total annualized cost (TAC) is considered the performance evaluation of the optimal design of WENS. The proposed WENS superstructure model is mathematically formulated into a mixed-integer nonlinear programming (MINLP) problem and then solved utilizing the GAMS optimization platform. The proposed superstructure model of WENS is expected to closely reflect the actual conditions. The optimization strategy is demonstrated via a case study on the liquid ammonia production process.

Keywords: Mathematical programming · Process optimization · Work exchange network · Energy integration · Process synthesis

1 Introduction

In recent years, water shortages due to abnormal weather, power outages caused by overloaded power plants, or oil prices fluctuating due to the international situation have also affected industrial production and manufacturing as energy demand increases day by day. Governments and companies have put forward solutions to this dilemma, i.e., China Steel Corporation's cogeneration plant recovers waste heat from the steelmaking process for power generation and sells steam to neighboring chemical plants for use through pipelines. It not only saves processing costs, but also creates new benefits for the neighboring chemical plants. Energy integration and optimization are also ways to reduce the total annual cost of the chemical plants. Based on the current research and

I. Tolj et al. (eds.), *Recent Advances in Mechanical Engineering*, Lecture Notes in Mechanical Engineering, https://doi.org/10.1007/978-981-19-0867-5_13

development, most of them focus on the two shared utility issues of heat and water and even integrate both at the same time for optimization research [1]. In the gas processes, natural gas liquefaction, oil refining, petrochemical, and liquid ammonia production processes, compression processes are often used and unavoidable. Note that, most of the previous literature put more attention on heat integration and less on work integration, although pressurization work costs more expensive than heat [2]. Therefore, there is a need to properly utilize high-pressure streams to save more operating costs for the chemical plants. This study addresses the large gap of assumptions in the state-of-the-art methods so that the work integration is implementable in actual situations. Mathematical programming and careful thermodynamics consideration are the primary tools in this study. The details are described in the following sections. A case study on the liquid ammonia production process is employed to demonstrate the model implementation.

2 WENS

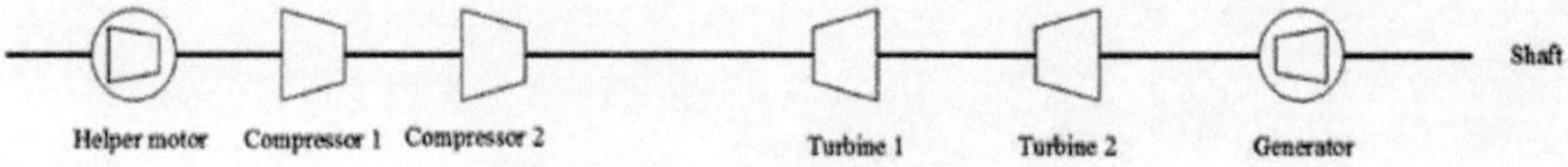

Fig. 1. Typical SSTC configuration [3]

A Work Exchange Network (WEN) is a combination of units usually found in Heat Exchange Network (HEN) and more. These units may consist of a network of 2-stream exchangers, utility heaters, utility coolers, single-shaft-turbine-compressor (SSTC) units (see Fig. 1), valves, turbines generating power, and compressors. Steam or electricity are the standard utilities to run a WEN properly. WEN considers the exchange of work and heat simultaneously. The temperatures and pressures are essential in WENS because the work of compressors and turbines is defined heavily by stream temperatures. For a turbine (compressor) to work efficiently, a high (low) inlet temperature is required. In WENS, without pressure-change equipment such as valves, turbines, and compressors, the streams at different pressures cannot be mixed unless the pressures are made equal. An SSTC consists of several compressor and turbine stages connected with a single shaft and operated at the same rotational speed. The turbines are the main driving force in WENS. Various compressors and turbines can interact simultaneously via a single shaft in a complex manner. The main challenge of stream matching is extremely difficult if all compressors and turbines cannot be operated in satisfactory regimes (away from limiting conditions such as surging and choking). The matching interaction of compressors and turbines is susceptible to the highly nonlinear functions of pressure, temperature, flow, compression/expansion ratio, etc. A complex optimization model is required to represent the highly nonlinear relationships [2]. Note that, even if the flow and inlet temperature are known constants, work duty in WENS forms a highly nonlinear function of pressure change.

There are several mathematical expressions to describe a proper WENS. The following are the main formulas implemented in this work:

- Superstructure logic constraints
- Mass, energy, and work balance
- Inlet and outlet temperature and pressure constraints
- Objective function (TAC)

The details of the constraints can be found elsewhere, with modifications to fit the computational requirements of this study.

3 Results and Discussion

In this study, a typical liquid ammonia production process is optimized by introducing SSTC to save TAC. The ammonia plants use water, air, and energy for production. Hydrocarbons are usually the primary energy source to produce hydrogen as the side product. Coal and electricity are also some of the alternative energy sources. Note that 77% of world ammonia capacity is made from natural gas due to the most efficient route, steam reforming of light hydrocarbons [4].

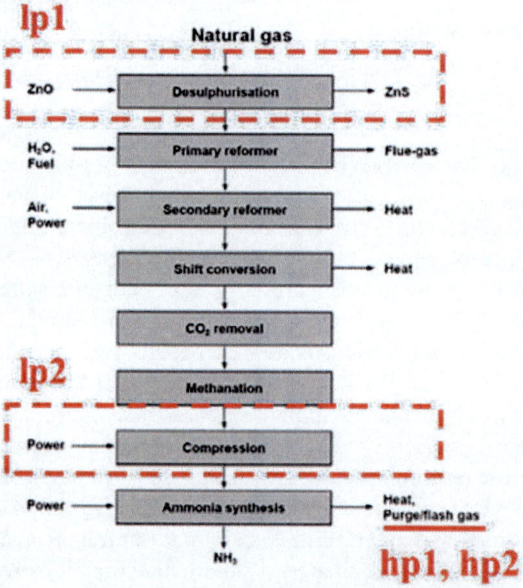

Fig. 2. Steam/air reforming process block diagram [4]

As shown in Fig. 2, the typical ammonia plants using natural gas as a feedstock include the following units [4]:

- Natural Gas Desulfurization
  Note that the natural gas may contain poison for the reformer catalyst in the form of sulfur at a maximum of 40 ppm by weight. Therefore, the desulfurization unit is

essential to reduce the sulfur content to about 5 ppm. The hydrogenation process is involved in producing hydrocarbons and hydrogen sulfide. Subsequently, the hydrogen sulfide is absorbed using zinc oxides.

- Reforming Unit
  There are two sections for this unit, the primary and secondary reforming. Hydrogen and carbon oxides are produced from the desulfurized hydrocarbon feed with steam in the primary reformer and hot air in the secondary reformer. Consequently, about 0.3 vol% CH_4 is contained in the reformed gas.
- Carbon Monoxide Conversion
  The central part of the CO contained in the reformed gas is catalytically converted to CO_2 in two catalytic stages with high and low temperatures in the CO-shift conversion.
- Carbon Dioxide Removal
  NH_3 is utilized to remove the carbon dioxide in converted gas. Therefore, ammonium hydrogen carbonate byproduct is generated in the CO_2 removal. The synthesis gas is the output with about 0.1 vol% CO_2.
- Methanation Unit
  To remove the small quantities of CO (0.1 vol%) and CO_2 (0.3 vol%), methane over a nickel catalyst is reacted accordingly. Note $CO + CO_2$ are poisons for the ammonia synthesis catalyst, and the content should be reduced to less than 10 ppm.
- Synthesis Unit
  A 300 bar synthesis gas is required for catalytic conversion of hydrogen and nitrogen to ammonia. Therefore, a centrifugal compressor is essential in this operation.

Table 1. Ammonia plant stream specifications

Stream	hp1	hp2	lp1	lp2
Components	Hydrogen (10%) Nitrogen (54.3684%) Methane (15.0822%) Argon (12.0457%) Ammonia (8.5037%)	Hydrogen (31.8546%) Nitrogen (13.146%) Methane (14.4036%) Argon (2.8%) Ammonia (37.7958%)	Nitrogen (79%) Oxygen (21%)	Hydrogen (17.0467%) Nitrogen (79.7893%) Methane (1.89784%) Argon (1.2661%)
Flow rate (kg/s)	0.59	0.59	7.5	7.5
Inlet temperature (K)	523	523	323	323
Inlet pressure (bar)	275	30	3	32
Outlet pressure (bar)	30	3.5	27	275
Min. temperature (K)	300	300	323	323
Max. temperature (K)	523	523	523	523

From these processes, two high-pressure streams (hp1 and hp2) and two low-pressure streams (lp1 and lp2) can be identified for process intensification with a work exchange network. The corresponding process conditions are listed in Table 1. Note that the corresponding streams are multi-component streams. Therefore, a suitable thermodynamics model is necessary to represent the stream's behavior at their operating condition ranges in terms of temperature and pressure.

Two sides of the high and low-pressure streams are described with the superstructures as shown in Figs. 3 and 4 as the basis to optimize the liquid ammonia production process. The corresponding binary variables enable all the possible interconnections of bypasses, valves, utility turbines, utility compressors, and SSTCs within the superstructures.

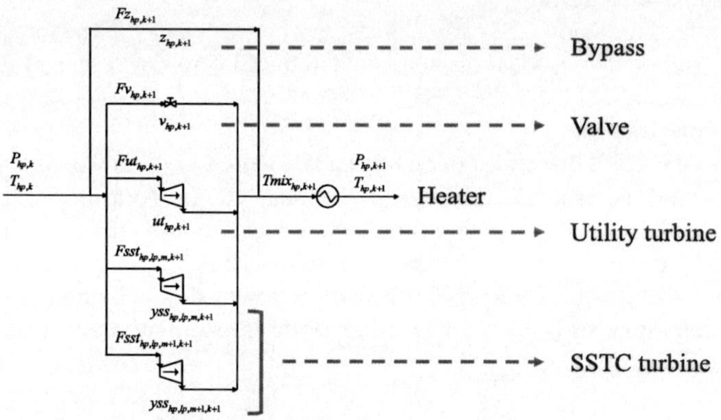

Fig. 3. Superstructure for high-pressure streams

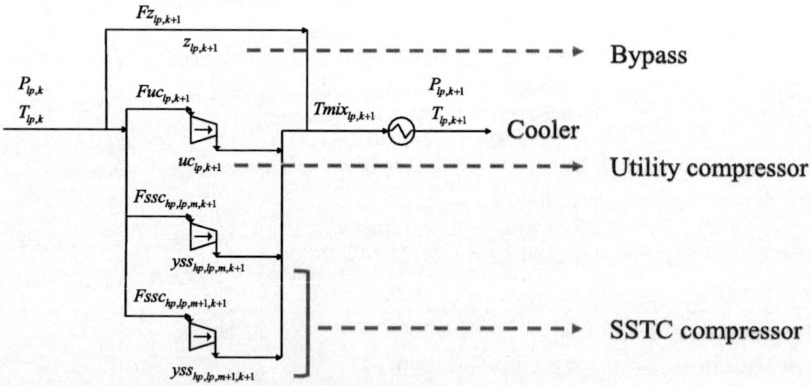

Fig. 4. Superstructure for low-pressure streams

An MINLP approach is formulated with the TAC as the objective function and the optimal configuration in Fig. 5. There are 4 SSTCs involved in the optimal design.

Moreover, the optimal configuration requires 4 helper motors to make up the power requirement of the subsequent SSTC compressors. Although more pieces of equipment are necessary with SSTC configuration, the costs are still significantly reduced, as shown in Fig. 5. A 6.12% reduction in TAC is made possible under WENS with SSTC compared to the original configuration with pure utility equipment (Base case).

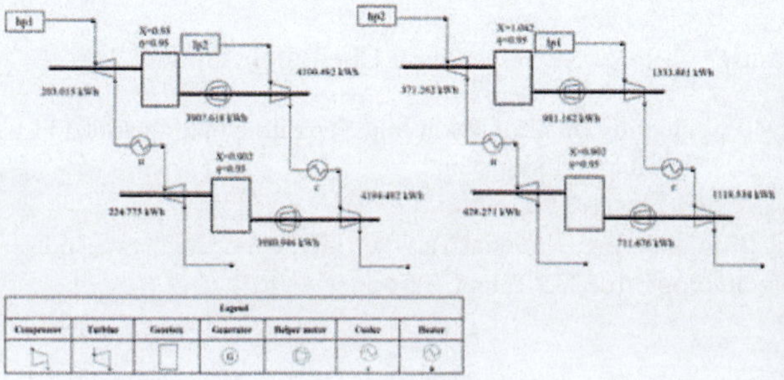

Cost	Base case	WENS
Capital cost ($/year)		
Valve	2,360	0
Utility equipment	960,000	0
SSTC unit	0	360,000
Heater and cooler	12,000	12,000
Helper motor	0	8,000
Total capital cost	974,360	380,000
Operating cost ($/year)		
Hot utility duty	0	99,795
Cold utility duty	142,358	143,090
Utility compressor	11,471,028	0
Helper motor	0	11,191,078
Total operating cost	11,613,386	11,433,963
TAC ($/year)	12,587,746	11,805,963

Fig. 5. Optimal WENS for the liquid ammonia production process and the cost breakdown

4 Conclusions

In this study, a novel WENS design strategy is generalized for work exchange network designers. A typical liquid ammonia production process is optimized, and the configuration is more practical in an industrial setup. The use of SSTCs is shown to be effective in reducing costs.

References

1. Cheng WH, Adi VSK (2018) Simultaneous optimization of non-isothermal design of water networks with regeneration and recycling. Process Integr Optim Sustain 2(3):183–203
2. Razib MS, Hasan MMF, Karimi IA (2012) Preliminary synthesis of work exchange networks. Comput Chem Eng 37:262–277
3. Yu H, Fu C, Gundersen T (2020) Work Exchange Networks (WENs) and Work and Heat Exchange Networks (WHENs)—a review of the current state-of-the-art. Indus Eng Chem Res 59(2):507–525
4. Aspen Technology Inc., Aspen Plus: Aspen Plus ammonia model. https://www.aspentech.com/en. Last accessed 15 Oct 2021

Study of the Microwave Heating Energy Effect on the Electrochemical Performance of GDCSr Electrospun Nanofiber Composite Cathode for Intermediate SOFC

Ika Dewi Wijayanti[1](✉), Putu Suwarta[1], and Chen-Chia Chou[2]

[1] Department of Mechanical Engineering, Institut Teknologi Sepuluh Nopember, 60111 Surabaya, Indonesia
dewi@me.its.ac.id

[2] Department of Mechanical Engineering, National Taiwan University of Science and Technology, 106335 Taipei, Taiwan

Abstract. The effect of microwave heating energy on the $Ce_{0.78}Gd_{0.2}Sr_{0.02}O_{2-\delta}$ (GDCSr) electrospun nanofiber composite cathode was studied. Two compositions containing 50 and 80 wt.% GDCSr nanofibers were mixed with LSCF6428 powder (L5G5 and L2G8). A temperature set of 900, 1000, and 1100 °C for 1 h were selected to perform microwave heating. The electrochemical performance was carried out at 500–800 °C. The phase composition and morphology of the composite cathode microstructure were analyzed by X-ray diffraction (XRD) and scanning electron microscopy (SEM). A porosity value close to 50% was achieved by the L5G5 composite cathode using the Archimedes method. As much 68.1 mA/cm^2 of I_0 and 1.6 Ω of resistance polarization were achieved by L5G5 cathode. The excellent collaboration between the unique properties of nanofibers and the selective heating properties of microwave energy enhances the electrochemical properties of composite cathode. Thus, a large number of triple boundary phase (TPB) sites can be effectively formed improving the transportation of oxygen ions into the electrolyte.

Keywords: Composite cathode · Intermediate SOFC · Microwave heating energy · Electrochemical performance · Electrospun nanofiber

1 Introduction

Cathode is one of the three main components of solid oxide fuel cell (SOFC) which plays important role in determining. However, the performance of SOFC is often limited by the slow kinetics of the oxygen reduction reaction at the cathode. Recently, alternative perovskite materials with mixed ionic and electronic conductivity (MIEC) have also been proposed as cathode materials for SOFC especially for operation at intermediate temperature, 500–800 °C. LSCF6428 is one of the more promising cathode materials for SOFC operation below 800 °C [1]. However, the ionic conductivity of an LSCF cathode drops rapidly as the temperature decreases. To overcome this problem, a composite

I. Tolj et al. (eds.), *Recent Advances in Mechanical Engineering*, Lecture Notes in Mechanical Engineering, https://doi.org/10.1007/978-981-19-0867-5_14

cathode can be considered as a promising material by providing a high density of triple phase boundary (TPB) while also having a contiguous ionically conducting phase which enhances the cell performance [2, 3].

A composite cathode containing $La_{0.6}Sr_{0.4}Co_{0.2}Fe_{0.8}O_{3-\delta}$ (LSCF) powder and $Ce_{0.78}Gd_{0.2}Sr_{0.02}O_{2-\delta}$ (GDCSr) electrospun nanofibers with GDCSr electrolytes which was conventionally fired at 1100 °C in air for 2 h were previously studied by our group [4]. An increase of 40% porosity was achieved by adding a second phase of 20 wt.% GDCSr nanofibers into the LSCF cathode. Furthermore, the exchange current density dramatically increased from 4.7 mA/cm^2 to 51.9 mA/cm^2, respectively, from LSCF to L8G2 alloy at 800 °C operating temperature. A reduction of the resistance polarization (Rp) from 23.4 to 2.1 Ω was reached by changing the cathode from LSCF to L8G2 alloy, respectively, representing the enhancement of the electrochemical properties by providing much amount of TPB. In this case, an increase of the sintering temperature of the composite cathode results in the enlargement of the grain, especially for the nanofibers causing a performance deterioration of the composite cathode [5, 6]. Therefore, a microwave heating energy is performed in this allowing the accelerating heating energy while maintaining the structure of the composite cathode.

Microwave heating is fundamentally different from conventional heating mainly in the heating methods, which facilitates a molecular level of rapid heating requiring less energy consumption, lower temperature, and shorter sintering time allowing an accelerated densification rate as a direct result of volumetric heating, grain growth control, uniform microstructure, and high density without the use of sintering aids [7]. Therefore, enhanced properties of materials can be produced by microwave heating techniques than by using conventional heating methods [8, 9]. However, the selective heating mechanism of microwave energy in certain materials causes the properties of the resulting material to be unpredictable because of the indirect relationship between the heating temperature and the resulting properties of the material.

2 Experimental Methods

2.1 Nanofiber Fabrication

$Gd(NO_3)_3.6H_2O$, $Ce(NO_3)0.6H_2O$, and $Sr(OH)_2.8H_2O$ with the typical purity of more than 99.5% were dissolved into the de-ionized water (DI water) by using magnetic stirrer. This mixture was then poured into the well-mixed solution containing PVP and ethanol. This polymer precursor solution was used for fabricating the electrospun nanofibers. The voltage of power supply was adjusted to 20 kV. The flow rate of injection pump was 1 ml/h and the working distance between the needle tip and drum was 8 cm.

2.2 LSCF Fabrication

La_2O_3, $SrCO_3$, Co_3O_4, and Fe_2O_3 powders with the typical purity of not less than 99.5% were mixed to produce LSCF6248 powder using oxide method. The slurry was then milled into the jar for 24 h then calcined in the furnace at 1100 °C for 3 h dwell time with 5 °C/min heating after drying 1 day.

2.3 Composite Cathode Fabrication

GDCSr nanofiber and LSCF6428 powder were mixed with a ratio of 50:50 and 80:20, respectively. Both were then poured into a 2-day well-immersed Terpineol and binder. Magnetic stirrer with a heating platform was used to convince the uniform mixing between LSCF powder and the GDCSr nanofibers. This cathode paste was then screen printed on to the GDCSr electrolyte with the surface area of 0.5 cm^2 and the thickness of 20 μm. Platinum paste was used as the counter and reference electrodes, while platinum wire was used as a current collector. This method can also be found in our previous work [4, 10].

2.4 Characterizations

Porosity Measurement. Archimedes method was used to measure the porosity.

Structural Characterization. Rigaku DMAX-B Japan was used to perform XRD measurement on the cathode with $\lambda = 1.5418$ Å ($CuK_{\alpha 1}$) and 5 deg/min scanning rate from 20° to 80° measurement range. A JEOL JSM-6500F Japan was used to characterize the microstructure and distribution of phases in the cathode.

Electrochemical Measurement. AC impedance analyzer (Solartron 1260 and 1287 UK) was used to measure the electrochemical performance of the composite cathodes at 500–800 °C range. A frequency range of 0.1–100 kHz and 100 mV voltage amplitude under alternating current was applied. In addition, Arrhenius diagram containing natural logarithmic of the resistance polarization (Rp) and the inverse of temperature was plotted to determine the activation energy of the cathodes [11, 12]. The exchange current density was also calculated and estimated from the Tafel measurements.

3 Results and Discussion

3.1 Porosity Data

Porosity measurements were performed on a green pellet containing screen printed of composite cathode paste. The porosity data for the alloys at 900, 1000, 1100 °C of microwave heating (MH) and 1100 °C of conventional heating (CH) can be seen in Fig. 1. At 900 °C, the porosity value for all cathodes is the highest compared to the other temperatures. In contrast to the CH, the porosity values for all cathodes are the lowest. Indeed, the porosity data obtained by 1100 °C MH is still higher than that of by 1100 °C CH. On the other hand, the addition of GDCSr nanofiber into LSCF powder increases the porosity of the cathodes. Further to this, the increase of GDCSr nanofiber composition to 80 wt.% results in the more increasing of the porosity data. These phenomena indicate that the addition of nanofiber contributes to maintain the porosity formation in the cathodes. Indeed, the selective heating properties of microwave energy play an important role to hinder the growth of the cathodes, thus the porosity formation becomes more pronounce when the composition of nanofiber increases. When it comes to the higher MH, the porosity data for all composites decreases equally. On the other hand, the porosity data of the studied alloys resulting from the CH shows the lower value compared to the MH. Performing the CH to the cathodes results in the lower porosity formation compared to the MH. In addition, the smallest porosity formations were achieved by the LSCF cathode using CH at 1100 °C [4].

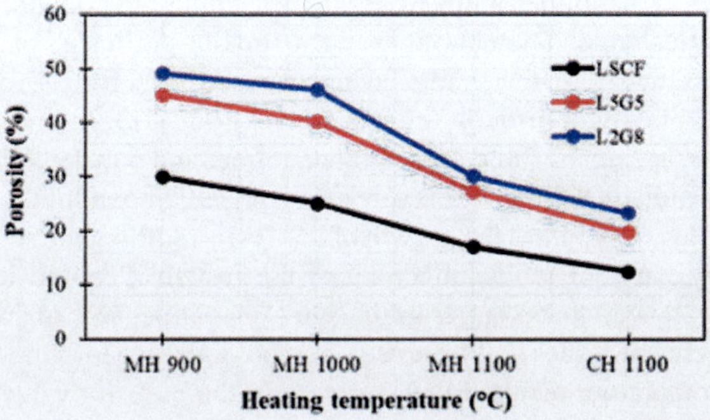

Fig. 1. Porosity data for LSCF, L5G5, and L2G8 cathode alloys measured by Archimedes method at 900, 1000, 1100 °C of microwave heating (MH) and 1100 °C of conventional heating (CH) temperatures.

3.2 Structural Characterizations

XRD pattern. Table 1 shows the calculated crystallite size from XRD patterns. It is shown that the crystallinity of the composite cathodes was improved. The existence of nanofiber can suppress the growth of the crystallite size. A decrease in crystallite size from 211.05 to 173.03 and then increase to 281.17 nm was obtained by LSCF, L5G5, and L2G8 alloys, respectively, at 1 h of 900° C. The similar trend was also observed when the temperature was increased to 1000 and 1100° C. Indeed, L5G5 alloy achieved the smallest crystallite size compared to the other alloys showing that the optimum chemical composition was obtained in this alloy.

Table 1. Crystallite size of the studied alloys at different microwave heating temperatures.

Temperature (°C)	Crystallite size (nm)		
	LSCF	L5G5	L2G8
900	211.05	173.03	281.17
1000	337.5	167.94	364.22
1100	506.53	356.76	467.63

SEM images. Microstructures of the studied alloys at three different MH can be seen in Table 2. Increasing the temperature from 900,1000, and to 1100 °C results in an increase in the grain size of the LSCF cathode from 846, 1563, and to 2282 nm, respectively, which can be seen in Table 3a–c. Similarly, the grain size of L5G5 cathode increases linearly with the temperature. However, the increase in the grain size of LSCF cathode is more significant compared to that of the L5G5 as seen in Table 3d–f. At 900 and 1000 °C, the GDCSr nanofiber and LSCF powder which are contained in L5G5 alloy are clearly distinguishable. The nanofiber has a smaller grain size of 340 and 470 nm compared to the powder of 630 and 796 nm. Indeed, at 1100 °C both of them are indistinguishable, making it difficult to measure the grain size. In line with L5G5 alloy, at all temperatures, as seen in Table 3g–i, both nanofiber and powder which are contained in L2G8 alloy are indistinguishable. The presence of GDCSr nanofibers in the composite cathodes contributed to suppress the growth of LSCF. In addition, the selective heating by microwave energy can be beneficial in hindering the growth of composite cathode alloys. A previous study showed that a larger grain size of nanofiber was obtained using CH in L8G2 alloy [4]. When it comes to the porosity, it is obviously seen in the SEM images that increasing the temperature resulted in a decrease in the porosity value due to the grain growth of nanofibers and powder containing in the composite cathodes. However, the significant grain growth of pure LSCF cathode causes lower porosity values compared

Table 2. SEM images of (a), (b), (c) for LSCF; (d), (e), (f) for L5G5; and (g), (h), (i) for L2G8 using microwave heating at 900, 1000, and 1100 °C for 1 h on top of electrolyte GDCSr.

Alloys	Microwave heating temperature (°C)		
	900	1000	1100
LSCF	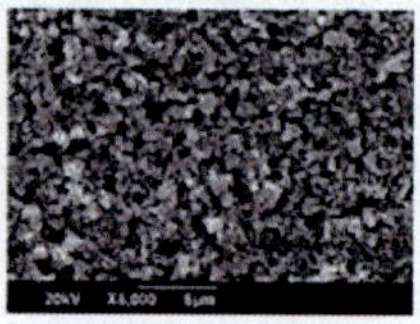 (a)	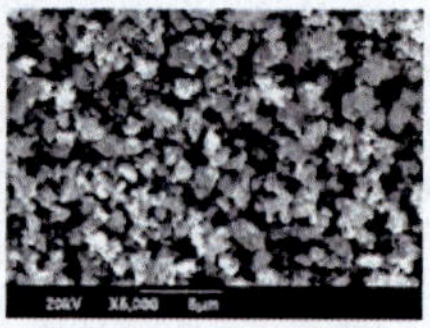 (b)	 (c)
L5G5	 (d)	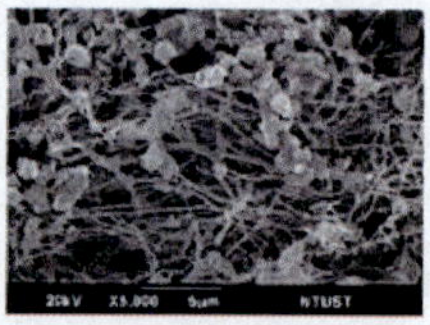 (e)	 (f)
L2G8	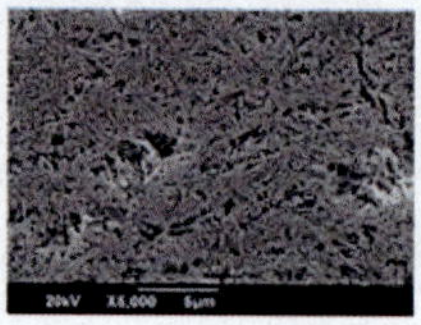 (g)	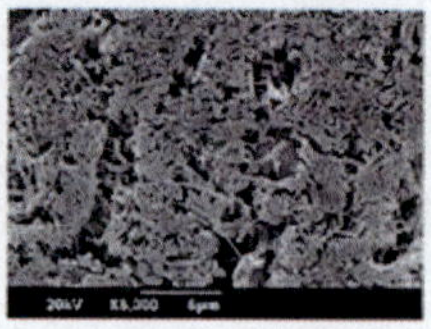 (h)	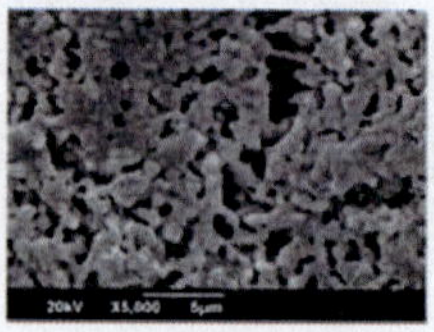 (i)

to other cathodes. This phenomenon indicates that the GDCSr nanofibers contribute to inhibit the growth of the composite cathode resulting in an increase in porosity. In addition to the increased porosity, an increased contact area, an enlarged surface area, and an improved three-phase boundary are achieved.

3.3 Electrochemical Performances

Electrochemical Impedance Spectroscopy (EIS). The equivalent circuit model as in [4] was then applied to fit the resulted Nyquist plot from EIS measurements. As seen in Table 3, at lower heating and operating temperatures, the Rp values were larger than at higher heating and operating temperatures showing that the charge transfer process is difficult to occur, thus the electrode reaction at the lower temperature is limited. Indeed, at higher operating temperatures, a decrease in the Rp value was achieved. In addition, L5G5 cathode obtained the smallest Rp value when the higher heating and operating temperatures were applied. These phenomena indicate that the presence of nanofiber in the cathode contributes to enhancement of the porosity formation representing the electron sites, which can easily break down the oxygen gas into the oxygen ions to continue the transportation to the electrolyte of the composite cathodes.

Table 3. Resistance polarization data from the resulted fitting for all the composite cathode electrodes using 900, 1000, and 1100 °C microwave heating temperatures at 500–800 °C operating temperatures.

Alloys	Resistance polarization (Ohm)											
	Microwave heating temperatures (°C)											
	900				1000				1100			
	Operating temperatures (°C)											
	500	600	700	800	500	600	700	800	500	600	700	800
LSCF	290.9	24.9	2.6	2.2	915.6	80.8	8.7	3.1	3370	192.8	20.7	7.4
L5G5	4791	424.6	45.9	7.7	884.1	48.1	4.7	1.6	1281.9	101.2	12.3	1.8
L2G8	6334.8	319	39.7	17.9	3402	112.3	26.8	11.9	2067.8	86.5	13.8	7.2

Activation Energy. Table 4 shows the calculated activation energy for all the studied alloys which clearly describes that the LSCF has the smallest activation energy compared to the other cathodes followed by L5G5 and L2G8 alloys.

In the case of the largest L5G5 activation energy, the available catalytic sites are abundant to increase the oxygen reduction reaction rate at the cathode. The transfer of oxygen ions from the cathode to the anode is faster, the charge transfer capacity of the anode is larger, so that the oxygen reduction rate is rapidly larger and the performance of SOFC is significantly higher. In fact, the addition of GDCSr nanofibers decreases the activation energy which can be related to the formation of higher porosity and poor association between GDCSr nanofibers and LSCF powder at the cathode.

Table 4. The activation energy of the studied alloys at 900, 1000, and 1100 °C microwave heating temperatures.

Composite cathodes	Activation energy (eV)		
	Microwave heating temperature (°C)		
	900	1000	1100
LSCF	1.127	1.130	1.134
L5G5	1.272	1.231	1.289
L2G8	1.312	1.306	1.294

***Exchange Current Density*.** The contribution of nanofiber in the catalytic activity of SOFC including the absorption of oxygen ions on the cathode surface, the reduction of electron sites in the triple phase boundary region, and the release of electron from reduction of hydrogen molecules on the anode surface can be determined using Tafel curve.

Table 5 shows I_0 value for all studied alloys using MH of 900, 1000, and 1100 °C at 500–800 °C operating temperatures. At 900 °C MH, LSCF achieved the highest I_0. However, at 1000 and 1100 °C, L5G5 shows the opposite. Indeed, an increase in wt.% of nanofiber into LSCF resulted in a significant decrease in the I_0, which especially occurred at the lower MH. Porosity becomes key parameter determining the performance of SOFC including I_0 and Rp. Further to this, poor connection between nanofiber and powder at the lower MH led to reduce the oxygen ion conduction and increase the Rp.

In fact, the higher number of nanofibers in the composite cathode absorbs insufficient amount of microwave energy. Thus, the growth of LSFC grains is inhibited and their shape remains. In other words, the structure seems to have smaller LSCF powder than the others.

Table 5. The exchange current density value (mA/cm^2) of the composite cathodes at microwave heating of 900, 1000, 1100 °C at 500–800 °C operating temperature.

Alloys	Exchange current density (mA/cm^2)											
	Microwave heating temperatures (°C)											
	900				1000				1100			
	Operating temperatures (°C)											
	500	600	700	800	500	600	700	800	500	600	700	800
LSCF	0.27	3.6	38.0	49.5	0.15	3.1	27.4	35.1	0.02	0.5	4.8	14.7
L5G5	0.016	0.257	4.8	14.8	0.086	1.097	11.4	68.1	0.061	0.876	8.0	60.5
L2G8	0.012	0.278	2.5	6.1	0.023	0.789	3.7	9.2	0.038	1.025	7.2	15.1

4 Conclusions

The effect of microwave heating energy on the electrochemical performance of GDCSr electrospun nanofiber composite cathode for intermediate SOFC was studied. Selective heating of microwave energy significantly affects the properties of the composite cathode. The existence of GDCSr nanofiber as the second phase in the LSCF alloy inhibits the growth of the composite cathode grains. These phenomena can be beneficial for the electrochemical performance of the cathodes especially when the amount of nanofiber is 50 wt.% and the microwave temperatures are 1000 and 1100 °C. As much 68.1 mA/cm^2 of I_0 and 1.6 Ω of Rp were achieved by L5G5 cathode. The excellent TPB connection between nanofiber and powder with the optimum amount of nanofiber resulted in the excellent performance of the composite cathodes.

Acknowledgements. Authors wish to acknowledge the Department of Mechanical Engineering NTUST for the assistance and the place for doing the experiments and the Department of Mechanical Engineering ITS for the support to finish this research.

References

1. Perry Murray E, Sever MJ, Barnett SA (2002) Electrochemical performance of (La, Sr)(Co, Fe)O3–(Ce, Gd)O3 composite cathodes. Solid State Ionics 148:27–34. https://doi.org/10.1016/S0167-2738(02)00102-9
2. Zhu H, Kee RJ (2017) Modeling protonic-ceramic fuel cells with porous composite electrodes in a button-cell configuration. J Electrochem Soc 164:F1400. https://doi.org/10.1149/2.0591713jes
3. Lu X, Heenan TMM, Bailey JJ, Li T, Li K, Brett DJL, Shearing PR (2017) Correlation between triple phase boundary and the microstructure of solid oxide fuel cell anodes: the role of composition, porosity and Ni densification. J Power Sources 365:210–219. https://doi.org/10.1016/j.jpowsour.2017.08.095
4. Wijayanti ID, Chou C-C (2015) (La, Sr)(Co, Fe) O3-(Ce, Gd)(Sr) O2 electrospun nanofiber composite cathode for solid oxide fuel cells. In: 2015 International conference on advanced mechatronics, intelligent manufacture, and industrial automation (ICAMIMIA), pp 194–197. IEEE (2015)
5. Aruna ST, Balaji LS, Kumar SS, Prakash BS (2017) Electrospinning in solid oxide fuel cells—a review. Renew Sustain Energy Rev 67:673–682. https://doi.org/10.1016/j.rser.2016.09.003
6. Ahn M, Cho J, Lee W (2019) One-step fabrication of composite nanofibers for solid oxide fuel cell electrodes. J Power Sources 434:226749. https://doi.org/10.1016/j.jpowsour.2019.226749
7. Fariñas JC, Moreno R, Pérez A, García MA, García-Hernández M, Salvador MD, Borrell A (2018) Microwave-assisted solution synthesis, microwave sintering and magnetic properties of cobalt ferrite. J Eur Ceram Soc 38:2360–2368. https://doi.org/10.1016/j.jeurceramsoc.2017.12.052
8. Ali SM, Anwar M, Somalu MR, Muchtar A (2017) Enhancement of the interfacial polarization resistance of La0. 6Sr0. 4Co0. 2Fe0. 8O3-δ cathode by microwave-assisted combustion method. Ceram Int 43:4647–4654

9. Jais AA, Ali SM, Anwar M, Somalu MR, Muchtar A, Isahak WNRW, Tan CY, Singh R, Brandon NP (2017) Enhanced ionic conductivity of scandia-ceria-stabilized-zirconia (10Sc1CeSZ) electrolyte synthesized by the microwave-assisted glycine nitrate process. Ceram Int 43:8119–8125
10. Hua C-H, Chou C-C (2015) Preparation of nanoscale composite LSCF/GDCS cathode materials by microwave sintering for intermediate-temperature SOFC applications. Ceram Int 41:S708–S712. https://doi.org/10.1016/j.ceramint.2015.03.128
11. Jensen F (1985) Activation energies and the Arrhenius equation. Qual Reliab Eng Int 1:13–17
12. Laidler KJ (1984) The development of the Arrhenius equation. J Chem Educ 61:494

Tricycle with an Electric Drive Mechanism for Post-stroke Rehabilitation

Rosadila Febritasari(✉) and I Made Londen Batan

Laboratory of Product Engineering, Department of Mechanical Engineering, Institut Teknologi Sepuluh Nopember, 60111 Surabaya, Indonesia
rosadilafeb@gmail.com

Abstract. The manual pedal tricycle is helpful in recovering the limb motor function for post-stroke patients. In certain conditions, they have to pedal the tricycle manually through hilly roads as hard as possible, whereas they still have muscle weakness and strongly discourage to do hard physical activities. Therefore, this research will be focused on developing the tricycle with an electric drive mechanism and finding the benefit for post-stroke mobility. Based on the ergonomics of the Indonesian body, an e-tricycle prototype is designed and manufactured in the delta concept. Assuming that the e-tricycle will be used on hilly or rough roads by 100 kg of rider's weight, so it requires a 500 W of electric motor power and a 48 V-14Ah lithium battery. Furthermore, the e-tricycle tested strength and stability based on SNI 7519:2009 and braking based on SNI 4404:2008 and SNI 1049:2008. The resulting test showed that the e-tricycle can move and brake well on flat and hilly roads at 13° slope angle with 18.9 km/h maximum speed. The e-tricycle is stable at 10° slope angle. The effect of road terrain and speed from electric support setting is significant on cycling time, heart rate, VO_2 max, and energy expenditure ($P < 0.05$). Cycling without electric support makes the leg muscles pedal harder when passing hilly roads. Therefore, the cyclists who are post-stroke patients are strongly recommended to use electric support so that the pedaling is lighter, the heart beats normally, and less energy expenditure. Thus, the second stroke attack does not relapse.

Keywords: Post-stroke patients · Rehabilitation · Electric motor · Battery · Electric tricycle

1 Introduction

Cycling using a tricycle is one of the post-stroke physical rehabilitation tools that has been proven to be useful in recovering limbs motor function [1], improving body fitness [2] and reducing their stress by doing leisure activity outdoor [3]. The cycling movement was applied in ITS-Tricycle (SERAITS) that was developed in the Product Design and Development Laboratory-ITS [4].

In general, stroke rehabilitation is conducted by the medical rehabilitation expert in a hospital or rehabilitation center. When rehabilitation is conducted outside of the hospital, such as in the park or public area, problems will come. For example, the stroke survivors

I. Tolj et al. (eds.), *Recent Advances in Mechanical Engineering*, Lecture Notes in Mechanical Engineering, https://doi.org/10.1007/978-981-19-0867-5_15

get tired easily to pedaling so they stop for a while for recovering their energy, and then continue cycling. In Indonesia, residential areas are usually flat and hilly terrain, another problem comes when the stroke survivors cycle using tricycle hardly through several hilly roads, whereas they have muscle weakness in their limbs which causes moving hardly and improperly [5]. Because of this condition, they are strongly discouraged to do vigorous physical activities [6]. They also can't pedal in a single full rotation, and usually the tricycle is equipped with pedal straps for safety so that their feet remain to stay on pedals when pedaling. Because of the stiff legs, the stroke survivors are only able to pedal in push-pull movements that means one strong foot push down the pedal and the other foot pulls back the other pedal [7].

To solve the problems, stroke rehabilitation experts mention that e-tricycle is innovative transportation with electric support modes and has various benefits. From the health aspect, e-tricycle can be adjusted for disabilities and chronic diseases rehabilitation programs, because it can be used to train leg muscles so that the muscles won't stiff again and can improve body fitness [8]. From the psychological aspect, cycling can motivate sedentary people and stroke survivors to perform physical activities to minimize stress and increase self-confidence [9]. Another study mention that electric support modes in e-bike provide pedal torque to reduce pedaling load which can be utilized to ease the stroke survivor's mobility in short durations and high speeds because it reduces energy expenditure and VO_2 max by 24% lower than conventional bike [9].

To help the post-stroke physical rehabilitation, an e-tricycle prototype was designed and made, which was built with an electric drive mechanism addition into a tricycle, besides pedaling manually. The mechanism has consisted of electrical components and supporting sensors. As electrical transportation and post-stroke physical rehabilitation tool, e-tricycle is tested according to the Indonesian National Standard (SNI) [10–12]. To find out the benefits of e-tricycle for stroke mobility, a driving test was conducted on a flat and hilly road with a certain speed from PAS to analyze the effect from cycling time, heart rate, VO_2 max, and energy expenditure responses.

The result of this study is an e-tricycle prototype as a physical rehabilitation tool and stroke mobility. The e-tricycle is expected to minimize the stress psychologically because while doing rehabilitation, they can also refresh their mind. This is very useful to speed up stroke recovery.

2 Methods

The research method that was used in this study was the application of design methods for manufacturing, which are divided into three, namely:

First: designing an electric tricycle, which begun with arranging list of requirements from the stroke survivor need to e-tricycle, then designing the frame, electric drive mechanism, and other accessories. The final design was realized by the manufacturing and assembly process to become a prototype electric tricycle.

Second: performance testing of e-tricycle which was conducted based on several standards, such as SNI 7519: 2009, is tricycle safety test [10], SNI 1049:2008 is bicycle safety requirements [11], and SNI 4404:2008 is vehicle braking test [12].

Third: a driving test is conducted to evaluate the e-tricycle movement ability on flat and hilly roads and also speed variations from PAS settings.

3 E-tricycle Prototype Design

3.1 List of Requirements

The design of electric tricycles is adjusted to stroke survivor needs, as seen in Table 1.

Table 1. List of requirements of e-tricycle

T/W	Customer needs	T/W	Customer needs
1. Function		4. Operational system	
T	The drive mechanism can be pedaled manually	T	Can be pedaled easily in flat and rough road
T	E-tricycle applies pedal system	T	Ease and comfort to ride
2. Safety		T	Electric motor powered pedaling
T	Maximum cyclist weight is 100 kg	W	Can move on a hilly road
T	Water-resistant wiring system	5. Manufacturing process	
T	E-tricycle operation can be done by on–off system easily	T	The tricycle frame can be manufactured
3. Geometry and dimension		T	Assembly process uses standard tools
T	E-tricycle has three wheels in delta concept	T	E-tricycle components can be assembled easily
T	Frame weight is less than 25 kg	6. Maintenance	
T	E-tricycle height is less than 1.5 m	T	Easy to maintain
T	The electrical component weight is less than 13 kg	T	The battery can be charged without removing the battery

* T = Term, W = Wish

3.2 Tricycle and Electric Drive Mechanism

Based on the ergonomics of the Indonesian body, an e-tricycle prototype is designed and manufactured in the delta concept in delta concept which has one wheel in front as a dead wheel and two rear wheels as a drive wheel [13]. Cyclist weight was assumed 100 kg and the e-tricycle was moved on the hilly or rough road, so the e-tricycle needs 500 W of motor power and 48 V-14Ah of Li-battery that can be seen in Fig. 1.

This mechanism is referred to as pedal and electric which indicates that the electric motor will turn on if the cyclist starts to pedal and gradually reduces the pedal power before reaching a certain speed. There is Pedal Assistance Sensor (PAS) as a motion detection sensor pedal (cadence) in the electric motor to reduce the pedaling load for cyclist who has leg muscle weakness so that they can adapt to pedaling movement. When pedaling movement is detected, the electric motor will turn on and give the torque motor to reach the speed that has been set by the cyclist.

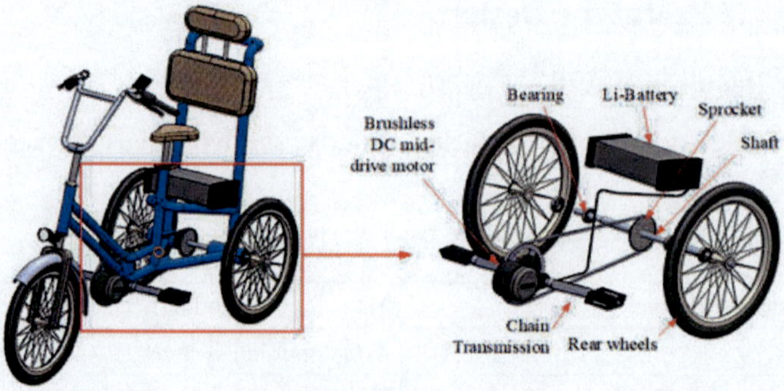

Fig. 1. Tricycle and electric drive mechanism design

3.3 E-Tricycle Prototype

Prototype electric tricycle has 1500 × 600 × 1100 mm^3 of dimension that can be seen in Fig. 2. The main components are the frame, 20-inch wheels, pedal safety, drive mechanism (electric motor), pedals, and transmission chain. While the main electronic components are: 500 W brushless DC mid-drive motor, 48 V-14 Ah Li-battery, throttle, controller, and LCD display. The supporting sensors inside the electric motor are speed detecting sensor, Pedal Assistance Sensor (PAS), and magnetic brake sensor.

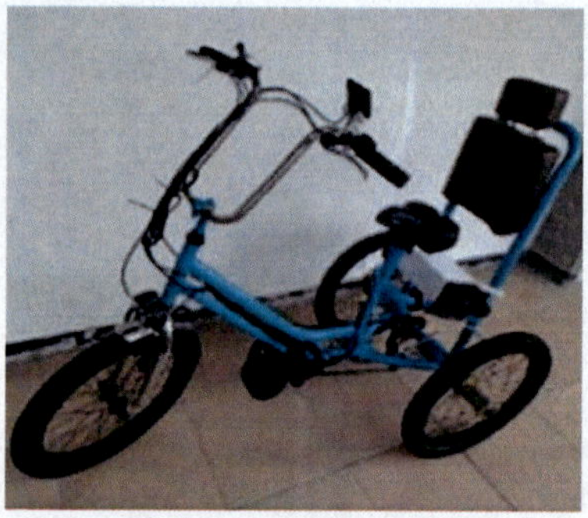

Fig. 2. E-tricycle prototype

4 Performance Test

4.1 Performance Test Procedure and Equipment

Performance test of e-tricycle refers to SNI 7519: 2009, that is, tricycle safety test, consists of static and dynamic strength test, dynamic test and stability test [10]. In static testing, the e-tricycle was given a load of 50 kg in stationary and moving condition. The e-tricycle was observed whether there was damage to the frame and its components. In stability testing, the e-tricycle is positioned in three directions (*downhill, uphill, side hill)* on 10° slope. It is expected that the bike will not roll over, overturn, or upside down. Dynamic testing conducted by e-tricycle moved straight, zig-zag, and u-turning in six level speeds using (PAS) on flat and hilly road (5°, 10°, and 13° of slope) [10].

Furthermore, vehicle brake testing as e-tricycle safety refers to SNI 4404:2008 and SNI 1049:2008. It was conducted using 70% of the e-tricycle maximum speed, namely, 15 km/h as initial speed in dry and wet terrain [12]. Braking distance must not exceed 7 m (dry conditions) and 5 m (wet conditions) [11].

Equipment for performance testing is an e-tricycle, speed detecting sensor, LCD display, stopwatch, and cone race.

4.2 Result of Performance Test

Static strength testing showed the e-tricycle can be pedaled manually and electrically, wheels can spin, and there is no damage to the backrest frame, post-seat frame, or rear axle, so it can be declared safe. In stability test, the results of stability testing showed no component damage, rolled over, upside down, and overturned so that it can be declared safe, as can be seen in Fig. 3. In dynamic test, the result showed the e-tricycle could move straight, U-turn, and zig-zag on flat roads. The e-tricycle was able to move 18.9 km/h of speed on uphill 13° of slope. And the result of brake test showed that the deceleration values were 2.42 m/s^2 and 2.88 m/s^2 and braking distances are 2.31 m and 2.83 m for dry and wet braking conditions, respectively.

(a) E-tricycle face to sidehill (b) E-tricycle face to uphill (c) E-tricycle face to downhill

Fig. 3. Stability test of e-tricycle

5 Driving Test

5.1 Driving Test Procedure and Equipments

Drive test of e-tricycle referred to the Roos Bulthuis' research [8] and was modified to factors and responses. The factors used in this study were road terrain (flat and uphill) and the speed variation of PAS on electric motors. The response parameters measured were cycling time, heart rate, VO_2 max, and energy expenditure.

The six participants were sedentary people (four men, two women, average age 30 ± 8 years, weight 68 ± 20 kg) and willing to be the object of research. The drive test procedure was conducted using e-tricycle in two different road terrains, such as flat and uphill road 5° of slope on 30 m distance.

E-tricycle speed was set in PAS 0, 1, and 5. In PAS 0, the electric motor does not provide pedal torque or the cyclists pedaled with their power at an average speed of 6 km/h. In PAS 1 and PAS 5, the electric motor provides torque on the paddle until the e-tricycle runs at an average speed of 8 km/h and 21 km/h, respectively. Each session changes, the participant's vital sign was checked and rested for 10 min.

Equipments for driving test are e-tricycle, speed detecting sensor, LCD display, stopwatch, and tensimeter.

5.2 Result of Driving Test

Quantifiable data was obtained from driving test and then the influence of both factors on responses by the two factors analysis of variance without interaction method was analyzed. P-value < 0.05 states a significant influence of statistical tests. All data were represented in the form of an average value of ± standard deviation (SD), as can be seen in Table 2 and the graphic can be seen in Fig. 4.

Table 2. Mean (±SD) data of cycling time (sec), heart rate (bpm), VO_2 max (ml/kg/min), and energy expenditure (kcal/min) in the flat and hilly road with acceleration based on PAS setting from 0 (6 km/h), 1 (8 km/h), and 5 (21 km/h)

Electric tricycle	Flat road						Uphill road						P-value
	PAS 0		PAS 1		PAS 5		PAS 0		PAS 1		PAS 5		
	Mean	SD	Mean	SD	Mean	SD	Mean	SD	Mean	SD	Mean	SD	
CT	21.41	5.77	16.00	3.95	11.85	3.89	22.22	5.73	18.22	2.70	15.88	2.106	< 0.05
HR	97.08	15.28	89.92	13.97	80.25	16.60	123.75	23.28	107.92	22.61	99.25	20.67	< 0.05
$VO_{2\ max}$	20.55	0.71	20.57	0.85	20.43	0.55	20.42	1.81	17.62	1.25	16.88	1.54	< 0.05
EE	4.21	0.99	3.72	0.82	3.19	0.79	6.53	2.19	5.12	1.82	4.46	1.45	< 0.05

* CT = cycling time; HR = Heart rate; VO_2 max; EE = Energy expenditure

All responses have a significant influence on road terrain and speed variations (P < 0.05). American College of Sports Medicine (ACSM) states muscle hypertrophy will effectively occur in moderate-intensity exercise which means the heart rate zone is in the range of 60–75% of the maximum heart rate [14]. A previous study from Jette et al. stated energy expenditure and VO_2 max are in the range of 5.0–7.4 kcal/min and 15.3–22.9 ml/kg/min (men) and 3.5–5.4 kcal/min and 12.6–19.8 ml/kg/min (women), respectively, for moderate exercise intensity [15].

Cycling time had a significant effect (P < 0.05) on factors. This means that cycling time is longer on the uphill road with PAS 0 than on the flat road as shown in Fig. 4a. This showed that the cyclist is able to go faster in a short time with PAS 1–5 on uphill road.

Heart rate had a significant effect on factors (P < 0.05). In Fig. 4b, cycling with PAS 0 caused the heart to beat faster than using PAS (1 and 5), as well as on the uphill road the heart rates are higher than on flat road. This means that a higher heart rate indicates the cyclist has performed moderate physical activity. As shown in Fig. 4b, the participant heart rate exceeds the lower limit of the heart rate zone in PAS 0 and on uphill road.

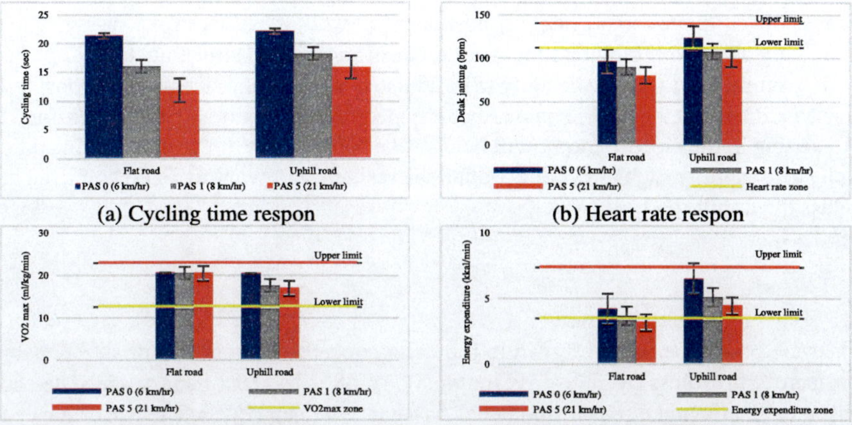

(a) Cycling time respon (b) Heart rate respon

(c) VO2 max respon (d) Energy expenditure respon

Fig. 4. Graph of responses to factors in the flat and hilly road with acceleration based on PAS setting from 0 (6 km/h), 1 (8 km/h), and 5 (21 km/h)

The oxygen capacity in the body is indicated by the VO_2max to determine the cyclist's fitness after cycling. VO_2 max respondents had a significant effect ($P < 0.05$) on factors (see Fig. 4c). This indicates that the fitness was smaller when cycling using PAS in e-tricycle which means that the fitness improvement occurs in a small capacity.

Energy expenditure showed pedaling energy when cycling. The participant's pedaling energy differed significantly in each condition. In Fig. 4d, the moderate-intensity exercises mostly occurred on uphill roads. This indicates the cyclist will spend smaller energy expenditure when using PAS so that they do not get tired quickly, or also they do not have to stop to wait for the energy recovering when they are tired.

5.3 Discussion of Test Drive

Our study showed that the heart is beating faster means the cyclist had performed moderate-intensity physical activity. Simons showed the average heart rate was higher significantly when cycling without PAS than using PAS in the highest mode [16]. It means that cycling with electric support is only able to achieve light-intensity exercise [8, 16]. Langford said heart rate and VO_2 max are lower on an electric bike than a conventional bike and on foot. Energy expenditure is also relatively stable on the uphill road due to electric support [9].

From a mobility aspect, the e-tricycle doesn't spend more energy expenditure to do mobility by utilizing PAS (1–5) on this electric motor. However, if the rider wants to focus on improving body fitness, muscle training, and cardiovascular training, they can use PAS in lower mode or just turn it off until at least moderate-intensity exercise [9]. Cycling without using PAS will make the body system, especially leg muscles work harder when passing the uphill road [5]. The heart beats faster and the energy expenditure is greater which indicates that the cyclist has performed a heavy physical activity [14]. In addition, stroke survivors are also not recommended to do the heavy physical activity because the cardiovascular system condition is not well. It causes stroke survivors to get tired easily as a result of a long recovery process and second stroke attack will be back.

Therefore, the e-tricycle design can be adjusted to the stroke survivors by considering their condition that is not the same as healthy people and requires special attention. The PAS setting in the electric motor will help stroke survivors in doing mobility in addition to physical therapy such as cycling around the residence or wanting to go home but have to pass the uphill road.

6 Conclusion

E-tricycle prototype was designed in the delta concept. The electric drive mechanism consists of a brushless DC mid-drive motor 500 W, 48 V-14Ah Li-battery, speed detecting sensor, magnetic brake sensor, and Pedal Assistance Sensor (PAS). The performance test results showed that there was no damage or fatigue on e-tricycle frame and components, the e-tricycle could move straight, U-turn, and zig-zag on a flat road and was able to move up 18.9 km/h of speed on uphill road with a slope of 13°, without upside down, overturned, and rolled over.

Based on statistical analysis, there was a significant influence between road terrain factors (flat and hilly) and speed of PAS settings (0, 1, and 5) to cycling time, heart rate, VO_2max, and energy expenditure response. E-tricycle with higher speeds in PAS 1–5 will affect cycling time shorter than PAS 0. The heart rate beats slower, VO_2max is lower, and also energy expenditure is smaller than PAS 0. The e-tricycle design can be aimed for stroke survivors and using the PAS setting (1–5) on this e-tricycle will ease them for doing mobility.

References

1. Febritasari R, Batan IML (2019) Implementation Fugl Meyer assessment of lower extremity method to develop a post-stroke rehabilitation procedure using ITS tricycle. IOP Conf Ser Mater Sci Eng 598(1)
2. Ofori EK, Frimpong E, Ademiluyi A, Olawale OA (2019) Ergometer cycling improves the ambulatory function and cardiovascular fitness of stroke patients—a randomized controlled trial. J Phys Ther Sci 31(3):211–216
3. Greenhalgh O et al (2019) An exploration of stroke survivors' perspectives on cycling and the use of electric bikes. Physiother Pract Res 40(2):117–126
4. Londen Batan IM, Sukma Lutiawan TNA, Salim LA (2019) Tricycle applications for physical therapy sufferers. IOP Conf Ser Mater Sci Eng 588(1)
5. Ada L, Canning CG, Low SL (2003) Stroke patients have selective muscle weakness in shortened range. Brain 126(3):724–731
6. Bohannon RW (2007) Muscle strength and muscle training after stroke. J Rehabil Med 39(1):14–20
7. THE STRAITSTIMES Homepage. https://www.straitstimes.com/lifestyle/home-design/cycling-made-easy. Accessed 23 Nov 2020
8. Bulthuis R, Tabak M, Schaake L, Hermens H (2021) Outdoor E-trike cycling: a low intensity physical activity. Assist Technol 00(00):1–8

9. Langford BC, Cherry CR, Bassett DR, Fitzhugh EC, Dhakal N (2017) Comparing physical activity of pedal-assist electric bikes with walking and conventional bicycles. J Transp Heal 6:463–473
10. SNI (2009) Keselamatan sepeda roda tiga. SNI 7519:2009, Indonesia
11. SNI (2008) Metoda pengereman kendaraan bermotor kategori L. Indonesia
12. SNI (2008) Sepeda – Syarat keselamatan, Indonesia
13. Batan IML (2012) Desain Produk, 1st edn. Guna Widya, Surabaya
14. Zuhl M (2020) Tips for monitoring aerobic exercise intensity. ACSM 1
15. Jette M, Sidney K, Blumchent G (1990) Metabolic equivalents (METS) in exercise testing, exercise prescription, and evaluation. Clin Cardiol 13:555–565
16. Simons M, Van Es E, Hendriksen I (2009) Electrically assisted cycling: A new mode for meeting physical activity guidelines?. Med Sci Sports Exerc 41(11):2097–2102

Design and Analysis of Bicycle Static as Physical Therapy Upper and Lower Body Exercise for Post-Stroke Patient

Luthfi Hakim[1](✉), Mohammad Muslimin[1], and Alfi Tranggono Agus Salim[2]

[1] Universitas Islam Majaphit, Mojokerto, Jawa Timur, Indonesia
luth@unim.ac.id

[2] Politeknik Negeri Madiun, Madiun, Jawa Timur, Indonesia

Abstract. Patient recovery after suffering a stroke can be done with therapy using the help of a physiotherapy device in the form of a stationary bicycle. The design of a stationary bicycle that can train the upper and lower sides of the patient's body simultaneously is packed in one tool in the form of a stationary bicycle. Validity and reliability tests were carried out on 20 respondents who had used the designed static bicycle aids. The results of these tests indicate that all questionnaire instruments for a patient response can be said to be valid. Cronbach's Alpha value is 0.716. Evaluation of every aspect of the static bicycle prototype was also carried out including the material aspect with a score of 79.4%, flexibility of 80.2%, function and safety aspects of 85.4%, and design aspect of 83.0%. Analysis of frame strength on a stationary bicycle is carried out theoretically and in a simulation. ST 37 steel material was chosen as a frame with a maximum shear stress of 235 MPa. Meanwhile, from the simulation results using commercial software, the maximum stress value is 114.50 MPa. This value is also compared with the theoretically obtained maximum stress value, the maximum stress value obtained is 103.76 MPa. This value shows that the results of testing the static bicycle frame are still in the safe category to use for a patient load of 100 kg.

Keywords: Bicycle static · Therapy · Frame strength · Upper and lower body exercise

1 Introduction

Stroke is characterized by morbidity and high levels of disability. It is not uncommon for stroke patients to become persons with disabilities so that they cannot live freely [1]. Besides, it has become a major cause of long-term disability, inability to walk unaided, and dependence on other people in everyday life [2]. In general, the cause of stroke is due to bleeding from blood vessels in the brain and blockage of the blood vessels [3, 4]. Of all stroke patients, 87% were ischemic, 10% intracerebral hemorrhage, and 3% were subarachnoid hemorrhage strokes [5]. The dominant risk factors for stroke are increasing age, coronary heart disease, diabetes mellitus, hypertension, and heart failure [6].

Rehabilitation practice was chosen as a way to minimize disability after a stroke. The duration of rehabilitation for post-stroke patients depends on the level of stroke

I. Tolj et al. (eds.), *Recent Advances in Mechanical Engineering*, Lecture Notes in Mechanical Engineering, https://doi.org/10.1007/978-981-19-0867-5_16

the patient has experienced. The average patient was hospitalized for 16 days, and then continued on an outpatient basis for several weeks. During the rehabilitation process, exercises are needed to help reduce the level of disability in the patient's limbs. One of the rehabilitation activities carried out by post-stroke patients is with the help of physiotherapy tools which are carried out by the help of medical personnel. Various physiotherapy tools have been developed by researchers.

The innovations that have been made by previous researchers to develop rehabilitation and care technology for post-stroke patients include high technology systems in the form of mechanical bicycles [7–9]. Cloud technology has been developed as a tool for post-stroke patient training [10], comparing Robot-Assisted Training (RAT) with Task-Specific Training (TST) to compare muscle fatigue activity and pain levels [11], and use game methods in rehabilitating post-patient patients. Stroke [12], the application of the PSS concept for the rehabilitation of post-stroke exoskeleton-based patients has been used in China to maximize medical utility and increase the novel value of the work. The three outcome scenarios obtained from the rehabilitative exoskeleton were developed from previous studies based on three levels of rehabilitation, namely, base level, middle level, and top level [13].

Therapy of stroke patients can be done through physiotherapy by experts, including by applying a stroke patient rehabilitation bicycle [2, 14], robotic technology to assist recovery [11], training patterns in the form of games [12], participants through herbal medicine [15], and others. However, what is commonly done by post-stroke patients is to use a therapeutic tool that can restore nerves that were previously unable to function properly. It is proven that there have been many innovations in equipment that have been developed by previous researchers.

Designing a mechanical bicycle that can be used as a rehabilitation aid for post-stroke patients has been developed previously [2, 14]. However, the tools that have been developed are limited to the rehabilitation of the legs (lower) or arms (upper), and the shape of the products produced does not match the anthropometry of the Indonesian people. Therefore, this study tries to develop a product that can rehabilitate the legs (lower) and arms (upper) of post-stroke patients. The size of the product can be adjusted to the user's posture by making it a portable form.

The steps taken to develop these bicycles include designing bicycle products based on post-stroke patient anthropometric data, using QFD attributes, and following up with analysis using HoQ (House of Quality) or weighting [16, 17]. However, this study will design a stationary bike for the therapy of post-stroke patients. Analysis of the strength of a static bicycle frame is carried out with the help of commercial software so that the results obtained can be used to state that the tool that has been developed is safe to use [14]. The results of the frame strength analysis with the help of the software are then compared with the theoretical analysis.

2 Research Methodology

The method used in this design is an analytical method to analyze the strength of a stationary bicycle frame for post-stroke patients with the help of commercial software. The research begins by collecting the information needed to design the stationary bike.

Then three-dimensional modeling was made, and the strength of the frame was analyzed using the help of commercial software. Information gathering is needed to determine the geometry of the bicycle to be designed. The determination of the bicycle geometry is based on the posture of an adult human with a height of 164–180 cm with a seat design that can be adjusted for the height. The patient's weight is assumed to be 100 kg.

The frame design is necessary to be investigated because it is used to support all bicycle components. Besides, the frame is also useful to maintain driving stability. The frame material used is ST37 steel with a maximum tension of 235 MPa. The frame design is shown in Fig. 1.

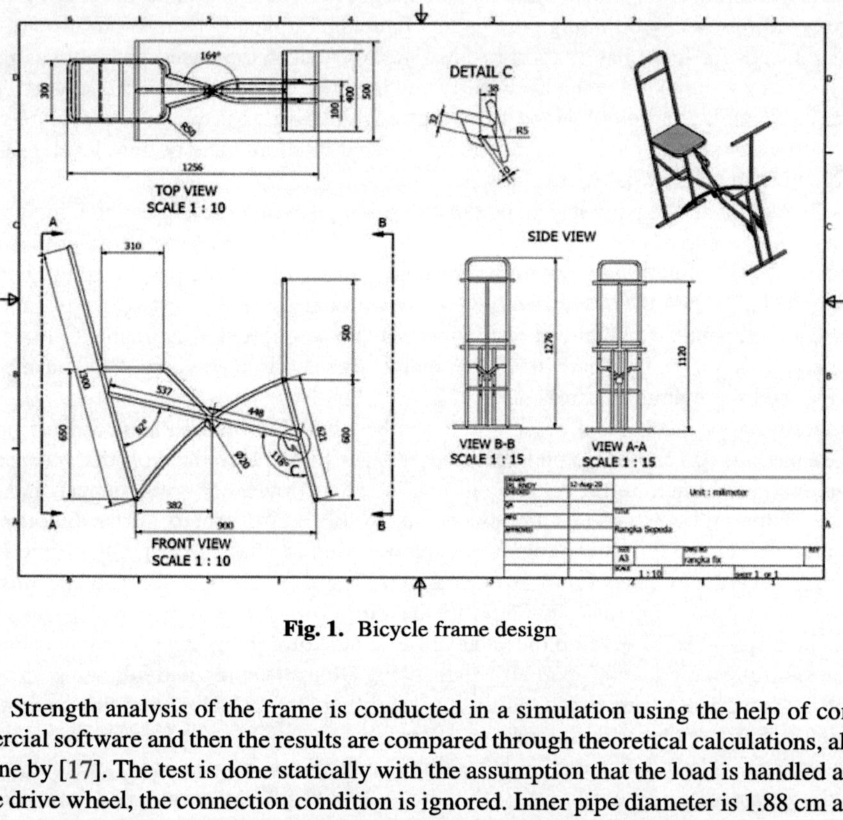

Fig. 1. Bicycle frame design

Strength analysis of the frame is conducted in a simulation using the help of commercial software and then the results are compared through theoretical calculations, also done by [17]. The test is done statically with the assumption that the load is handled and the drive wheel, the connection condition is ignored. Inner pipe diameter is 1.88 cm and outer pipe diameter is 2.0 cm. In general, the material specifications for ST37 can be found in Table 1.

Theoretically, the frame strength is calculated based on the static loading. The force on each part of the frame is analyzed, and then the greatest force is divided by the smallest area value to obtain the maximum stress experienced by the bicycle frame, using Eqs. (1) and (2), respectively.

$$A_{min} = \pi\left(\frac{D_o}{2}\right)^2 - \pi\left(\frac{D_i}{2}\right)^2 \tag{1}$$

Table 1. Steel specifications ST37

Parameter	Constant
Mass density	7.85 g/cm^3
Yield strength	235 MPa
Ultimate tensile strength	340 MPa
Young's modulus	217 Gpa

$$S_{max} = \frac{F}{A_{min}} \tag{2}$$

If the allowable strength does not exceed the induced stress within the material, then the design is considered safe, and the process can be continued at the manufacturing stage of each component. These components are then assembled and tested directly on 20 post-stroke patients to get responses regarding the quality of the bicycle that has been designed. The quality is determined by several aspects including material aspects, flexibility aspects, function and safety aspects, and design aspects.

3 Results and Discussion

The theoretical force calculation on each static bicycle frame component has been used to obtain the maximum stress value. To get the maximum stress value on a static bicycle frame, a force analysis has been carried out against certain points. As seen in Fig. 2, analysis of the reactions that occur at points A, B, C, D, and E is carried out to obtain the maximum stress value that occurs in each part of the frame. The force on each part of the frame is analyzed by dividing the force component into x- and y-coordinates. To obtain the maximum stress force on each rod, the result of the force obtained is divided by the value of the smallest area of the frame. The greatest force is obtained in the frame section of AE and DE, which is 3791 N. This value is then divided by the minimum area value which is 0.365×10^{-4} m^2. So that the maximum stress value is 103.67 MPa.

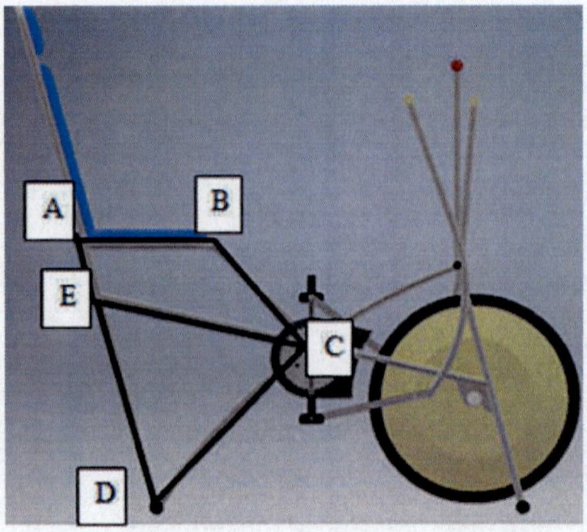

Fig. 2. Component style on the bicycle frame

The results of the theoretical calculations are then compared with the maximum stress values obtained by simulation using commercial software. The simulation results show that the maximum von Mises stress obtained is 114.5 MPa. The stress is still below the yield strength value of the material so that permanent deformation will not occur (Table 2).

Table 2. Forces acting on the bicycle frame

Frame	Great style	Type
AB	981	Press
BC	1387	Press
CD	1571	Press
AE	3791	Press
DE	3791	Press

A part from the yield strength data obtained from the simulation results, displacement values were also obtained from the simulation. With the patient loading evenly of 100 kg, the displacement that occurs in the frame of 0.4113 mm changes from the initial frame with a maximum criterion of 1.5 mm. So that the frame is still in the safe category to use. As shown in Fig. 4, the maximum displacement is indicated by a red area while the minimum displacement is indicated by a blue area.

The results of the theoretical calculations are then compared with the maximum stress values obtained by simulation using commercial software. The following is the comparison of the maximum stress that occurs on a bicycle frame, as seen in Fig. 3. The figure shows that the theoretical maximum stress value obtained is 103.76 MPa while the maximum stress obtained by simulation is 114.5 MPa.

The stress value obtained shows that there is a difference in the value obtained between the two of 10.74 MPa with an error percentage value of 9.37, as seen in Table 3. The value of both stresses is still below the allowable strength of ST37, thus the bike frame designed is still in the safe category for use with a patient load of 100 kg. Likewise, comparisons have been made with similar studies to validate the results.

Table 3. Maximum stress ratio

Comparison	Simulation	Theoretic	Difference	Error (%)
Researcher	114.50 MPa	103.76 MPa	10.74 MPa	9.37
Imran (2018) [17]	26.66 MPa	27.40 MPa	0.74 MPa	2.70

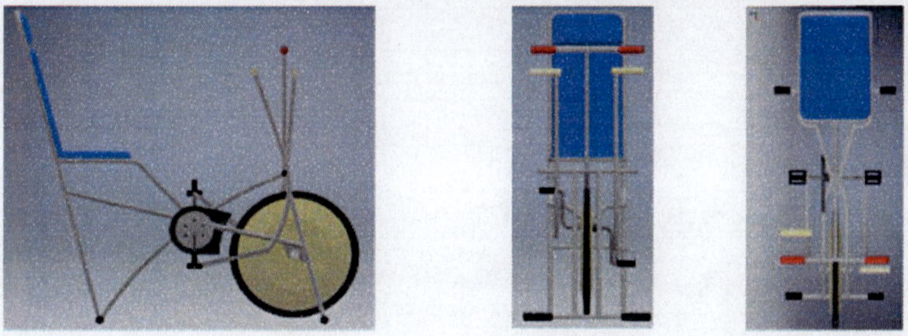

(a) (b) (c)

Fig. 3. Bicycle static display of post-stroke patients seen from **a** side view, **b** front look, **c** top view

In detail, the appearance of bicycle designs for post-stroke patients can be seen in Figs. 3 and 4. The static bicycle for post-stroke patients has a length of 1256 mm, a width of 300 mm, and a height of 1276 mm, with a bicycle mass of 20 kg.

Patient response data were obtained from questionnaires to 20 post-stroke patients undergoing rehabilitation in several hospitals. Patients are scattered in several hospitals, including patients at the HASANAH's hospital in Mojokerto City. The data obtained were then tested for the level of validity and reliability.

The results of the validity test of the patient response questionnaire can be said to be valid if the r-count is greater than the r-table. For 20 samples, the r-table test is 0.444, so it can be said that all questionnaire instruments for the patient's response can be said to be valid. Cronbach's Alpha value is 0.716. Each aspect is assessed based on assessment indicators in the range of 1 to 5. The maximum score for each aspect of the 20 questions is 500. So that each aspect can be assessed in the range of 0 to 100. In detail, the value of each aspect can be seen in Table 4.

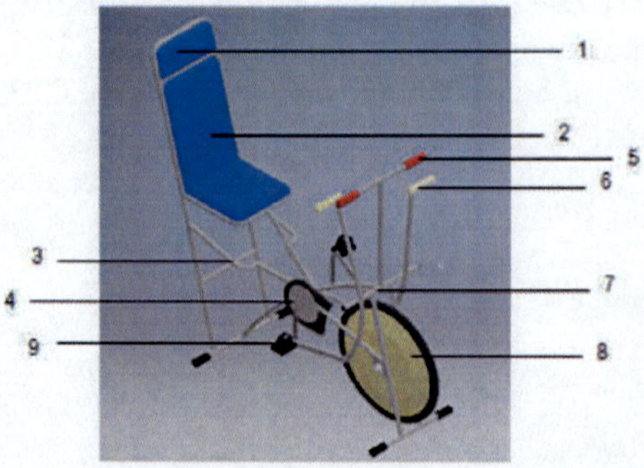

Fig. 4. Static bicycles for post-stroke patients

Table 4. Scores on each aspect

Assessment aspects	Score acquisition	Percentage
Material aspects	397	79.4
Flexibility aspects	401	80.2
Functions and safety aspects	427	85.4
Design aspects	415	83.0

4 Conclusion

The design and analysis of a stationary bicycle for post-stroke patients has paid attention to ergonomic theory so that the product to be produced is adjusted to the patient's height between 164 and 180 cm. The strength of the skeleton is analyzed theoretically and in a simulation. Theoretically, the maximum stress value of the frame is 103.76 MPa, while simulated maximum stress value is 114.50 MPa. This value is still carried by the allowable stress value of the ST37 material as the frame material used, so the frame design is still in the safe category with a patient load of 100 kg. From the results of the validity and reliability test, it is said to be valid and reliable. The results of these tests indicate that all questionnaire instruments for the patient response can be said to be valid. Cronbach's Alpha value is 0.716. This value shows that the patient's response questionnaire meets the criteria reliably, namely, $r\_Count > r\_Table$ is $0.716 > 0.444$. Evaluation of every aspect of the static bicycle prototype was also carried out including the material aspect with a score percentage of 79.4%, flexibility aspect of 80.2%, function and safety aspect of 85.4%, and design aspect of 83.0%.

Acknowledgements. The researchers express their gratitude to the Directorate of Research and Community Service, Ministry of Education and Culture (DRPM-DIKTI) with number SK 8/E1/KPT/2020 dated January 24, 2020, and contract agreement number 187/SP2H/LT/DRPM/2020 on March 9, 2020, and the contract between the researcher and LP4MP of the Islamic University of Majapahit no.02/LP4MP/UNIM/K/IV/2020 on April 8, 2020.

References

1. Asakawa T, Zong L, Wang L, Xia Y, Namba H (2017) Unmet challenges for rehabilitation after stroke in China. Lancet 390(10090):121–122. https://doi.org/10.1016/S0140-6736(17)31584-2
2. Barbosa D, Santos CP, Martins M (2015) The application of cycling and cycling combined with feedback in the rehabilitation of stroke patients: a review. J Stroke Cerebrovasc Dis 24(2):253–273. https://doi.org/10.1016/j.jstrokecerebrovasdis.2014.09.006
3. van Vliet P, Wulf G (2006) Extrinsic feedback for motor learning after stroke: what is the evidence? Disabil Rehabil 28(13–14):831–840. https://doi.org/10.1080/09638280500534937
4. Derrickson H (2013) Principles of anatomy and physiology, 14th edn by Tortora GJ, Bryan, pp 12–14
5. Virani SS et al. (2020) Heart disease and stroke statistics—2020 update: a report from the American Heart Association
6. Ghani L, Mihardja LK, Delima D (2016) Faktor Risiko Dominan Penderita Stroke di Indonesia. Bul Penelit Kesehat 44(1). https://doi.org/10.22435/bpk.v44i1.4949.49-58
7. Dewi DS, Rakhmawati A, Batan IML, Wessiani NA (2019) Product design for post-stroke rehabilitation bicycle with Kansei engineering approach. IOP Conf Ser Mater Sci Eng 598(1). https://doi.org/10.1088/1757-899X/598/1/012087
8. Batan IML, Rodika, Riva'I M (2015) Three wheel bike as physical therapy equipment for post-stroke patient. Appl Mech Mater 776:337–342. https://doi.org/10.4028/www.scientific.net/amm.776.337
9. Londen Batan IM, Sukma Lutiawan TNA, Salim LA (2019) Tricycle applications for physical therapy sufferers. IOP Conf Ser Mater Sci Eng 588(1). https://doi.org/10.1088/1757-899X/588/1/012034
10. Hossain MS, Hoda M, Muhammad G, Almogren A, Alamri A (2018) Cloud-supported framework for patients in post-stroke disability rehabilitation. Telemat Informat 35(4):826–836. https://doi.org/10.1016/j.tele.2017.12.001
11. Shahar N, Schwartz I, Portnoy S (2019) Differences in muscle activity and fatigue of the upper limb between task-specific training and robot assisted training among individuals post stroke. J Biomech 89:28–33. https://doi.org/10.1016/j.jbiomech.2019.04.005
12. Trombetta M, Bazzanello Henrique PP, Brum MR, Colussi EL, De Marchi ACB, Rieder R (2017) Motion Rehab AVE 3D: A VR-based exergame for post-stroke rehabilitation. Comput Meth Progr Biomed 151:15–20. https://doi.org/10.1016/j.cmpb.2017.08.008
13. Xiao Y, Zhao K, Ma Z-X, Li X, Qiu Y-P (2017) Integrated medical rehabilitation delivery in China. Chronic Dis Transl Med 3(2):75–81. https://doi.org/10.1016/j.cdtm.2017.02.003
14. Sholeh A, Ivianda Juniani A, Novrita Devi Y (2018) Analisis dan Perancangan Sepeda Statis untuk Rehabilitasi Penderita Stroke. Politek. Perkapalan Negeri Surabaya, pp 11–16
15. Kwon CY, Lee B, Chung SY, Kim JW (2019) Herbal medicine for post-stroke anxiety: a systematic review and meta-analysis of randomized controlled trials. Complement Ther Clin Pract 35:237–252. https://doi.org/10.1016/j.ctcp.2019.02.015
16. Homkhiew C, Ratanawilai T, Pochana K (2012) Application of a quality function deployment technique to design and develop furniture products. Songklanakarin J Sci Technol 34(6):663–668
17. Marcielo D, Anton, Irawan AP (2015) Perancangan Dan Analisis Kekuatan Konstruksi Dan Powertrain Pada Prototype Hand-Crank Cycle (Sepeda Engkol Tangan) 11(1):69–79

The Number of Slots and Electromagnetic Core Effects on Eddy Current Brake Performance: A Computational Approach

Wahyu Dwi Aprian[1], Ubaidillah[1](✉), Muhammad Nizam[2], Mufti Reza Aulia Putra[1], and Saiful Amri Mazlan[3]

[1] Department of Mechanical Engineering, Universitas Sebelas Maret, Surakarta, Indonesia
ubaidillah_ft@staff.uns.ac.id
[2] Department of Electrical Engineering, Universitas Sebelas Maret, Surakarta, Indonesia
[3] Engineering Materials and Structures (eMast) iKohza, Malaysia-Japan International Institute of Technology (MJIIT), Universiti Teknologi Malaysia, Jalan Sultan Yahya Petra, 54100 Kuala Lumpur, Malaysia

Abstract. This study discusses Eddy Current Brake (ECB) by using electromagnet as the source of magnetism. The purpose of this study was to investigate the relationship between braking torque and variations in the number of half-circle slots on ECB. The method applied in this study was the Finite Element Method (FEM), where the simulated ECB modeling was varied by 10, 20, and 30 slots using a change in the number of electromagnetic field sources. The results of this study obtained a description of the braking torque with the largest value in the variation of 10 slots using four electromagnets, namely, 59.18 Nm.

Keywords: Eddy current brake · Half-circle slotted · Finite element method (FEM) · Electromagnet

1 Introduction

Eddy current brake is a type of electric braking that utilizes eddy currents to generate braking forces [1]. Eddy currents are induced currents that occur due to changes in the magnetic field in a conductor. The ECB braking system does not require friction, so these brakes have the advantage of being easy to maintain compared to conventional brakes. This is because the ECB brake has no wear rate, does not make any noise, and has a fast response [2]. Research on ECB using three different materials, namely, aluminum, copper, and brass as the disk conductor material has been carried out by Krishna. The study resulted in the data that showed aluminum has less spread, and it saves the cost and time required for a higher percentage of speed [3]. Another research on the effect of the aluminum conductor disk surface on the increase in torque has been carried out by Prayoga. The research resulted in the data that showed the half-circle slotted conductor disk has the highest torque [4]. Prayoga's research can still be developed by varying the number of slots on the ECB conductor disk. This development has never been studied

I. Tolj et al. (eds.), *Recent Advances in Mechanical Engineering*, Lecture Notes in Mechanical Engineering, https://doi.org/10.1007/978-981-19-0867-5_17

before. By varying the number of slots, it is possible to influence the performance of the ECB, and therefore it is necessary to research the effect of adding variations in the number of slots to the performance of the ECB.

Many types of research in the field of ECB have been carried out, including variations in air gaps, conductor materials, and the shape of the conductor contour disk. Simulation is a process of imitating a real system and its surroundings or state of affairs [5]. In this study, a magnetic field simulation of an electromagnet as a stator and a disk conductor as a rotor was carried out by varying the number of slots of the half-circle type which had never been developed before. This research was expected to improve the performance of the ECB.

2 Methodology

This research used the Finite Element Method (FEM) to model the performance of the ECB. The data generated in the simulation process was in the form of a trend curve that described the effects between the parameters and the performance of the ECB. Several ECB parameters were developed to be varied to know their effects on the performance of the ECB. These parameters were variations in the number of half-circle type slots and the number of electromagnets applied magnetic sources. The variations in the number of slots were 10, 20, and 30 slots with the addition of magnetic sources, at a speed of 300 rpm until it reached one full rotation. The use of a rotating speed of 300 refers to the speed of light vehicles and the selection of the number of slots refers to the study of Prayoga [4]. The design to be simulated in the study is shown in Fig. 1.

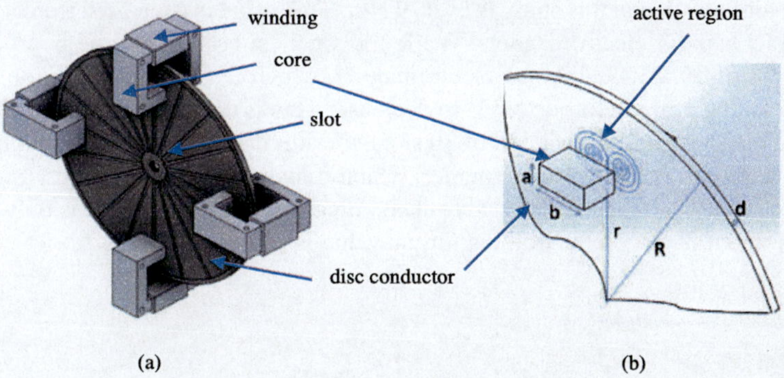

Fig. 1. **a** 3D model ECB; **b** Eddy current design variable [6]

This simulation research used several parameters that have been set up based on previous researches. It facilitated the design validation process so that the resulting design became valid and the resulting data could be accounted for. The design parameters used in the study are shown in Table 1.

Table 1. ECB design parameters [6]

Variable	Unit	Value
Current (i)	A	20
Number of a coil (*N*)	–	360
Length of pole shoe (a)	mm	30
Width of pole shoe (b)	mm	12.5
The total length of winding core (l)	mm	248
Distance from center to pole Shoe center (r)	mm	83.5
Air gap (t)	mm	0.5
Thick disk (d)	mm	5
Radius disk brake (r)	mm	120
Relative permeability aluminum (μ_{al})	$H \cdot m^{-1}$	1.26×10^{-6}
Relative permeability iron (μ_{fe})	$H \cdot m^{-1}$	6.3×10^{-3}
Aluminum conductivity (α)	$S \cdot m^{-1}$	$36,9 \times 10^{6}$

3 Result and Discussion

3.1 Variations in the Number of Slots

The trend of each electromagnet variation is different for each additional slot. The highest braking torque value in this study is 59,179 Nm. This value is generated from a variation of 10 slots using 4 electromagnets. While the smallest braking torque is produced by a variation of 30 slots using one electromagnet, which is 15,085 Nm. In the variation of 30 slots, the braking torque tends to decrease. This is due to the distance differences between the slots. In the variation of 30 slots, the slot distance is getting smaller so that the active region is also getting smaller, which makes the braking torque to decrease. Variations in the number of slots have no significant effect or the effect is only less than 1 Nm or about 3–5% on the braking torque value is shown in Fig. 2.

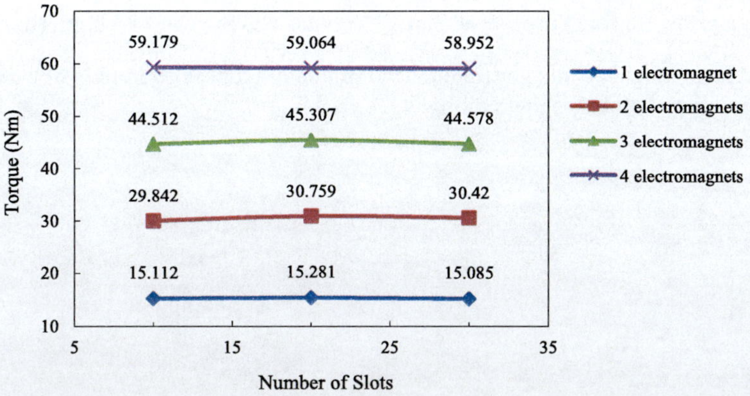

Fig. 2. The relationship between variations in the number of slots and braking torque

3.2 Variation in the Number of Electromagnets

The effect of the addition of electromagnets is very significant on the performance of the ECB. It was explained by the theory that the more electromagnets used in ECB, the higher the magnetic field produced [7]. The ECB has many parameters that can affect its performance. These parameters affect the torque generated by the ECB. In this paper, to improve the performance of the ECB, variations were made using the Finite Element Method (FEM) by varying the number of slots and the number of electromagnets. The electromagnetic addition curve is shown in Fig. 3.

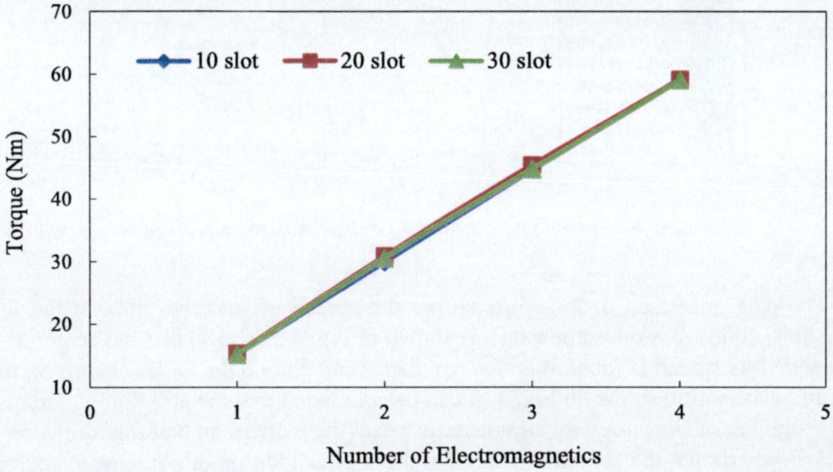

Fig. 3. The relationship between variations in the number of electromagnets and braking torque

3.3 Variations in the Number of Slot Effects for the Spread of Magnetic Flux

The spread of magnetic flux generated in the simulation for each variation of the addition of the number of slots is shown in Fig. 4.

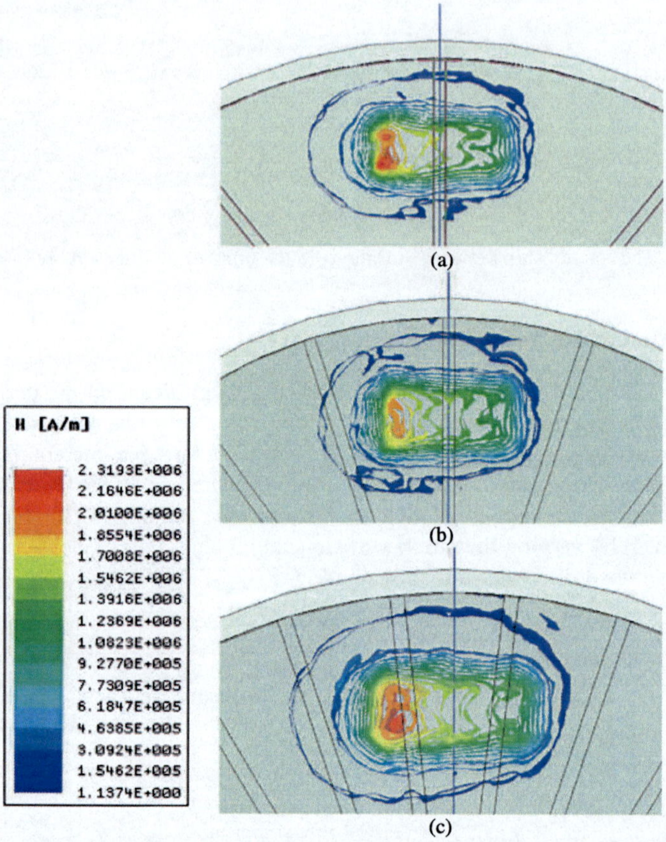

Fig. 4. Magnetic flux spread at **a** 10 slots, **b** 20 slots, **c** 30 slots

The data generated in the study shows the spread of magnetic flux in the three variations almost has the same value as shown in Fig. 4. It shows that the center of the magnetic flux spread is focused on the conductor and coil on the ECB. Therefore, from the flux distribution shown in Fig. 4, it can be concluded that the addition of variations in the number of slots does not significantly affect the increase in braking torque on the ECB. Furthermore, the spread of magnetic flux to the addition of a magnetic source is shown in Fig. 5.

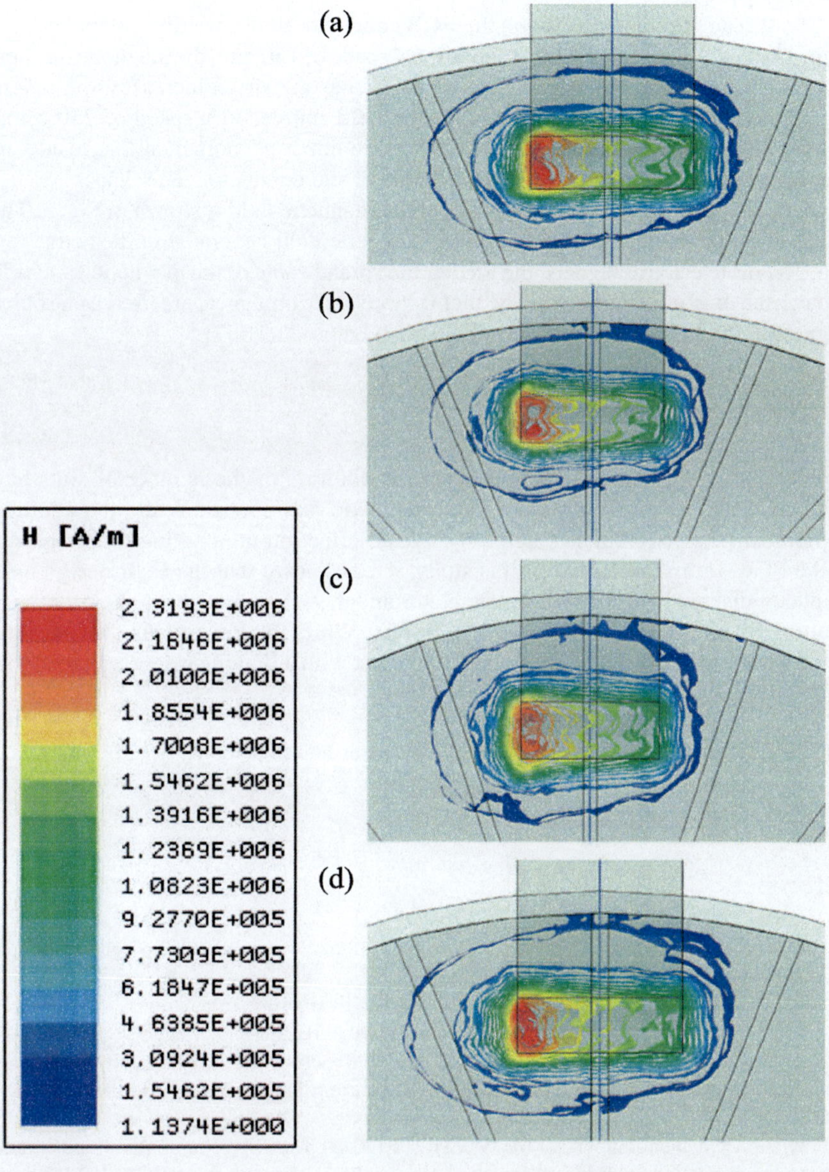

Fig. 5. Magnetic flux spread at **a** one electromagnet, **b** two electromagnets, **c** three electromagnets, **d** four electromagnets

The spread of magnetic flux on the ECB conductor with variations in the number of magnets shows different trends. At a rotational speed of 150 rpm, the magnetic flux spread is concentrated in the magnetic field source, and as the speed increases, the magnetic flux spreads further away from the magnetic field source. At a speed of 750 rpm, the braking torque decreases. This is due to the skin effect phenomenon that results in an increase of current spread at the farthest part of the conductor. The dispersion of the magnetic flux concentrated at the source of the magnetic field is shown in Fig. 5. This is indicated by the tendency of the stronger magnetic field generated in the active region area. The more electromagnets, the greater the spread value of the magnetic spread. This is under the existing theory, namely, that if more electromagnets are used as a source of magnets in the ECB, more magnetic fields will be produced [7].

4 Conclusion

From this study, it can be concluded that the addition of the number of slots has no significant effect on braking torque and magnetic flux spread. While the addition in the number of electromagnets has a significant effect on the braking torque produced by the ECB. From the results of this study, it was known that the ECB design using a conductor disk with 10 slots using four electromagnets produced the highest torque. The resulting braking torque value was 59,179 Nm. While the lowest value was generated by the variation of 30 slots using 1 electromagnet with a braking torque value of 15,085 Nm.

Acknowledgements. This paper was funded by Riset Inovatif Produktif (RISPRO), Lembaga Pengelola Dana Pendidikan (LPDP) 2021.

References

1. Lequesne B (1997) Eddy-current machines with permanent magnets and solid rotors. IEEE Trans Ind Appl J 33(5):1289–1294
2. Shin HJ, Choi JY, Cho HW, Jang SM (2013) Analytical torque calculations and experimental testing of permanent magnet axial eddy current brake. IEEE Trans Magn J 49(7):4152–4155
3. Anantha Krishna GL, Sathish Kumar KM (2018) Experimental investigation of influence of various parameters on permanent magnet Eddy current braking system. Mater Today Proc J 5(1):2575–2581
4. Prayoga AR, Ubaidillah Nizam M, Waloyo HT (2019) The influence of aluminum conductor shape modification on Eddy-current brake using finite element method. In: ICEVT 2019—Proceeding 6th International conference on electrical vehicle technology 2019, pp 146–150
5. Putra MRA, Nizam M, Tjahjana DDDP, Aziz M, Prabowo AR (2020) Application of multiple unipolar axial eddy current brakes for lightweight electric vehicle braking. Appl Sci 10(13)
6. Waloyo HT, Ubaidillah U, Tjahjana DDDP, Nizam M, Aziz M (2020) A novel approach on the unipolar axial type eddy current brake model considering the skin effect. Energies 13(7):1–15
7. Putra MRA, Nizam M, Tjahjana DDDP, Arfandi M, Ubaidillah, Choirunisa I (2019) Design study in single disk axial eddy current brake. In: Proceeding—2018 5th International conference on electrical vehicle technology ICEVT 2018, pp 158–160

The Bulb Concept in Drag Reduction on Pentamaran

Wiwin Sulistyawati(✉) and Purwo Joko Suranto

Department of Naval Architecture Engineering, UPN Veteran Jakarta, Jakarta 12450, Indonesia
wiwinsulistyawati@upnvj.ac.id

Abstract. Many computational and experimental studies have supported the research of multi-hull and their advances in decreasing interference. The bulb concept as a retrofit on multi-hull of several studies has proven particularly suitable for drag reduction of interference factors. This study examines the hydrodynamic of pentamaran with bulbs and varying degrees of rotation. Experiments were carried out to identify interference wave fluctuations between the main hull and the outrigger. This approach is based on Michell's Thin Ship theory, and it looks at the highest wave profile that happens from each speed at the x/L and y/B places of the pentamaran. Wave fluctuation investigations provide precise bulb placement to break up interference. Then further analysis is done by Computational Fluid Dynamic (CFD) to determine drag components from variation changes in the angle of the bulb. The results show that the total drag coefficient of a pentamaran with bulbs can be reduced by up to 16.78%, with the most significant interference reduction reaching 29.03%. The analysis and investigations results are presented in graphs generated by computerised tools and visual captures of experiments.

Keywords: Pentamaran · Bulbs · Interference · Experiment · Thin ship theory · CFD

1 Introduction

Efficiency and the development of innovations in energy control have become and will continue to become increasingly crucial to the heightened awareness of the need to reduce carbon emissions. Efficiency on ships requires measures that increase hydrodynamic efficiency to reduce drag. The hydrodynamic efficiency of vessels is generally achieved by changing the hull's shape and optimisation. In multi-hull, hull shape design, inter-hull position configuration, hull-symmetrical configuration, and complexity optimisation are required to improve overall efficiency and hydrodynamic performance as in research [1–3]. The retrofit concept with a centre bulb on a multi-hull is considered adequate to reduce inter-demi-hull interference. The concept of the centre bulb in catamarans was investigated by [4, 5] at its impact on drag and efficiency. Another study [6] found that employing a centre bulb reduced waves by up to 13% in experimental results and 15% in numerical data, both at Fr 0.6. Meanwhile, [7], using a centre bulb in a catamaran, will minimise total drag by 25.76% at Fr 0.47.

I. Tolj et al. (eds.), *Recent Advances in Mechanical Engineering*, Lecture Notes in Mechanical Engineering, https://doi.org/10.1007/978-981-19-0867-5_18

This study investigates a pentamaran with a bulb between the hulls to reduce interference and total drag. This work is a continuation of the first author's work on the pentamaran in [8]. Experiments carried out the research to identify most of the interference between the hull and total drag. A longitudinal-lateral wave cutting method is also utilised to investigate wave fluctuations, using the Michell Thin Ship approach with the simple tool, "Michlet". The determination of longitudinal-lateral wave cutting points is determined and simulated against variations in speed. Measurement of the drag component of the pentamaran also uses Computational Fluid Dynamic (CFD) ANSYS. The initial results of the experiment, thin ship theory, and CFD were compared, then the changes were analysed in interference that occurred to variations in changes in bulb angle and speed.

2 Wave Interference

2.1 Thin Ship Theory

The wave interference on a multi-hull can be considered an interrelated function of the hull's shape, the outrigger's variation position, the outrigger separation distance, and the speed. The study focuses on the interference and the effects on the multi-hull in the research of [9–12]. According to the thin ship theory in [13, 14], the linear superposition of a multi-hull wave pattern is a wave spectrum that combines the hull shape's factor and the hull placement effect. The final effect produces a wave interference $F(\theta)$ by combining the wave amplitude $A_j(\theta)$ of the number of hulls j at (xj, yj) and the total displacement of the multi-hull σ_j.

$$F(\theta) = \sum_{\sigma=1}^{N} \sigma_j e^{ik(\theta)[xj\cos\theta + yj\sin\theta]} \tag{1}$$

The complex amplitude resulting from several hulls j or free wave spectrum $k_O = g/V^2$, g for gravity and the path of k-integration passes above the pole at $k = k_0 \sec^2\theta$. For multi-hull, then amplitude function $A(\theta)$ *is*

$$A(\theta) = \sum_{j=1}^{N} A_j(\theta) e^{ik(\theta)[xj\cos\theta + yj\sin\theta]} \tag{2}$$

$A(\theta)$ is related to the wave pattern or Kochin function; wave number $k(\theta) = k_0 \sec^2\theta$; and a function of phase $\omega = (x, y, \theta) = x\cos\theta + y\sin\theta$. Then $A(\theta)$ can be identified from the wave height $\zeta(x, y)$ at the point (x, y) on the longitudinal section along one side of the wave pattern. The wave of the multi-hull $\zeta(x, y)$ is expressed as

$$\zeta(x, y) = \sum_{j=1}^{N} Rw \int_{-\pi/2}^{\pi/2} A_j e^{-ik(\theta)[(x-xj)\cos\theta + (z-zj)\sin\theta]} d\theta \tag{3}$$

then the wave drag, *Rw*:

$$Rw = \frac{\pi}{2}\rho U^2 \int_{-\pi/2}^{\pi/2} |A(\theta)|^2 \cos^3\theta d\theta \tag{4}$$

The setting of an appropriate outrigger is to reduce interference and wave drag. It can prevent the highest wave energy concentration from occurring as a result of wave encounters between hulls.

2.2 Experimental

Interference, I_F, results from hull interactions and is assumed by the ratio of *N*-drag of each hull and total drag of multi-hull. In calm water, the determination of interference, I_F, is expressed as

$$I_F = \frac{R_{T_{multihull}} - \sum R_{T_{N-hull}}}{R_{T_{multihull}}} \tag{5}$$

According to the International Towing Tank Conference (ITTC), the measurement of total drag on towing-tank experiments on calm-water ship drag is as follows:

$$R_T = (1+k)R_F + R_w \tag{6}$$

The Prohaska method was used to determine the pentamaran form factor *(k + 1)* using a low-speed experiment (0.15–0.29 m/s). The total drag coefficient is defined as follows:

$$C_T = {R_T}/{0.5\rho V^2 S} = (1+k)C_F + C_W \tag{7}$$

where **ρ**—water density of the, V—ship speed, S—wet surface area of hull, and C_F—friction drag coefficient that is evaluated by ITTC 1957 correlation line:

$$C_F = {0.075}/{(\log_{10} Rn - 2)^2} = {R_F}/{0.5\rho V^2 S} \tag{8}$$

3 Interference Investigation

3.1 Identified by Experimental

The investigation was conducted on a pentamaran with outriggers configuration, as shown in Fig. 1, with parameters given in Table 1. Minimum drag was achieved using Michell's Thin Ship optimisation in [8] with a stagger of 0.36*Lpp* (*Lpp* = length of perpendicular of the main hull) and clearances of 1.35*Bml* (*Bml* = breadth moulded of the main hull) for the forward outriggers, and 1.5*Bml* for the back outriggers. Furthermore, the occurrence of wave interference and the wave fluctuations were investigated

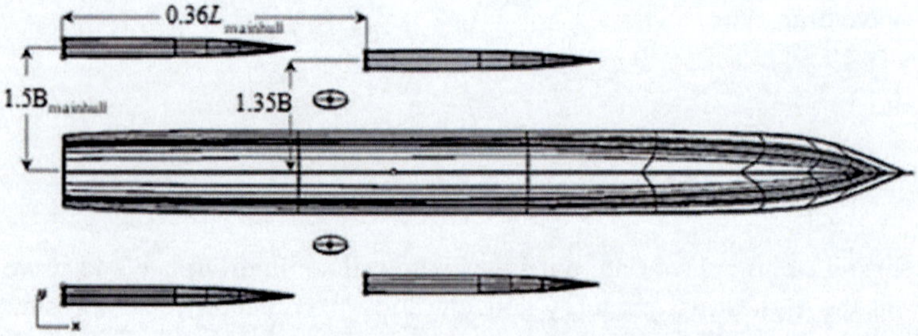

Fig. 1. Pentamaran configuration with bulb

Table 1. The main parameter of pentamaran

Main dimension	Main hull	Outriggers
Total Length (mm)	1500	414
Breadth (mm)	150	30
Draft (mm)	24	12
Height (mm)	90	78
Wetted area (mm^2)	23.63×10^4	20.06×10^3
Volume displ. (mm^3)	23.13×10^5	13.80×10^4
Deadrise β (deg)	20	35

with a wave-cut method by the simple tool, "Michlet". The convergence is obtained by varying the n-theta value, which calculates the wave drag on the ship's hull. This value is predetermined and increased until a variation of less than 10% is achieved.

Figure 2 depicts the investigation of wave patterns indicating interference with the experiment. The experiment range of model speeds 0.58–2.30 m/s corresponding to Froude number, Fr, is 0.1–0.7. The experiments were carried out in the towing tank at Institut Teknologi Sepuluh Nopember Surabaya, which measures 50 m long by 3 m wide and has a water depth of 2 m. Figures 3 and 4 show a comparison of drag components C_T-C_W between the results of experimental data and a calculation based on thin-ship theory. Figures 5 and 6 show a comparison of the experimental results and the Michell tool with Computational Fluid Dynamics (CFD) on the drag component.

Comparing the experiment, Michell's tool and CFD (Figs. 3, 4, 5 and 6), the results of Michell's tool and experiment data on total drag coefficient, C_T, show a consistent trend line but not on wave drag coefficient, C_W. The mean total drag coefficient CT had a deviation of 2.953%. The wave drag coefficient C_W is significant, about 9.049%, especially in Fr range 0.4–0.55. However, the comparison between CFD and experiments deviates up to >29%. It is most likely due to the effects of trim and sinkage, which are not accounted for in CFD. It agrees with the findings of other studies [14, 15] that comparing Michell calculations and experiment data reveals a consistent trend, particularly at high speeds, but provides a significant deviation on CFD.

Fig. 2. Interference investigation by experiment on medium speeds 1.92 and 2.30 m/s

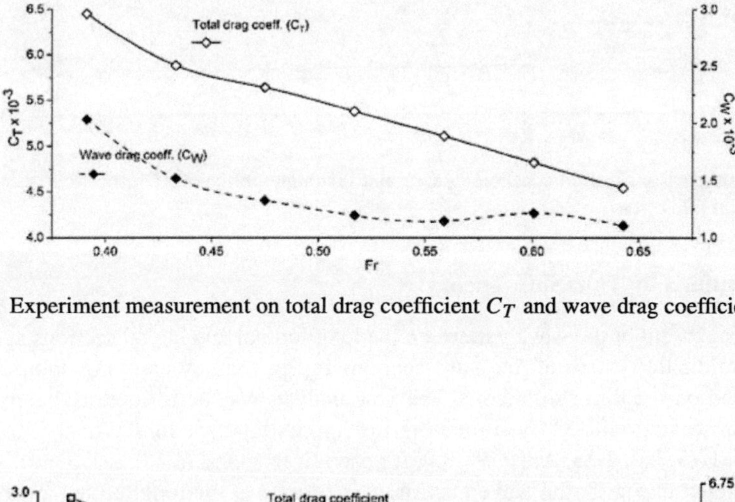

Fig. 3. Experiment measurement on total drag coefficient C_T and wave drag coefficient, C_W

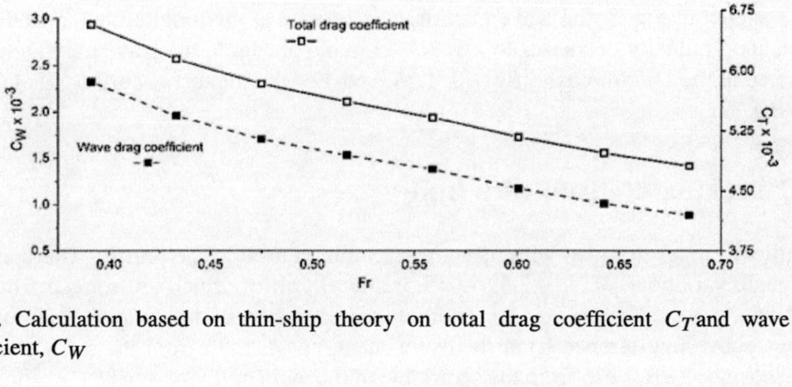

Fig. 4. Calculation based on thin-ship theory on total drag coefficient C_T and wave drag coefficient, C_W

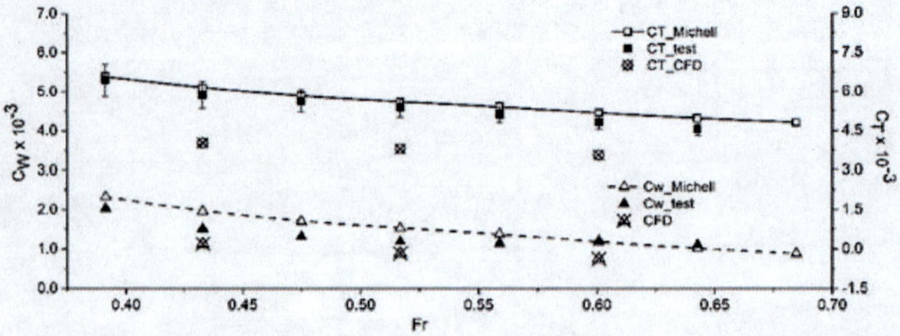

Fig. 5. Comparison total drag coefficient C_T and wave drag coefficient, C_W, between experiment, Michell and CFD

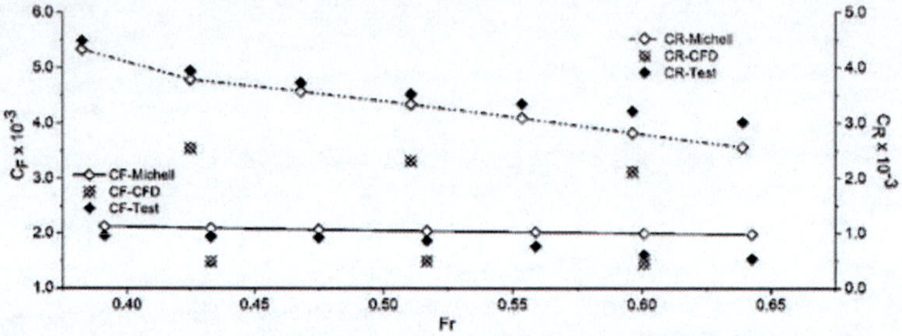

Fig. 6. Comparison friction coefficient, C_F, and residual coefficient, C_R, between experiment, Michell and CFD

3.2 Identified by Thin Ship Theory

The measurement of the wave pattern on the longitudinal and lateral sections according to the coordinate system of the wave sections in Fig. 7 is by applying computational tools based on the thin ship theory. The first analysis was performed to determine the maximum wave produced by simulating the longitudinal sections x/L: 0.5, 0.75, 1.0, 1.25, 1.5, 1.75, 2.0, 2.25, 2.5, 2.75, 3.0, for each Fr from 0.4 to 1.0.

As shown in Fig. 8, the wave maximum is formed at the longitudinal cut x/L 1.25 Fr, 0.6 and gradually decreases to x/L 3.5. Looking at Fig. 9, the wave maximum was generated at the longitudinal, x/L, 1.0, 1.25, and 1.5; the highest occurs at x/L 1.25 on the y/B 1.77.

4 Conceptual Retrofit with Bulb

The ellipse model from [6] with 59 mm × 15 mm is used in the retrofit. There are six bulb angle variations: 0, 1.0, 2.5, 5.0, 7.5, and 10, all of which are rotated from the bulb's centre, as shown in Fig. 10. The bulb's assumption on the longitudinal position is constructive interference from the experiment, at x/L 0.33 from the stern. And the lateral position, y/B 0.275 from the centreline of the main hull. The vertical position was determined by investigating the highest wave occurrence from the wave-cut method,

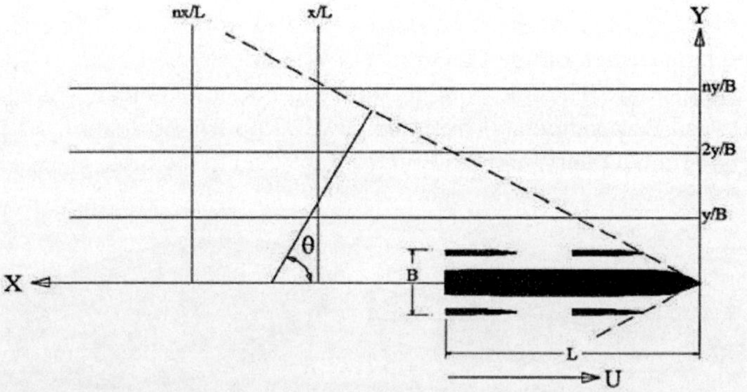

Fig. 7. The wave-cut coordinate system on longitudinal x/L and lateral y/B

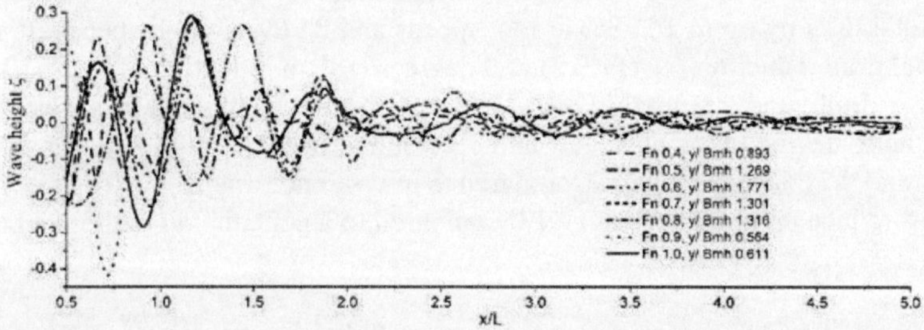

Fig. 8. Wave profile result in longitudinal wave-cut simulation by applying computational tools based on the thin ship theory

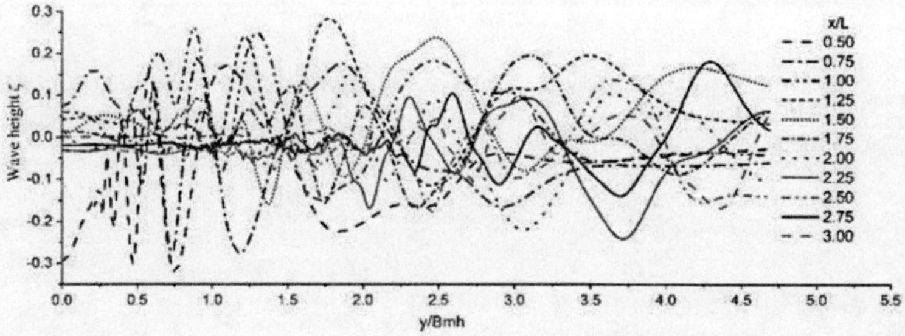

Fig. 9. Wave profile result on lateral wave-cut simulation by applying computational tools based on the thin ship theory

measured from the height of the waterline on the z-axis, z/T 0.44. The effect of retrofit position and bulb degree on the CFD process was investigated using ANSYS and the ITTC 2011 guidelines. The inlet, outlet, side, and bottom boundaries are set at 1 Lpp (Length between Perpendiculars) from the bow, 2 Lpp from the stern, 1 Lpp from the keel to the bottom boundary, and 0.5 Lpp from the keel to the top.

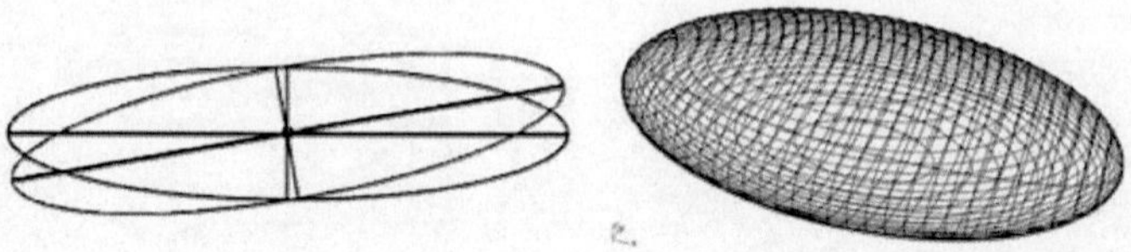

Fig. 10. Angle rotation setting of the bulb as a retrofit (Danışman 2014)

Figures 11 and 12 show the results of CFD calculations on total drag performed with CFX-Post to determine total drag, R_T, and interference, I_F, using Eq. (1). The simulation study at six angles of the bulb set up on a pentamaran found that 1.0° of bulb angle can reduce total drag by up to 16.78% at low speeds and 2.96% at high speeds. It shows a 0.5% total drag reduction C_T (Fr 0.2) at 0 degrees bulb, a reduction of up to 10.4% (Fr 0.2) at 2.5° bulb, and a reduction of 5.47% at 2.5° bulb. Compared to the pentamaran without bulb, degree bulbs of 5, 7.5, and 10 resulted in total drag increases of 7.35%, 11.64%, and 9.21%, respectively. Compared to interference drag, I_F, 1.0 degrees of bulb angle can reduce up to 29.03% at Fr 1.0 compared to a pentamaran without a bulb.

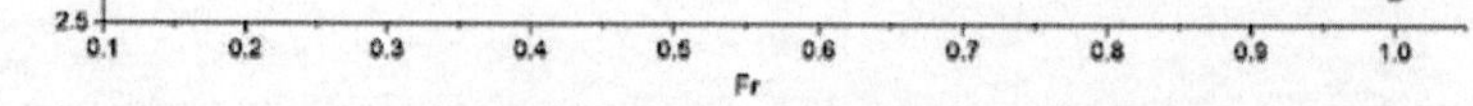

Fig. 11. Comparison of total drag coefficient, C_T, by applying Computational Fluid Dynamic (CFD)

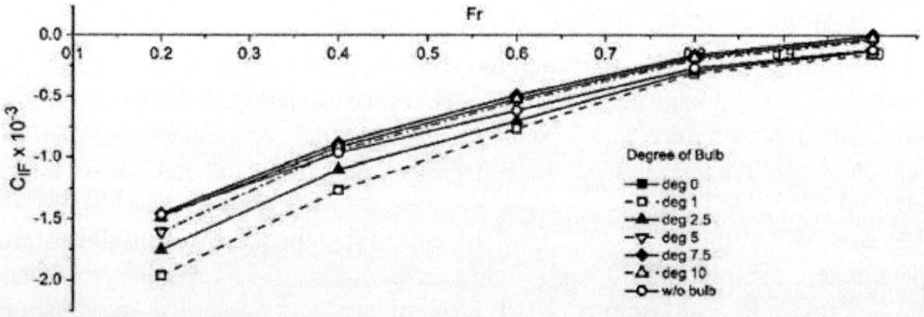

Fig. 12. Comparison calculation on wave interference coefficient, C_{IF}, by applying Computational Fluid Dynamic (CFD)

5 Conclusion

This research has identified wave interference through experiments and visually analysed the wave patterns formed between the hulls. Thin ship theory was also used to calculate drag components to determine the wave spectrum's fluctuation point by wave-cut in longitudinal and lateral directions. These methods make it easier to assume that the bulb is in the proper position. The comparison of drag coefficient components based on computation and experiments reveals poor agreement below Fr 0.5. It arose as a result of the difficulty in determining the proper form factor of the test model. Comparing experiment results, thin ship theory, and CFD on a pentamaran without the bulb confirms that CFD can be used to analyse the bulb further. The results of the analysis of bulb 1.0 degrees bring the best angle than the others on the total drag reduced up to 16.78% at Fr 0.6. Furthermore, the interference is reduced by 29.03% at Fr 1.0 compared to the pentamaran without a bulb. This study can be extended to optimising the bulb position on the pentamaran numerically and by comparison using the towing tank test.

Acknowledgements. UPN Veteran Jakarta has supported this work according to the Chancellor's decision No. 751/UN61.0/HK.02/2021 about the independent research of 2021. The authors also wish to thank Dr Leo Lazaukas of the University of Adelaide for the Michlet tool.

References

1. Yeung RW, Wan WH (2008) Multihull and surface-effect ship configuration design: a framework for powering minimization. J Offshore Mech Arctic Eng 130(3):31005(1–9)
2. Yanuar KTW, Widjaja MP (2017) Energy efficiency of high speed tetramaran ship model with minimum resistance configuration. Int J Mech Eng Robot Res 6(4):263–267
3. Sulistyawati WY, Pamitran AS (2020) Michell investigation of the significant influence on the hydrodynamic of a warp-chine pentamaran. J Ship Prod Des 36(3):202–212
4. Bruzzone D, Grasso A, Zotti I (2008) Nonlinear seakeeping analysis of catamarans with central bulb. In: 6th international conference on high performance marine vehicles. Carlo Bertorello-editor Univ. di Napoli Federico II, ISBN: 9788890117497, Naples, pp 47–61
5. Zotti I (2007) Medium speed catamaran with large central bulbs: experimental investigation on resistance and vertical motions. In: International conference on marine research and transportation ICMRT, pp 167–174

6. Danışman DB (2014) Reduction of demi-hull wave interference resistance in fast displacement catamarans utilising an optimised centerbulb concept. Ocean Eng 91:227–234
7. Samuel S, Kim, DJ, Iqbal M, Bahatmaka A, Prabowo AR (2018) Modification of traditional catamaran to reduce total resistance: configuration of Centerbulb. In: 17th international congress of the international maritime association of the mediterranean (IMAM 2017). Lisbon, pp 541–546
8. Sulistyawati WY, Pamitran AS (2020) Michell's thin ship theory in optimisation of warp-chine on pentamaran configuration. J Appl Fluid Mech 13(3)
9. Zaghi S, Broglia R, Di Mascio A (2010) Experimental and numerical investigations on fast catamarans interference effects. J Hydrodyn Ser B 22(5):528–533
10. Broglia R, Jacob B, Zaghi S, Stern F, Olivieri A (2014) Experimental investigation of interference effects for high-speed catamarans. Ocean Eng 76:75–85
11. Castiglione T, He W, Stern F, Bova S (2014) URANS simulations of catamaran interference in shallow water. J Mar Sci Technol 19(1):33–51
12. He W, Castiglione T, Kandasamy M, Stern F (2015) Numerical analysis of the interference effects on resistance, sinkage and trim of a fast catamaran. J Mar Sci Technol 20(2):292–308
13. Tuck EO (2003) Computation and minimisation of ship waves. In: Hill JM, Moore R (eds) SIAM proceedings in applied mathematics, vol 116; SIAM 116(1):1–16
14. Tuck EO, Lazauskas L (2008) Drag on a ship and Michell's integral. In: 22nd international congress of theoretical and applied mechanics (XXII ICTAM). Adelaide, South Australia, pp 1–2
15. Lin D, Prasanta SK, Hamid H (2020) Application of Michell's integral to high-speed round-bilge hull forms. J Ship Prod Des 36(3):189–201

PV On-Grid Charging Station for Electric Scooter: Design and Fabrication

Arif Indro Sultoni(✉), Aneke Rintiasti, Aan Anto Suhartono, and Zainal Panutup Aji

Balai Riset Dan Standardisasi Industri, 60244 Surabaya, Indonesia
arif.indro.sultoni@gmail.com

Abstract. Charging station is the main infrastructure of electric vehicles. It is needed if vehicle voyages at long distances. Commonly, electric scooter was charged at home or office. So it cannot cruise at long range. Solar energy is a proper renewable energy can be applied to electric vehicles due to gas emission. In this paper, charging station for electric scooter is designed and developed. Constant Current-Constant Voltage (CC–CV) is applied as a charging method. The station is powered by a 1.2 KWP on-grid photovoltaic array to reduce electric consumption. Photovoltaic array is constructed on the rooftop of the charging station. The charging station is equipped with friendly user interface human–machine communication for the billing system. Billing communication to GSM server is done by using SIM808 and internal communication by using CAN-bus. Experimental result shows photovoltaic contributes 33% of charging power requirement in clear weather. It needs 50 min time charging from 53 to 90% of battery SOC. Sequence billing process needs 3.05 min.

Keywords: Photovoltaic · On-grid · CC-CV · Charging station

1 Introduction

Electric vehicle is developed to substitute internal combustion engine because of its efficiency and no emission to ambient. The invention of energy storage and electric drive motor including their controllers is boosting the development of electric vehicles. Many countries already obligate the use of electric vehicles by regulation or incentive. In Indonesia, regulation for boosting electric vehicle implementation was stated in Presidential Regulation 55/2019 [1].

Motorcycle is the most favorite vehicle in Indonesia. The selling of motorcycles increased significantly from 2003 until 2020 before the pandemic. At the end of 2019, it was recorded that 112.771.136 units of motorcycles were used in Indonesia [2]. There is a unique behavior for the use of motorcycle in Indonesia. Indonesian people ride motorcycle for miles and go to another district. It is different from other countries that use motorcycle only in a city or district. So they need a charging station for electric motorcycle/scooter spread at city, suburb or village.

I. Tolj et al. (eds.), *Recent Advances in Mechanical Engineering*, Lecture Notes in Mechanical Engineering, https://doi.org/10.1007/978-981-19-0867-5_19

Many control strategies for charging station were purposed. Non-isolated Bi-directional DC-DC converter for the PHEV Charging station was developed by Sureshkumar et al. [3]. This paper compares three circuits of converters: Buck-Boost Cascade Converter, Cuk Converter and Half-Bridge DC–DC Converter. Simulation shows that Half-Bridge DC–DC Converter is better than Buck-Boost and Cuk because of the smaller number of passive components, lower switch current stress and higher efficiency. González et al. purpose control strategy for power converter of charging station with active function [4]. Simulation tests are conducted using a 100 kW power electronic converter model under different load and grid conditions. Results show the currents injected/demanded by the charger following the set points provided by the charger energy management system with active, reactive, harmonic and imbalance control that are presented under distorted and unbalanced low voltage grid conditions. Constant current/constant voltage controller for power-efficient charger was developed by Al Hassan [5].

Photovoltaic was commonly used to reduce the electric consumption of vehicle charging stations. Fancis proposed solar-powered EV smart charging station [5]. During winter, photovoltaic reduced 88% electric demand for charging. Panatarani designed and developed charging station equipped with RFID [6]. Renewable energy form PV and wind turbine supply the charger and the excess of energy stored to battery. Result shows it only charges 2.4 kW of power by current 24 A. All the above papers discuss charging stations for electric cars. It is rare that papers that discuss charging stations for scooters or electric motorcycles. One of them was a paper titled Charging Station for Electric Motor Cycle Powered by Solar Energy that was developed by Siriwattanapong and Chantharasenawong [7]. The charging station uses off-grid photovoltaic and is not yet equipped with a billing system.

On line metering EV charging station infrastructure was proposed by Lam et al. [8]. They develop third party customer service platform that facilitated a single bill issued by a customer. Experiment and evaluation of their proposed system show that the throughput achieved is about 5 Mbps at 10 ms end-to-end delay in power line communication. Mierau et al. proposed REST software architecture to integrate mobility service providers, distribution network operators and utilities to central stakeholders [9].

This research aimed to support the Presidential Regulation by introducing a prototype of PV on-grid charging station for electric scooters with design and local content. The prototype is equipped with a friendly user interface for the e-money self-billing system.

2 Method

Design of charging station is developed by identifying a common scooter battery pack that existed in Indonesia. Electrical source from grid and photovoltaic is combined as on-grid type connection. Topology of the proposed system is depicted in Fig. 1. AC–DC converter, charger controller and billing-monitoring are designed and fabricated in this research.

Photovoltaic array with 1.2 KWP is installed on rooftop. DC output from the array is optimized and converted to AC current by the inverter. AC current from inverter and grid is converted to DC current by AC–DC converter and then controlled to charge electric scooter. Controller is interfaced with billing and monitoring system. Charging station realization is shown in Fig. 2.

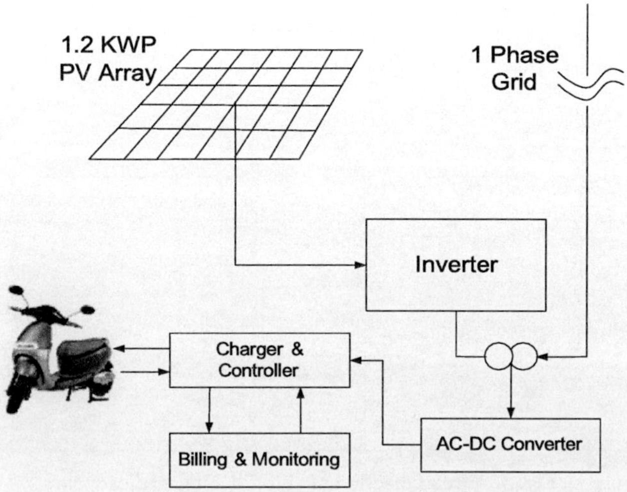

Fig. 1. Topology of proposed system

Fig. 2. Picture of charging station

The prototype of the charging station is installed at Balai Riset dan Standardisasi Industri Surabaya that appeared on Google Maps as a public charging station so the customers find the location easily.

2.1 *Charger and Controller*

Charge and controller is designed suitable for electric scooter that is commonly outstanding in Indonesia. Constant Current-Constant Voltage (CC–CV) method is applied to control battery charging. CC–CV diagram is depicted in Fig. 3.

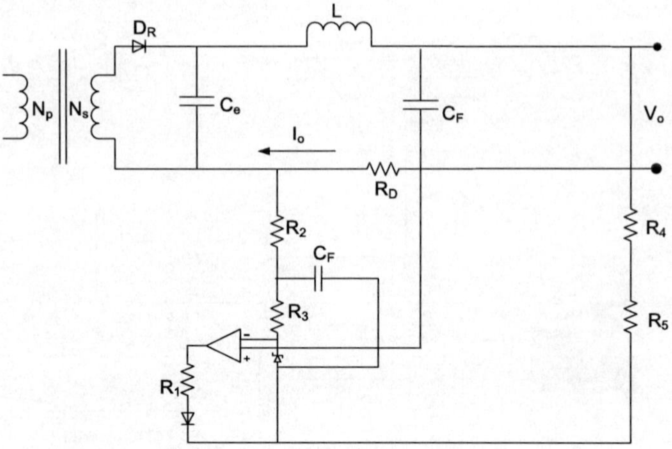

Fig. 3. Circuit diagram of current/voltage adjustment

Current Control (CC) is actuated by varying $f \cdot t_{on}$ with pulse wide modulation as

$$I = I_o\left(\frac{1 - f.t_{on}}{2}\right) \tag{1}$$

where f = switching frequency, t_{on} = on period of power MOSFET.

V_o is output charge voltage controlled by frequency modulation $f \cdot t_{on}$ that is written as

$$V_o = \frac{V_{DC\_max}}{N}\left(\frac{f.t_{on}}{1 - f.t_{on}}\right) \tag{2}$$

where $V_{DC\_max:}$ maximum DC supply voltage; and N = Primary/Secondary turn ratio. If V_r is given as reference voltage, the voltage detected at R_d is V_d, so voltage passing to U_4 is.

$$V_z = \frac{(R_2 + R_3)}{R_2} V_r \tag{3}$$

Constant voltage control (CV) actuated by giving R_5 complies with

$$R_5 = \frac{V_z.R_4}{V_o - V_z} \tag{4}$$

STM 32F2 series is used as microcontroller units to control Pulse Wide Modulation (PWM) as shown in Fig. 4.

Pin PE3 for input on/off switch, PE 9 and PE 14 are for adjusting Pulse Width Modulation of CC–CV and PA4 for adjusting voltage output. On the left side, current and voltage sensor is attached and also an OLED display to standalone monitoring.

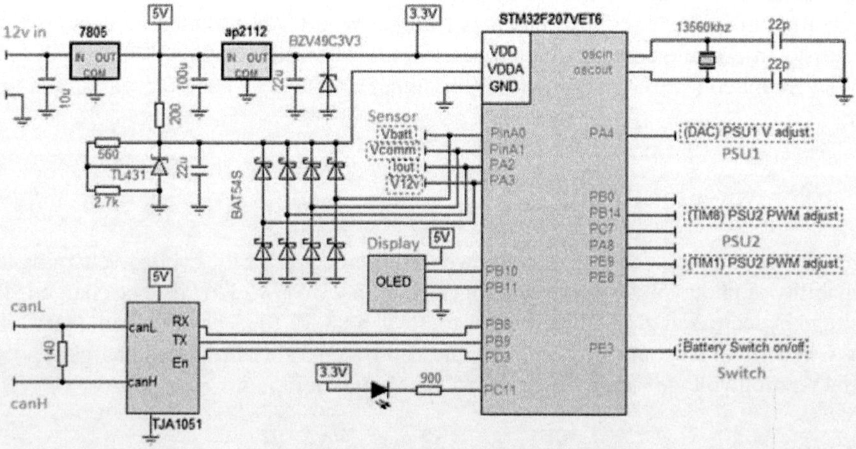

Fig. 4. Circuit diagram of MCU

2.2 *Communication of Billing and Monitoring System*

Billing and monitoring is designed friendly to customers with an Android base user interface. Block diagram of billing and monitoring is described in Fig. 5.

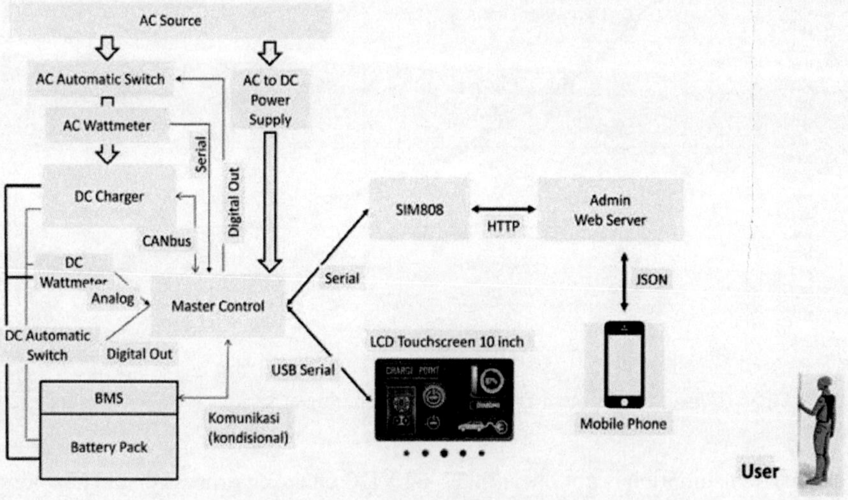

Fig. 5. Block diagram of billing and monitoring system

Customer can order how much energy is needed to buy by the LCD touch screen. Command from a customer is informed to cloud by SIM808 and then transmitted to master control. Master control instructs Charger with CAN-bus communication. Customer get information on charging process via LCD Screen including: Battery SOC (%), energy supplied (W.h), charging time (hh:mm:ss), voltage (V) and current of charging (A).

3 Experimental Result

Experiment is conducted to observe power requirement at various voltage charging and capability of photovoltaic power-sharing. Battery pack 64 V-23AH is to be charged. The voltage is decreased gradually from 64 to 35 V by 1 V. Total power (Grid + PV) and PV is recorded. At sunny day photovoltaic contribute 33% of required charging power at 64V rating voltage, and 6.7% at 34 Volt as shown in Fig. 6.

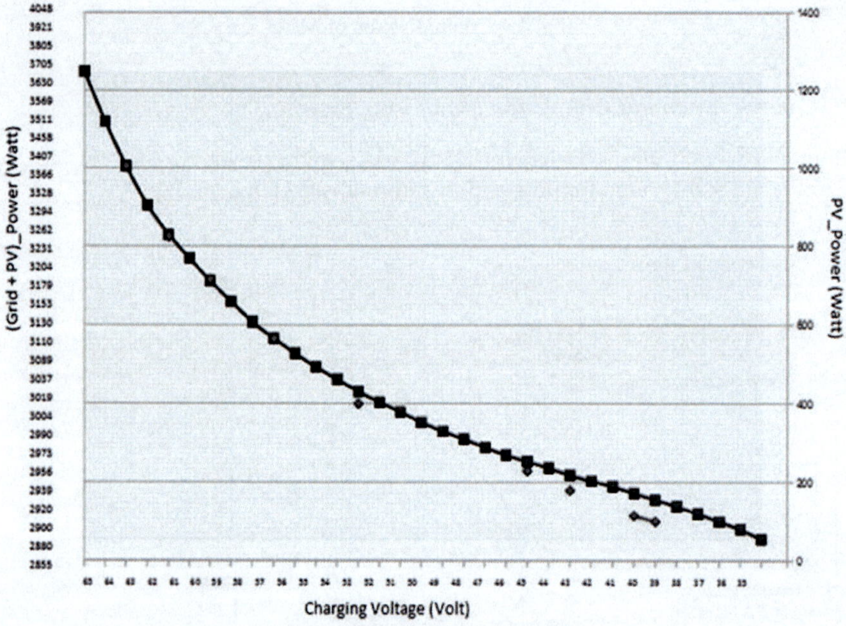

Fig. 6. Power requirement and photovoltaic sharing at various voltage charging

Power consumption is not linear in 57–64 VDC charging process due to heat losses.

Relation between battery SOC and required charging time is shown in Fig. 7. From 35 V (53% SOC) to 59 V (90% SOC), it just needs 50 min time to charge. From 59 to 63 V, charging speed is decreased and then goes constant until full charge. It needs 180 min from 53% SOC until full charge.

Time sequence for billing process is shown by Table 1. Total time required for Authorization and Identification, Set-point and Validation, Payment, Cloud Record and Data Logging is 183 s or 3.05 min.

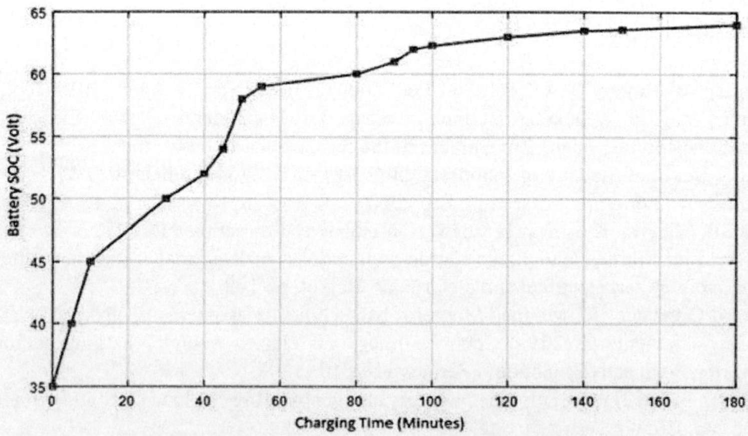

Fig. 7. Relation between battery SOC and charging time

Table 1. Time sequences of billing process

No. Sequence	Process	Time
1	Authorization and identification	45 s
2	Set-point and validation	30 s
3	Payment	57 s
4	Cloud record	38 s
5	Data logging	13 s
	Total time	183c

4 Conclusion

Charging station for electric scooter powered by on-grid photovoltaic array is designed and fabricated. The design and fabrication include AC–DC converter, charger controller and billing communication system. On-grid photovoltaic array with 1.2 KW of peak power is installed on the rooftop of the charging station. Experiment is carried out to charge a 64 V–23AH battery pack of an electric scooter. Result shows photovoltaic are capable to share 33% of total power charging consumption at 64 V rating voltage. To charge a battery from 53% SOC to 90% SOC, it needs 50 min and to achieve 100%, it needs 180 min. It needs 3.05 min to execute self-service billing by e-money.

Acknowledgements. The authors are very grateful to The Chairman of Agency for Industrial Standardization and Service Policy—Ministry of Industry of Indonesia that supports the funds of the research, and the anonymous reviewers for their valuable comments which significantly improve the presentation of this paper. Also, the authors honor The Chairman of Balai Riset dan Standardisasi Industri Surabaya that supports our activity as long as conducting the research.

References

1. President of Republic of Indonesia. Peraturan Presiden No. 55 Tahun 2019 Tentang Percepatan Program Kendaraan Bermotor Listrik Berbasis Baterai (Battery Electric Vehicle) Untuk Transportasi Jalan; Government of Indonesia: Jakarta Pusat, Indonesia, 2019.
2. Central Berau of Statistic of Indonesia. https://www.bps.go.id/indicator/17/57/1/jumlah-kendaraan-bermotor.html. Acessed 27 Feb 2020
3. Tank SB, Manavar K, Adroja N (2015) Non-isolated Bi-directional DC_DC converter for plug in hybrid electric vehicle charge station applications. In: National conference on emerging trand in computer, electrical and electronics. Rajkot, pp 1–9
4. Barrero-González F, Milanés-Montero MI, González-Romera E, Romero-Cadaval E, Roncero-Clemente C (2019) Control strategy for electric vehicle charging station power converters with active function. Energies 12(20):1–18
5. Al Hasan F (2012) Power efficiency battery charger by using constant current/constant voltage controller. Circuit Syst 3(2):180–186
6. Francis N (2018) Solar powered EV smart charging station. In: E-mobility power system integration symposium. Stockholm
7. Panatarani C, Murtaddo D, Maulana DW, Irawan S, Joni IM (2016) Design and development of electric vehicle charging station equipped with RFID. In: 2nd Padjadjaran international physics symposium. Bandung, pp 030007–1–030007–4
8. Siriwattanapong A, Chantharasenawong C (2017) Electric motorcycle charging station powered by solar energy. In: 8th TSME-international conference on mechanical engineering. Bangkok
9. Lam KL, Ko KT, Tung HY, Tung HC, Lee WC, Tsang KF, Lai LL (2011) Advanced metering infrastructure for electric vehicle charging. Smart Grid Renewe Energy 2(4):312–323
10. Mierau M, Fey S, Kohrs R, Wittwer C (2013) Communication solutions for a cloud-based charging management system for a fleet of shared use electric vehicle. World Electric Vehicle J 4(6):1125–1135

Robust Parameter Design to Optimize Kerf Width and Material Removal Rate on Plasma Cutting of Stainless Steel 316

Wima Azis Khanafi, Yanuar Rohmat Aji Pradana, and Suprayitno(✉)

Department of Mechanical Engineering, Faculty of Engineering, Universitas Negeri Malang, Malang, Indonesia
suprayitno@um.ac.id

Abstract. Plasma cutting is an unconventional machining that is widely used in industry because of its productive efficiency. The desired cutting quality is kerf width and material removal rate (MRR). Those cutting qualities on plasma cutting are influenced by the parameters of electrical current (I), cutting speed (Cs), stand of distance (SoD), and gas pressure (P). In this research, material SS 316 was used on plasma cutting using several different parameter combinations. Taguchi robust parameter design is adopted as an optimization strategy, and fractional orthogonal array L_{18} is chosen as the experimental design. Signal-to-noise ratio (S/N) robustness index is adopted as the quality measure of the cutting. The optimization strategy found that I = 60A, Cs = 1500 mm/min, SoD = 2 mm, and P = 5.5 bar give the optimum in MRR with 13.60498 in S/N or improve 0.1955 compared from initial design, while I = 50A, Cs = 1500 mm/min, SoD = 3 mm, and P = 5 bar give the optimum in kerf width with −8.0701 in S/N or improve 1,46,595 compared from initial design.

Keywords: Robust Parameter Design · Plasma Cutting · Kerf Width · MRR · Taguchi Method

1 Introduction

Plasma cutting is widely used because its efficiency is 2.5 times higher than autogenous cutting [1]. SS-316 is a material that is often cut by plasma cutting, because SS-316 has low machinability when cut by conventional machining [1, 2]. In this research, material SS-316 is chosen because this material is often used on important machine parts like in chemical and petrochemical processing, food and beverage processing, medical, marine, water treatment, and pharmaceutical processing [3].

In cutting processes, the productivity can be indicated by MRR, and for cutting quality it can be seen in the kerf width quality [4, 5]. In this study, the main objective is to obtain high MRR and small kerf width. This objective can be reached by finding the right combination of each parameter in the plasma cutting. Ramakrishnan et al. (2018) studied surface roughness, kerf width, and HAZ on plasma cutting by finding the best combination of electrical current, cutting speed, stand of distance, and gas pressure [1],

I. Tolj et al. (eds.), *Recent Advances in Mechanical Engineering*, Lecture Notes in Mechanical Engineering, https://doi.org/10.1007/978-981-19-0867-5_20

while Sharma et al. (2017) conducted research using the Taguchi method to find the best combination of current, cutting speed, stand of distance, and gas pressure on MRR [10].

In this research, four parameters of plasma cutting are chosen: electrical current (I), cutting speed (Cs), stand of distance (SoD), and gas pressure (P) [6–8]. Taguchi method as one of the popular and efficient methods in robust parameter design is selected as the optimization strategy. Taguchi introduced the S/N ratio as the quality measure of robustness [9]. Then, Taguchi formulated three different S/N ratios according to the objective.

- S/N ratio Smaller the Better (STB).

$$S/N_{\mathrm{STB}} = -10 \cdot \log_{10}\left(\frac{1}{n}\sum_{i=1}^{n} y_i^2\right) \tag{1}$$

- S/N ratio *Nominal the Best* (NTB).

$$S/N_{\mathrm{NTB}} = -10 \cdot \log_{10}\left(\frac{1}{n}\sum_{i=1}^{n} (y_i - m)^2\right) \tag{2}$$

- S/N ratio Larger the Better (LTB).

$$S/N_{\mathrm{LTB}} = -10 \cdot \log_{10}\left(\frac{1}{n}\sum_{i=1}^{n} \frac{1}{y_i^2}\right) \tag{3}$$

2 Methodology

2.1 Optimization Problem Formulation

The objective of this research is to find parameter combination of plasma cutting that produces a maximum MRR and minimum kerf width. The optimization problem formulation in this study is as follows:

Objective: Max. (MRR) and
Min. (Kerf width)
Parameters: Electrical Current (I)
Cutting Speed (Cs)
Stand of Distance (SoD)
Gas Pressure (P)

The range of each parameter is shown in Table 1. To accommodate response fluctuation due to the noise factor effect, three replications of each parameter combination are taken to estimate the mean and standard deviation. LTB S/N ratio formulation is used to maximize MRR, and LTB S/N ratio formulation is to minimize kerf width.

Table 1. Parameters and levels

Parameters	Level 1	Level 2	Level 3
Current (I), (A)	50	55	60
Cutting Speed (Cs), (mm/min)	1000	1250	1500
Stand of Distance (SoD), (mm)	1	2	3
Gas Pressure (P), (bar)	5	5.5	6

2.2 Design of Experiment

Lower bond and upper bond of each plasma cutting parameter are defined as lower and upper levels, respectively. The parameter setting that is practically used has been set as middle level. All four plasma cutting parameters and each level are shown in Tabel 1 as follows:

For four 3-level factors, L_9 fractional factorial is the minimum number of experiment that can be used as the experimental design, but preliminary results show that L_9 design failed to capture the optimum. It was probably due to interaction and/or non-linear effects. L_{18} fractional factorial array is then adopted as the design of experiment, as interaction effects are distributed evenly on each column [12] (Fig. 1).

Fig. 1. Plasma cutting MQ-1325

The experiment was conducted at PT. Sinar Prima Indonesia located in Lowokwaru, Malang, Indonesia. In this work, we use Pioneer Plasma Cutting MQ-1325. Three replications are carried out to estimate the average response and its fluctuation due to noise factor effects. The MRR is calculated using Eq. (4).

$$MRR = \frac{m_{wa} - m_{wb}}{\rho_w . t} \tag{4}$$

where MRR is material removal rate (cm^3/min), m_{wa} is the initial mass of workpiece (gram), m_{wb} is final mass of workpiece (gram), ρ_w is the density of workpiece (gram/cm^3), and t is workpiece machining time (min) [11], while kerf width is calculated using Eq. (5) [1, 2].

$$K_{Avrg} = \frac{K_t + K_b}{2} \tag{5}$$

where K_{Avrg} is the average kerf width (mm), Kt is top kerf width (mm), and Kb is bottom kerf width (mm). Material SS-316 with 35 × 20 × 3 mm of its dimension was used in this experiment.

3 Results and Discussion

3.1 Measurement Results for MRR and Kerf Width

Physical experiment was conducted with three replication samples for each 18 total different parameter combinations. The overall value of the MRR and kerf width experimental data can be seen in Table 2.

Table 2. Complete data of MRR and kerf width L_{18}

I	Cs	SoD	P	EXP	MRR			Kerf width		
					Mean	Stdev	S/N	Mean	Stdev	S/N
1	1	1	1	**1**	2,842	0,329	8,902	2,650	0,339	−8,535
1	2	2	2	**2**	3,134	0,153	9,892	3,008	0,386	−9,637
1	3	3	3	**3**	3,413	0,390	10,497	2,762	0,718	−9,107
2	1	1	2	**4**	3,377	0,310	10,462	3,010	0,567	−9,723
2	2	2	3	**5**	3,903	0,110	11,818	3,485	0,156	−10,853
2	3	3	1	**6**	2,824	0,458	8,687	2,503	0,613	**−8,223**
3	1	2	1	**7**	3,362	0,755	9,921	4,030	0,474	−12,166
3	2	3	2	**8**	3,958	0,112	11,939	3,120	0,092	−9,887
3	3	1	3	**9**	3,866	0,136	11,728	3,082	0,520	−9,898
1	1	3	3	**10**	3,195	0,361	9,927	3,353	0,163	−10,520
1	2	1	1	**11**	3,193	0,192	10,038	2,855	0,205	−9,134
1	3	2	2	**12**	3,472	0,244	10,747	3,172	0,276	−10,058
2	1	2	3	**13**	3,812	0,345	11,518	3,255	0,193	−10,266
2	2	3	1	**14**	3,278	0,383	10,137	2,923	0,458	−9,423
2	3	1	2	**15**	4,259	0,487	12,420	2,797	0,352	−9,001
3	1	3	2	**16**	4,697	0,217	**13,409**	2,983	0,294	−9,536
3	2	1	3	**17**	3,989	0,285	11,952	3,425	0,573	−10,813
3	3	2	1	**18**	4,382	0,300	12,772	2,853	0,482	−9,229

Parameter combination of plasma cutting as shown in experimental 16 on Table 2 is parameter setting that was usually used in practice, then called initial design. It has the best S/N in MRR among the 18 samples. For kerf width, experimental sample 6 is better than the initial design as shown in Table 2.

Based on this experimental data, L_{18}, an analysis of mean, was conducted to estimate the main effect of each level parameter. The main effect was then plotted to see the mean response and obtain the optimum combination level.

3.2 Effect Plot

The effect plot is obtained from the calculation of the analysis of the mean based on the data in Table 2 both for MRR and kerf width. The plot effect of MRR can be seen in Figs. 2 and 3 for kerf width.

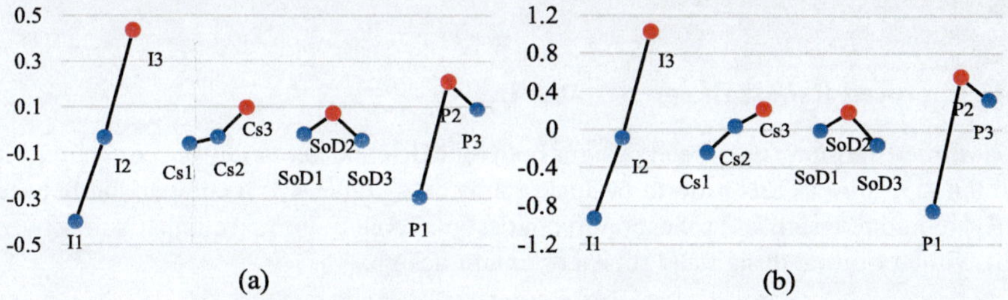

Fig. 2. Effect plot **a** mean MRR, **b** S/N ratio MRR

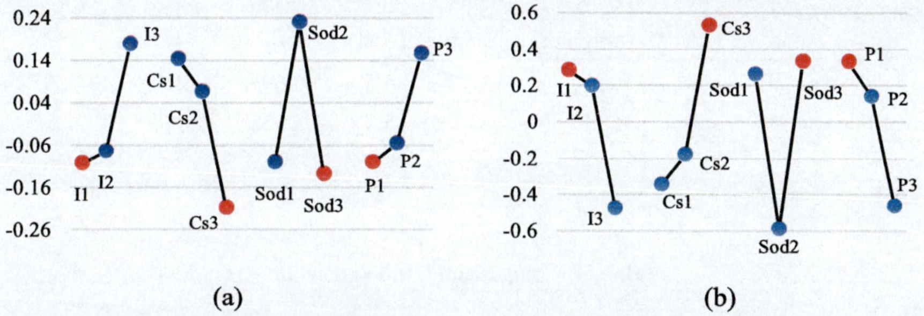

Fig. 3. Effect plot **a** mean kerf width, **b** S/N ratio kerf width

Figures 2 and 3 show the mean response S/N ratio (a) for MRR and (b) for kerf width, respectively. As shown in Fig. 2b, the optimum MRR (robust design MRR) can be predicted by a combination of parameters current level 3 (60 A), cutting speed level 3 (1500 mm/min), stand of distance level 2 (2 mm), and gas pressure level 2 (5.5 bars). The optimum kerf width (robust design kerf width) can be predicted with a combination of current level 1 (50 A), cutting speed level 3 (1500 mm/min), stand of distance level 3 (3 mm), and gas pressure level 1 (5 bars), shown in Fig. 3b.

3.3 Verification Robust Design

Additional experiments were conducted to verify the predicted optimum both for MRR and kerf width. The optimum parameter level for MRR is 3–3–2–2, and for kerf width it is 1–3–3–1. The verification results of these two optimum conditions are shown in Table 3.

Table 3. Verification results of predicted optimum of MRR and kerf width

Opt design	I (A)	Cs (mm/min)	SoD (mm)	P (bar)	Replicates			Mean	Stdv	S/N
					1	2	3			
MRR	60	1500	2	5,5	6,142	4,145	4,909	5,065	1,008	13,605
Kerf	50	1500	3	5	2,675	2,76	2,08	2,505	0,37	−8,07

3.4 Improved Robust Design to Initial Design

The verified optimum parameter designs both for MRR and kerf width are compared with the initial design in S/N ratio to evaluate robustness. Table 4 is a comparison between MRR of initial design and robust optimum design. Table 5 shows a comparison between kerf width of initial design and robust optimum design.

Table 4. Comparison result MRR

Design parameters	I (A)	Cs (mm/min)	SoD (mm)	P (bar)	Mean	Stdev	S/N Ratio
Initial design	60	1000	3	5.5	4.697	0,217	13.4095
Robust design	60	1500	2	5.5	5.065	1,008	13.6050
Gain							0.1955

Table 5. Comparison result kerf width

Design parameters	I (A)	Cs (mm/min)	SoD (mm)	P (bar)	Mean	Stdev	S/N Ratio
Initial design	60	1000	3	5,5	2,983	0,29	−9,5361
Robust design	50	1500	3	5	2,505	0,37	−8,0701
Gain							1,46,595

As shown in Table 4, robust optimum design in MRR improves S/N ratio by 0.1955 compared to initial design. Even though the standard deviation is worse, the mean MRR improves, so the overall S/N ratio is better than the initial design. A similar condition in kerf width, the robust optimum design slightly decreases in standard deviation, but improves in mean kerf width. The S/N ratio increases by 1.466 compared to the initial design.

3.5 Analysis of Variance (ANOVA)

Analysis of variance was carried out to determine the contribution of each parameter to the output (MRR and kerf width). The main effect of each parameter was considered by ANOVA, but not for interaction effect among parameters, as was done with the previous Taguchi method. ANOVA for S/N MRR can be seen in Table 6 and S/N kerf width in Table 7.

Table 6. ANOVA S/N MRR

S/N ratio	SS	DoF	MS	Fo	Ft	C (%)	Rank
I	11,521	2	5,760	4,960	3,17	38,78	1
Cs	0,622	2	0,311	0,268	3,17	2,09	3
SoD	0,360	2	0,180	0,155	3,17	1,21	4
P	6,755	2	3,377	2,908	3,17	22,74	2
Error	10,451	9	1,161			35,18	
Total	29,708	17				100	

Table 7. ANOVA S/N Kerf Width

S/N ratio	SS	DoF	MS	Fo	Ft	C (%)	Rank
I	2,0638	2	1,0319	1,8829	3,17	13,98	3
Cs	2,5688	2	1,2844	2,3436	3,17	17,40	2
SoD	3,1468	2	1,5734	2,8710	3,17	21,32	1
P	2,0485	2	1,0243	1,8690	3,17	13,88	4
Error	4,9323	9	0,1096			33,42	
Total	14,7601	17				100%	

As can be seen in Table 6, the current parameter (I) has the highest effect on MRR, namely 38.78%, followed by gas pressure (P) 22.74%, cutting speed (Cs) 2.09%, and the last position is the stand of distance (SoD) of 1.21%. Based on Table 7, it is known that the stand of distance (SoD) parameter has the highest effect on the kerf width of 21.32%, followed by the cutting speed (Cs) parameter of 17.40%, the current (I) parameter of 13.98%, and the smallest effect is gas pressure (P) with a value of 13.88%. The results of the ANOVA confirm the main effect plot for S/N ratio of the Taguchi method in Figs. 2 and 3.

4 Conclusion

The conclusion from the optimization of plasma cutting parameters is that a robust design for maximum MRR is obtained at a combination of 3–3–2–2 where the current is 60 A, cutting speed is 1500 mm/min, stand of distance is 2 mm, and gas pressure is 5.5 bar. Robust design for minimum kerf width is obtained by a combination of 1–3–3–1 where current is 50 A, cutting speed is 1500 mm/min, stand of distance is 3 mm, and gas pressure is 5 bar. The enhancement robust design against the initial design for the MRR is 0.36785 cm^3/min and S/N ratio is 0.19550. Robust design improvement for kerf width is −0.48 mm and S/N ratio is 0.80720. The contribution of each parameter to MRR is current (I) 38.78%, gas pressure (P) 22.74%, cutting speed (Cs) 2.09%, and stand of distance (SoD) 1.21%. The contribution of each parameter to the kerf width is stand of distance (SoD) 21.32%, cutting speed (Cs) 17.40%, current (I) 13.98%, and gas pressure (P) 13.88%.

Acknowledgements. This publication is a part of research work that is fully funded by PNBP UM 2021 under contract No. 5.3.890/UN32.14.1/LT/2021.

References

1. Ramakrishnan H, Balasundaram R, Karthikeyan N, Ganesh N (2018) Experimental investigation of cut quality characteristics on SS321 using plasma arc cutting. J Brazilian Soc Mech Sci Eng 8:1–11
2. Bhowmick S, Basu J, Majumdar G, Bandyopadhyay A (2018) Experimental study of plasma arc cutting of AISI 304 stainless steel. Mater. Today Proc 5(2):4541–4550
3. Company SS (2014) Specification sheet: alloy 316/316L, 1
4. Masoudi S, Mirabdolahi M, Dayyani M (2018) Development of an intelligent model to optimize heat-affected zone, kerf, and roughness in 309 stainless steel plasma cutting by using experimental results. Mater Manuf Process 00(00):1–12
5. Kumar KR, Sreebalaji VS, Ganesan V (2015) Analysis of Kerf width and cutting speed characteristics of Aluminium/Tungsten carbide composites
6. Parthiban A, Prasath JP, Vivek P, Pugazhenthi R (2018) Experimental investigation of plasma arc cutting for stainless steel sheet. Int J Mech Prod Eng Res Dev 8(1):907–914
7. Patel P, Soni S, Kotkunde N, Khanna N (2018) Study the effect of process parameters in plasma arc cutting on Quard-400 material using analysis of variance. Mater. Today Proc 5(2):6023–6029
8. Yang SYS (2001) An expert system for plasma cutting process quality prediction and optimal parameter suggestion
9. Yu J, Chang C, Suprayitno (2017) Robust design optimisation via surrogate network model and soft outer array design. Int J Prod Res 7543:1–15
10. Sharma S, Gupta M, Kumar R, Bindra NS (2017) Experimental analysis and optimization of process parameters in plasma Arc cutting machine of EN-45A material using Taguchi and ANOVA method. 121006(7):1394–1398
11. Jahan MP, Green B (2014) Varieties, and applications, vol 11, Elsevier
12. Phadke MS (1989) Quality engineering using robust design. Prentice-Hall PTR, Inc

Effects of Taper Pin and Two-Step Pin on Fatigue Strength in Micro Friction Stir Spot Welding of AA1100

Ario Sunar Baskoro(✉), Gatot Prayogo, Mohammad Azwar Amat, Muhammad Shadani Pahlevi, and Rafi Ahmad Eshandra

Department of Mechanical Engineering, Faculty of Engineering, Universitas Indonesia, Depok, Indonesia
ario@eng.ui.ac.id

Abstract. Micro Friction Stir Spot Welding (mFSSW) is a welding method on a thin sheet of metal. This welding method is frequently used because of its advantages compared to the other conventional welding method such as RSW. Micro FSSW produced better performance in aluminum alloy with a strong welding joint and longer fatigue life. This research intends to discover the effect of tools geometry on welding joints produced by the mFSSW method using a base material of aluminum alloy 1100 in the form of repetitive load through the dynamic fatigue test. In this research, three different tools were used, where each tool had different pin and shoulder dimensions. The fatigue test was performed on five different loads in each tool. The testing result would be processed and analyzed to create an S–N diagram and failure mode analysis for each specimen to discover which tools geometry has the best endurance against repetitive load. The results showed that Tool 3 (two-step pin), which had a small outer pin diameter and a large shoulder diameter compared to other tools, produced an S–N diagram that fitted the basic theory and had the smallest statistical error. The failure mode generated from all tests was in the form of a nugget pullout.

Keywords: Micro friction stir spot welding (mFSSW) · Tools geometry · AA1100 thin plate · Fatigue test · Failure mode

1 Introduction

Friction stir welding (FSW) is a welding method invented by The Welding Institute in 1991 which utilizes friction to generate heat [1]. One of the developments of this method is called friction stir spot welding (FSSW), developed and patented by Mazda Motor Corporation and Kawasaki Heavy Industry. The weld tools only rotate in one place after they are plunged into the workpiece. This method produced a strong welding joint and longer fatigue than other conventional welding techniques such as resistance spot welding (RSW) [2].

For a decade, some researchers are looking for the possibility of joining a thin plate of aluminum alloy with a thickness of less than 1 mm, such as Scialpi et al. [3]

I. Tolj et al. (eds.), *Recent Advances in Mechanical Engineering*, Lecture Notes in Mechanical Engineering, https://doi.org/10.1007/978-981-19-0867-5_21

began to join AA2024-T3 to AA6082-T6 sheets with a thickness of 800 μm using butt configuration. Bagheri et al. [4] modified the FSSW process with vibration to join 2.0 and 0.7 mm AA5083 aluminum alloy sheets. Baskoro et al. [5–7] also successfully joined a thin plate of aluminum alloy 1100 with a sheet thickness of 400 μm in a lap joint configuration using micro friction stir welding (mFSW) and micro friction stir spot welding (mFSSW). According to Wang et al. [8], mFSSW is one of the emerging technologies that could be used for soldering iron. This technology would benefit the assembly of the circuits of electronic devices and medical items, and micro-devices that are susceptible to overheating. Meanwhile, the technologies nowadays and in the future focusing on the micro-based technologies, mFSSW, could excel in producing those.

Previous studies on joining aluminum alloy 1100 using mFSW and mFSSW generally focused on mechanical properties and microstructural characterization. In the meantime, it is also important to know the joint's fatigue life that is produced by mFSSW of thin aluminum sheets in the form of repetitive load through the dynamic fatigue test. The selection of the design of experiment (DoE) in this study was based on the results of a previous study done by Baskoro et al. [5–7].

2 Materials and Method

The materials used in this study were AA1100 sheets with a thickness of 0.42 mm, and cut specimens of 75 mm × 25 mm. Micro friction stir spot welding was performed using an overlap joint at a 25 × 25 mm center. Meanwhile, the specification of the tools used is a High-Strength Steel (HSS) with the specification as shown in Table 1. Figure 1 shows the welding contour of each tool on the welded materials.

Table 1. Tools dimensions

Code name	Name	Dimension (mm)			
		Shoulder diameter	Outer pin diameter	Inner pin diameter	Pin height
Tool 1	Small taper pin	4.099	1.811	2.515	0.629
Tool 2	Medium taper pin	4.124	2.288	2.942	0.604
Tool 3	Two-step pin	4.987	1.572	1.572	1.572

The mFSSW process used a mini-drill machine attached to a CNC vertical milling machine. The initial rotational tool speed that was used was 33,000 rpm. However, because of the welding process, the rotational speed was not constant over time, and it decelerated. Meanwhile, the fatigue test machine that is being used is Hung Ta HT-9711 Dynamic Fatigue Testing Machine as shown in Fig. 2.

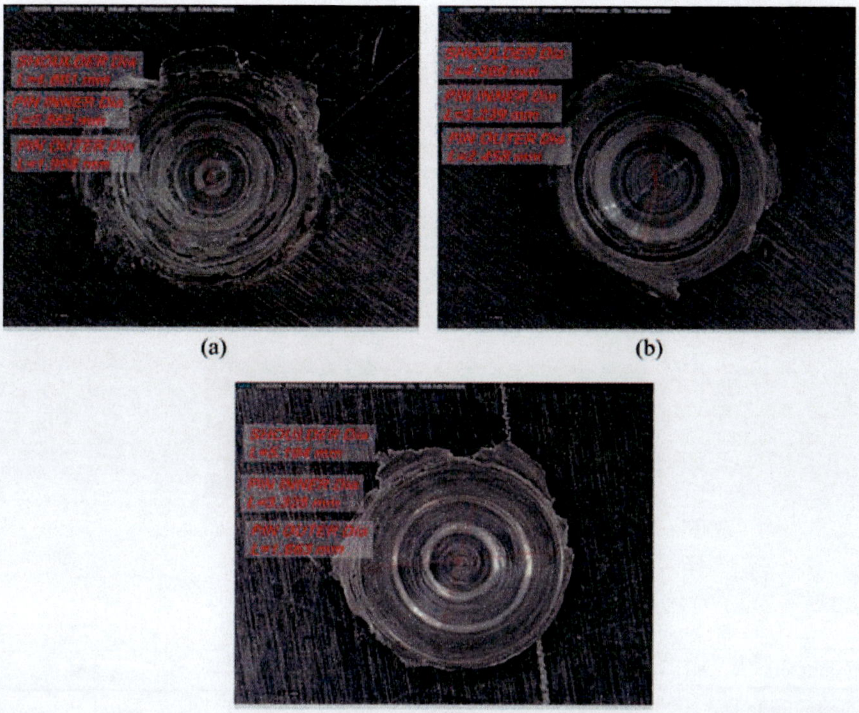

(a) (b) (c)

Fig. 1. Welding contour of each tool: **a** tool 1 (small taper pin), **b** tool 2 (medium taper pin), and **c** tool 3 (two-step pin)

The force ratio and frequency that were being used in this test were 0.1 and 10 Hz. From that parameter, we should know the maximum load and preload required to be input into the machine by using Eq. (1). Five maximum loads of 200, 188, 175, 163, and 150 N were used for this test.

$$ForceRatio(RP) = \frac{Pmin}{Pmax} = \frac{Preload - Amplitude}{Preload + Amplitude} \tag{1}$$

After some calculation, the following Table 2 will summarize the parameter inputted in the fatigue test machine.

Each parameter will be tested one time, while the S–N diagram will be created using linear regression. The plunge depth in the mFSSW process was set to be ±600 micron, where each of the samples would be tested by each of the previously stated parameters. The complete list of the plunge depth from each specimen is shown in Table 3.

Fig. 2. Hung Ta HT-9711 dynamic fatigue testing machine

Table 2. Fatigue test parameters

Parameters	1	2	3	4	5
Maximum load (N)	200	188	175	163	150
Preload (N)	110	103.4	96.25	89.65	82.5
Amplitude	90	84.6	78.75	73.35	67.5
Minimum load (N)	20	18.8	17.5	16.3	15

Table 3. The plunge depth of each sample

Number	Maximum load applied (N)	Plunge depth (micron)		
		Tool 1 (small taper pin)	Tool 2 (medium taper pin)	Tool 3 (two-step pin)
1	200	600	590	620
2	188	610	630	620
3	175	570	580	600
4	163	570	620	620
5	150	680	570	650

3 Results and Discussion

3.1 Fatigue Life Analysis

Table 4 shows the complete results of fatigue tests for each tool. And Fig. 3. shows the S–N diagram for each tool. The results show that the trends tend to be inclined, where the smaller maximum load that is being applied for the test would result in higher cycles to failure. However, each datum per tool does not represent the trends that are created. To see which tools produced better data, we calculated the error for each tool using their R that is produced with the help of Microsoft Excel. Each tool produced R values: 0.4347 for Tool 1, 0.3016 for Tool 2, and 0.8033 for Tool 3. From those values themselves, it is shown that Tool 3 produces the best data compared to other tools. Table 1 shows that Tool 1, a two-step pin, has a smaller pin outer diameter and larger shoulder diameter than the other tools.

Table 4. Complete result of Fatigue Tests for each Tool

Number	Maximum load applied (N)	Number of cycles to failure
Tool 1 (small taper pin)		
1	200	7,100
2	188	55,200
3	175	11,750
4	163	33,300
5	150	78,000
Tool 2 (medium taper pin)		
1	200	5,875
2	188	30,100
3	175	3,750
4	163	37,600
5	150	37,500
Tool 3 (two-step pin)		
1	200	9,100
2	188	43,700
3	175	119,465
4	163	201,750
5	150	335,847

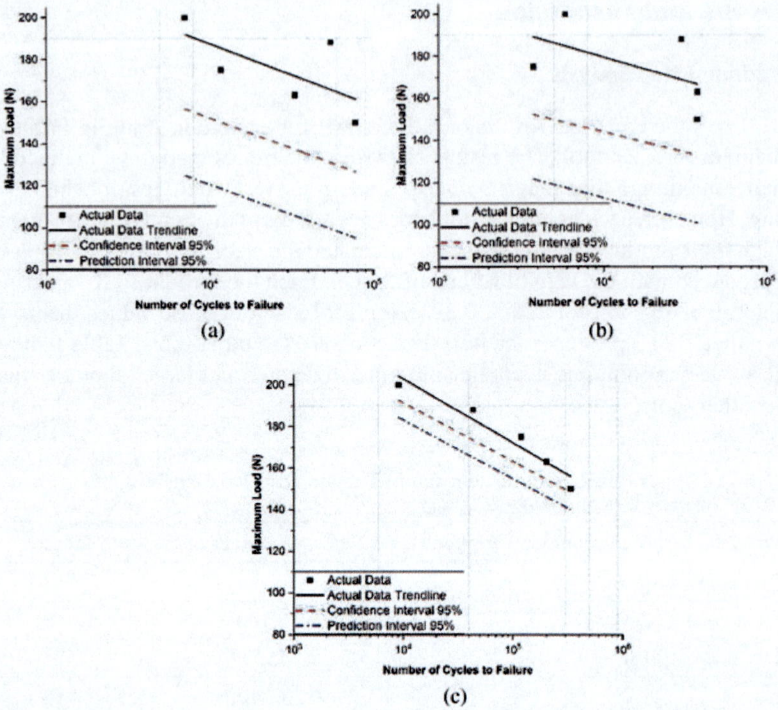

Fig. 3. S–N Diagram: **a** tool 1 (small taper pin), **b** tool 2 (medium taper pin), and **c** tool 3 (two-step pin)

3.2 Failure Type

After the fatigue testing of all specimens, it shows that nugget pullouts happen to all the specimens. It is shown by the attached weld nugget on one half of the specimen, resulting in a hole in another side of the specimen as shown in Fig. 4.

4 Conclusions

1. Tools with a small pin outer diameter and large diameter shoulder produce welded joints with better strength, characterized by the number of cycles leading to failure compared to other tools.
2. Tool 1 produces the most minimal statistical error test data, where the data generated is under the basic theory where the smaller the load tested, the more cycles generated.
3. The failure mode resulting from this test is in the form of a nugget pullout.

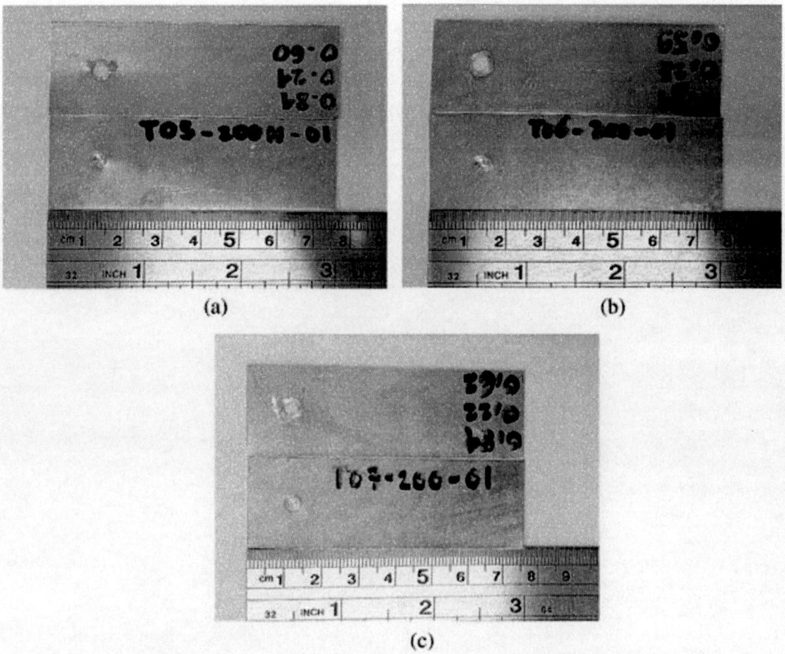

Fig. 4. The typical failure for each tool: **a** tool 1 (small taper pin), **b** tool 2 (medium taper pin), and **c** tool 3 (two-step pin)

Acknowledgements. This work is supported by the PUTI Q1 program of the Directorate of Research and Development Universitas Indonesia 2020 with contract number NKB-4002/UN2.RST/HKP.05.00/2020.

References

1. Thomas WM, Nicholas ED (1997) Friction stir welding for the transportation industries. Mater Des 18:269–273
2. Yuan W, Mishra RS, Webb S, Chen YL, Carlson B, Herling DR, Grant GJ (2011) Effect of tool design and process parameters on properties of Al alloy 6016 friction stir spot welds. J Mater Process Technol 211(6):972–977
3. Scialpi, A., De Giorgi,, M., De Filippis, L. A. C., Nobile, R., Panella, F. W. (2008) Mechanical analysis of ultra-thin friction stir welding joined sheets with dissimilar and similar materials. Mater Des 29(5):928–936
4. Bagheri B, Abbasi M, Abdollahzadeh A, Omidvar H (2020) Advanced approach to modify friction stir spot welding process. Met Mater Int 26:1562–1573
5. Baskoro AS, Nugroho AAD, Rahayu D, Suwarsono, Kiswanto G, Winarto W (2013) Effects of welding parameters in micro friction stir lap welding of aluminum A1100. Adv Mater Res 789:356–359
6. Baskoro AS, Suwarsono, Kiswanto G, Winarto (2014) Effects of high-speed tool rotation in micro friction stir spot welding of Aluminum A1100. Appl Mech Mater 493:739–742

7. Baskoro AS, Hadisiswojo S, Kiswanto G, Winarto, Amat MA, Chen ZW (2020) Influence of welding parameters on macrostructural and thermomechanical properties in micro friction stir spot welded under high-speed tool rotation. Int J Adv Manufact Technol 106(1–2):163–175
8. Wang DA, Chao CW, Lin PC, Uan JY (2010) Mechanical characterization of friction stir spot microwelds. J Mater Process Technol 210(14):1942–1948

The Strength Analysis of Dovetail Joint on 3D Printed Part Polylactic Acid Using Computer Simulation

Arya Mulyana Kusumah, Redyarsa Dharma Bintara(✉), and Suprayitno

State University of Malang, Jl. Semarang No.5 Sumbersari, Malang, Indonesia
redyarsa.dharma.ft@um.ac.id

Abstract. 3D printing is one of the manufacturing technologies that can be used for making a prototype part. Fused deposition modeling (FDM) is one 3D printing technology that has economic advantages in the use of materials and a variety of machine choices. Currently, the working capacity of 3D printing machines has many kinds but 235 × 235 × 250 mm of the working area is seldom seen on the market. Its working area becomes a problem if it is used to print a big model that has oversize than the working area of the machine. Therefore, the big model can be segmented and printed separately to solve the current problem. Furthermore, all segmented part is assembled to get the desired 3D printed part. The 3D assembled part is connected by a joint structure. So, the purpose of this study is to investigate the characteristic of the dovetail joint on a 3D model assembled to tensile load on polylactic acid (PLA) material using computer simulation. The ASTM D3039 standard is applied to get the mechanical properties of PLA material that is printed by a 3D printer machine. Furthermore, the data obtained is made as an input in simulation software. The maximum shear stress and factor safety are two output parameter that is evaluated to define the best parameter combination. The results showed that the shear stress minimum is 18.805 MPa and the factor safety of the model is 0.56983. It occurred on variation models w_1 25 mm, w_3 6 mm, and t 30 mm. The significant result is due to the increase in joint area, angle of joint, and bending moment.

Keywords: 3D Printing · Dovetail Joint · Simulation

1 Introduction

Additive manufacturing is a technology that significantly evolved in the 3rd revolution industry. It is often well-known as rapid prototyping or 3D printing technology. The 3D printing technology can create a product by depositing material layer by layer [1]. Several types of 3D printing machines are already available in the market, such as direct metal sintering, fused deposition modeling (FDM), stereolithography, and selective laser sintering (SLS) [2]. The FDM machine is commonly found in the market because it has economic advantages in terms of materials used and machine. Currently, most economic of the FDM 3D printers has a working area of 235 × 235 × 250 mm.

I. Tolj et al. (eds.), *Recent Advances in Mechanical Engineering*, Lecture Notes in Mechanical Engineering, https://doi.org/10.1007/978-981-19-0867-5_22

The economic FDM 3D printer has limitation to manufacture a large dimension product. Therefore, the 3D model can be segmented and manufactured part by part to solve the current limitation of machine. Furthermore, all of 3D printed part is assembled together to get the desired product. However, the joint structure of assembled part needs to be designed and analyzed to get the robust design. Therefore, the purpose of the study is to analyze the strength of the joint structure with different thickness and width of the joint using computational simulation approach.

2 Method

2.1 Materials

Polylactic acid (PLA) is a polymer material that is commonly used as raw material for 3D printing machines. It is a strong material for several applications, but it still depends on the printing direction deposit layer [3]. PLA materials are often used because it has biodegradable thermoplastic properties and strong mechanical properties for various applications [4]. Several applications of 3D printing technology can be seen in the making of prosthetic hands, action figures, and so on [5]. In this study, PLA material has a specific property such as melt temperature (198–205 °C) and the original filament diameter is 1.75 mm since the material is used as a raw material for a 3D printer machine.

2.2 Experiment Procedure

The ASTM D3039 standard was obeyed as a tensile test of polymer, especially 3D printed part. The 3D model of the tensile test specimen was designed by using CAD software (Autodesk Inventor) which had a dimension of 250 × 25 × 2.75 mm. Furthermore, the 3D model was simulated on the 3D printing software (Cura) which is the parameter according to Table 1 [6]. Then the 3D model was manufactured by a 3D printing machine (Creality Ender 3). Meanwhile, the tensile test was performed by a universal tensile test machine (UTM) with a maximum load capacity of 5000 kg for getting the mechanical properties of the 3D printed part. The mechanical properties of the specimen were made as initial information for the simulation process.

2.3 Modeling and Simulation

The structure of the joint model was designed using CAD software. Furthermore, each model was simulated using finite element software (FEM). In the package of software, the mechanical properties of the model were inputted according to Table 2. Then, the structure of the joint parameter as shown in Fig. 1a was varied. Meanwhile, Fig. 1b is the meshing shape in which the hexagon type was applied to the model. The parameter design was varied into three different types such as width w_1 (25,50,75 mm), w_3 (6,7,8 mm), and thickness (t) (30 mm). Furthermore, the H_1 and H_2 was kept at 5 mm, whereas the length (l) of the model was 170 mm. Figure 2 shows the boundary condition of the simulation model. The displacement of the model was set according to the x, y, and z of global coordinate system ($x, y, z = 0$) on the thick model (see Fig. 2b). Furthermore, the

Table 1. Printing parameters of the tensile test specimen

Parameter	Value
Layer height (mm)	0.2
Wall thickness (mm)	0.8
Speed while extruding (mm/s)	60
Infill density (%)	100
Infill pattern	Lines
Printing temperature (°C)	205
Build plate temperature (°C)	60

uniaxial loading was applied to the model with respect to *y*-axis of the global coordinate system (see Fig. 2b). The contact surface was placed on the two faces that were shown in Fig. 2a. The contact surface was defined as a bonded type.

Table 2. Mechanical properties of PLA

Parameter	Value
Density (g/m3)	1.25
Young modulus (GPa)	1.52
Force @ peak (N)	3095.82
Strain (%)	4.3
Tensile strength (MPa)	45.528
Tensile Yield Strength (MPa)	39.8
Poisson's ratio	0.36

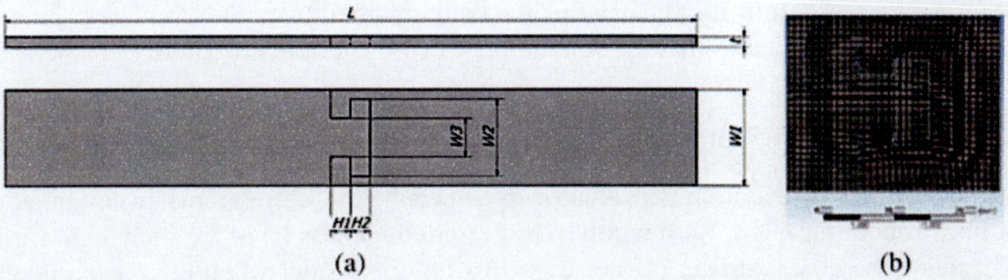

Fig. 1. Geometry of model test **a** parameter study and **b** element shape of the model test

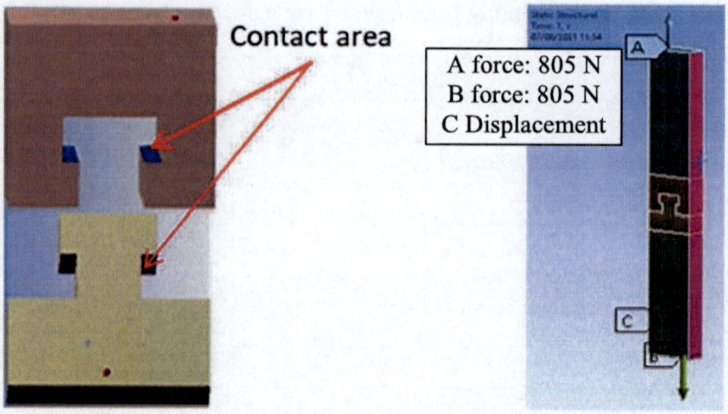

(a) (b)

Fig. 2. Boundary condition set to the model test **a** contact area on the dovetail joint and **b** direction of uniaxial loading

3 Result and Discussion

3.1 Validation Result

The tensile test is conducted to investigate the mechanical properties of the 3D printed part. Table 2 shows the average mechanical properties of the 3D printed part that was calculated with five repeated testings. The several datas such density (1.25 g/m_3) [7], yield stress (39.8 MPa) [7], and poisson ratio (0.36) [8] are also supported from the previous study.

Furthermore, the mechanical properties data that is recorded in Table 2 is made as input in simulation software for the validation process. The 3D model of the specimen test is loaded to the uniaxial tensile load; refer Table 2 (3095.82 N). Figure 3 shows the tensile test result by using computer simulation. Based on the result, the maximum stress of the model is 43.47 MPa. If it is compared with the yield stress of PLA material in Table 2 (45.528 MPa), the maximum stress of the simulation test tends to close the experimental test result. It indicates that the data obtained from the simulation process is correct. It also indicates that the simulation process is correct or valid.

3.2 The Stress Distribution of Dovetail Joint to Width and Thickness

The shear stress distribution is evaluated to determine the best parameter combination for joint structure design. Both width (w_1, w_3) and thickness (t) of the model are varied to evaluate the characteristic of the stress distribution model when the tensile load is applied. The tensile load is carried out in the direction of the y-axis with respect to the global coordinate system. Figure 4 shows an example of shear stress distribution on the design with variations w_1 (25 mm), w_3 (6 mm), and t (30 mm).

The shear stress distribution is observed around the connection model which depicts the highest stress concentration (see Fig. 4). Furthermore, the maximum shear stress models are recorded and analyzed for getting the best design. Moreover, the factor safety

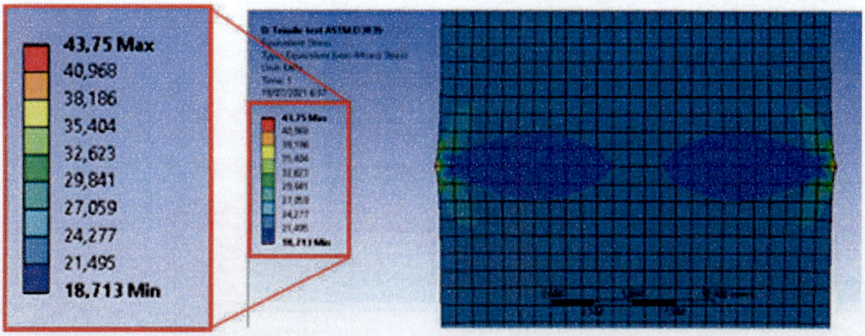

Fig. 3. Stress maximum on the 3D model of tensile test

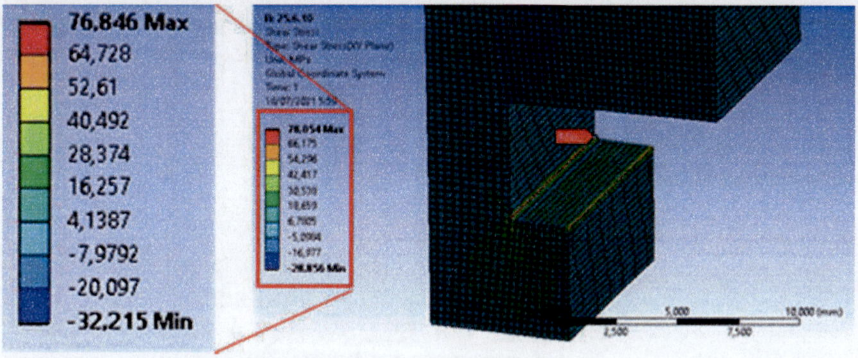

Fig. 4. Maximum shear stress on dovetail joint

(*F.S*) is also evaluated to determine whether the model is safe or fractures. Figure 5 shows the effect of parameter variation on shear stress maximum for all variation models. The results show that the maximum stress occurs around the connection model with the highest stress concentration and *F.S* is 25.282 MPa and 0.4597, respectively (w_1 75 mm; w_3 7 mm; t 30 mm). It is caused by increasing the bending moment on the model. The force acts on the *y*-axis direction of the models. The increase of w_1 can also increase the distance (d) of the contact point to the boundary of the geometry model (see Fig. 6). Furthermore, the bending moment can also increase the stress concentration around the joint area. The lowest shear stress occurs on the model which is w_1 (25 mm), w_3 (6 mm) and t (30 mm) with stress concentration and *F.S* is 18.805 MPa and 0.5698 respectively. It has a short distance (d) if it is compared to the rest of the variation. The small distance of w_3 also gives the large area of contact area. Therefore, the shear stress can distribute more uniformly.

$$\tau = \frac{F}{A} \tag{1}$$

$$M = F.d \tag{2}$$

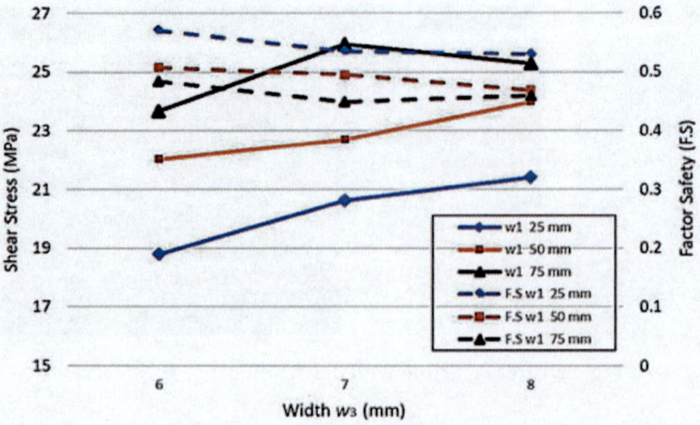

Fig. 5. Shear stress maximum to width and thickness of parameter set

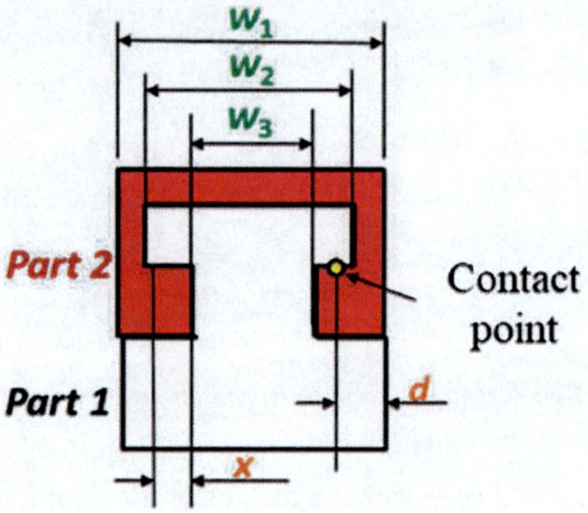

Fig. 6. Parameter design on contact region of the model

Equation (1) shows the correlation of shear stress (τ), force (F), and area (A). The shear stress is affected by the distribution of forces acting on the cross-sectional area. While Eq. (2) is the bending moment formula. M is represented as the bending moment and d is the distance from the contact point to right side of model.

4 Conclusion

Both width and thickness joint structure of the model can affect the effectiveness of shear stress distribution. Based on the result, the shear stress distribution can be caused by the differences in contact area. Moreover, the increase in width can also increase the bending moment of the model. The bending moment can increase the stress concentration so that the model tends to fail on the sharp area. The best design occurs on the variation model w_1 25 mm, w_3 6 mm, and thickness 30 mm have minimum values in shear stress

(18.805 MPa) and maximum of safety factor (0.56983). The significant result is due to an increase in the joint area, angle of joint, and bending moment.

References

1. Soares JB, Finamor J, Silva P, Roldo L (2018) Analysis of the influence of polylactic acid (PLA) colour on FDM 3D printing temperature and part finishing. Rapid Prototyp J 24(8):1305–1316
2. Bintara RD, Aminnudin A, Prasetiyo D, Arbianto F (2019) The characteristic of overhang object to material usage on FDM 3D printing technology. J Mech Eng Sci Technol 3(1):35–41
3. Sin LT, Tueen BS (2019) Polylactic acid a practical guide for the processing, manufacturing, and applications of PLA, 2nd edn. William Andrew, USA
4. Kovan V, Altan G, Topa ES (2017) Effect of layer thickness and print orientation on strength of 3D printed and adhesively bonded single lap joints. J Mech Sci Technol 31(5):2197–2201
5. Oropallo W, Piegl LA (2016) Ten challenges in 3D printing. Eng Comput 32(1):135–148
6. Liu W, Zhou J, Ma J, Wang (2018) Fabrication of PLA filaments and its printable performance. IOP Conf Ser Mater Sci Eng 275(1):0–6
7. Farah S, Anderson G, Langer R (2016) Physical and mechanical properties of PLA, and their functions in widespread applications A comprehensive review. Adv Drug Deliv Rev 107:367–392
8. MatWebHomepage. http://www.matweb.com/search/OverviewofmaterialsforPolylacticAcidBiopolymer. Accesed 18 July 2021

Simulation of Cooling Temperature and Injection Time on the Quality of the Toothbrush Head Product Produced by Injection Molding Machine

Achmad Wildan Alfikhrul Haqqi, Andoko(✉), and Yanuar Rohmat Aji Pradana

State University of Malang, Malang 65145, Indonesia
andoko.ft@um.ac.id

Abstract. Defects in injection molding machine products are analyzed through simulation. The simulation aims to analyze the effect of cooling temperature and injection time on the quality of the injection molded toothbrush head product. Product quality includes quality prediction, weld line, and short shot. The simulation method begins with product design, then determines the simulation parameters including coolant temperature (16, 17, and 18 °C) and injection time (2.0, 2.2, and 2.4 s). ANSYS was run to obtain mold temperature and simulate product quality using AutoDesk Moldflow Advise. Simulation results at a cooling temperature of 16 °C and an injection time of 2.0 s produce a mold temperature of 41.98 °C which has a percentage of product quality reaching 99.6%. Product defects are found in the form of bubbles and shrinkage caused by the temperature of the molten plastic being too high and the cooling temperature too low.

Keywords: Cooling temperature · Injection molding · Injection time · Product quality · Simulation · Toothbrush head

1 Introduction

Plastic is one of the chemical polymers that is widely used today, one of which is in the toothbrush head. Plastic has advantages over other polymer materials including light, strong, transparent, waterproof, and relatively cheaper [1]. One of the techniques widely used in the plastic forming process is injection molding [2, 3]. Injection molding is the manufacturing process of producing a component by injecting liquid material into a mold. Injection molding has advantages at high production speed, high tolerance, and little remaining material.

Factors that affect the quality of injection molded products include cooling temperature and injection time. Both are determinants of the productivity of the injection molding process, which depends on the shape and dimensions of the mold [4]. If the mold is too hot, then the molten plastic is slow in the cooling process, resulting in defects in the product. If the mold is too cold, then the product will solidify faster, and the shape of the product is not perfect. Defects in the product can be minimized by adjusting the

I. Tolj et al. (eds.), *Recent Advances in Mechanical Engineering*, Lecture Notes in Mechanical Engineering, https://doi.org/10.1007/978-981-19-0867-5_23

cooling temperature and cooling time optimally. So, the required cooling temperature setting includes an injection time of between 70% and 80% of one injection system cycle and the temperature difference should not exceed 25% [5]. One of the methods that can be used to determine the right parameters is simulation. Simulation is a solution to engineering problems using numerical calculations by modeling the conditions according to the actual situation [6]. Simulation has advantages in terms of cost, time, and effectiveness compared to experimental testing [7]. In this paper, the simulation aims to analyze the effect of cooling temperature and injection time on the quality of the injection molded toothbrush head product.

2 Material and Method

2.1 Material

The material used in this simulation is polystyrene plastic with molecule structure $SC_6H_5CH = CH_2$. Polystyrene is a good electrical insulator and low dielectric. Its tensile strength reaches about 8000 psi. The chemical of polystyrene is a long-chain hydrocarbon with every other carbon connected to the phenyl group (the name given to the aromatic ring of benzene, when bound to a complex carbon substituent). Chemical properties of polystyrene Material type weight: 0.96862 g/cm^3, Thermal conductivity: 0.15 W/m-C, Specific heat: 2100 J/Kg-C [8].

2.2 Method

The simulation procedure performed to analyze the cooling temperature and injection time was performed with ANSYS and Autodesk Moldflow Adviser. Ansys is used to determine the average temperature of molding whose data is used on Autodesk Moldflow Adviser to determine the quality of the varied products.

Data mold temperature of 42.83 °C is used to get product quality using Autodesk Moldflow Adviser software with an injection time of 2.4 s. The meshing node used is 100 with a mold temperature of 16, 17, 18 and a melting temperature of 230 °C to obtain the mold temperature to be analyzed using Autodesk Moldflow Adviser with the condition of the plastic point limit injected into the mold.

3 Result and Discussion

3.1 Quality Prediction

The results of the simulation mold temperature are presented in Table 1, mold temperature increases as the coolant temperature increases. Minimum and maximum limits of mold temperature used to mold thermoplastics.

Figure 1 shows the stream of thermal cavity molding. The cavity has the highest temperature at the gate location or the inlet of liquid plastic. The red region indicates the heat transfer of liquid plastic into the cavity for the process of forming toothbrush head product, the light blue region shows the average temperature of the cavity and maintains temperature stability during the product formation process.

Table 1. Temperature results

No.	Cooling temperature (°C)	Melt temperature (°C)	Mold temperature (°C)
1	16	230	41.07
2	17	230	41.98
3	18	230	42.83

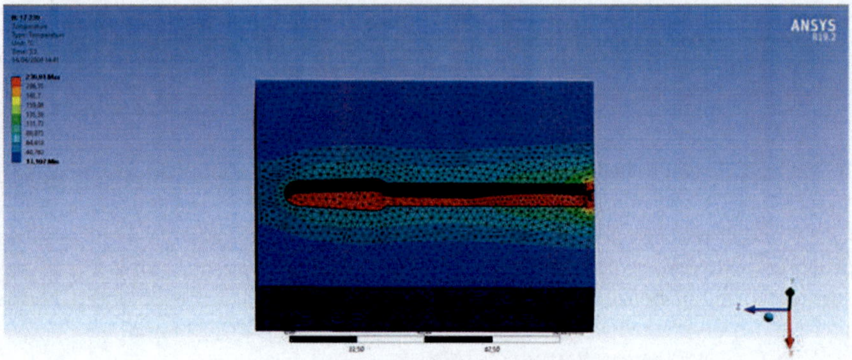

(a)

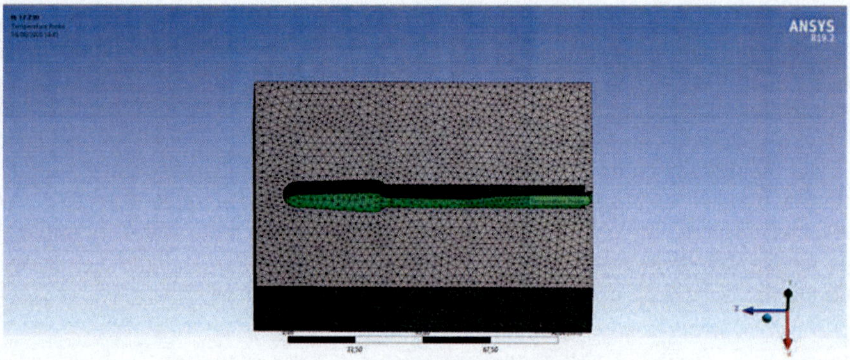

(b)

Fig. 1. **a** Contour plot of liquid plastic heat flow. **b** Probe temperature in cavity

Figure 1a shows the results of the temperature analysis when the cavity is given a cooling temperature of 24 °C. This result obtained a minimum temperature of 24,177 °C and an average cavity temperature of 41.7 °C with a maximum temperature of 200 °C. The recommended temperature should be in the specified range of thermoplastic materials used, at higher temperatures, cavity mold will be easier to fill due to lower plastic melt viscosity. However, high temperatures are less desirable due to longer cycle times which add to the cost [9].

Figure 1b shows the temperature of the probe in cavity molding used to obtain the temperature result in the product cavity. Graph of changes in mold temperature at various cooling temperatures is presented in Fig. 2. Cooling temperature or melt temperature has a significant influence on the mold temperature, and the magnitude of the mold temperature is proportional to the increase of both parameters.

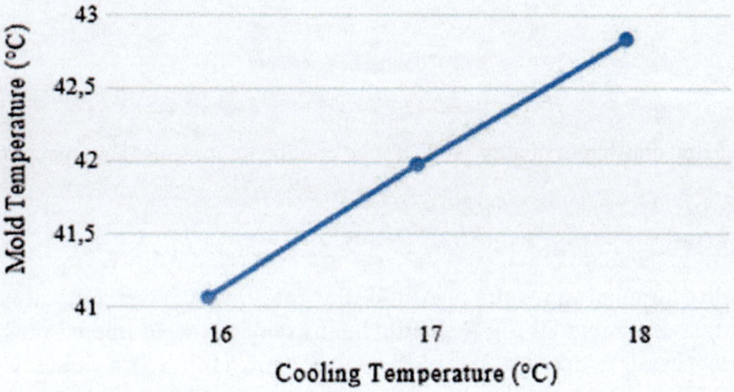

Fig. 2. Relationship of cooling temperature to mold temperature

3.2 Effect of Injection Time and Mold Temperature on Plastic Product Quality

Simulation results from Autodesk Moldflow Adviser software in the form of product quality percentage prediction are presented in Fig. 3. The product analysis process that has been carried out obtains the highest product quality prediction percentage of 99.6% and the smallest 95.80%. The highest percentage is at the cooling temperature of 18 °C and injection time of 2.4 s with a mold temperature of 41.7 °C producing a quality prediction percentage of 77.1%, while the smallest percentage at a cooling temperature of 16 °C and injection time of 2.2 s with mold temperature of 47.3 °C produces a quality prediction of 95.8%.

The difference in product quality analysis results, in Fig. 3, shows the parameters of cooling temperature and injection time along with the size of the mold temperature that equally affects the results of good product quality. The graphic explains that the cooling temperature and injection time is too fast. This condition causes short shot defects to become more dominant because the liquid plastic injected into the mold undergoes a faster cooling process before the liquid plastic fills the mold volume. On the other hand, if the cooling temperature and injection time are too long, the bubbles will be more dominant and the shrinkage volume will increase, therefore, the selection of optimal parameter setup is needed to produce good product quality [10].

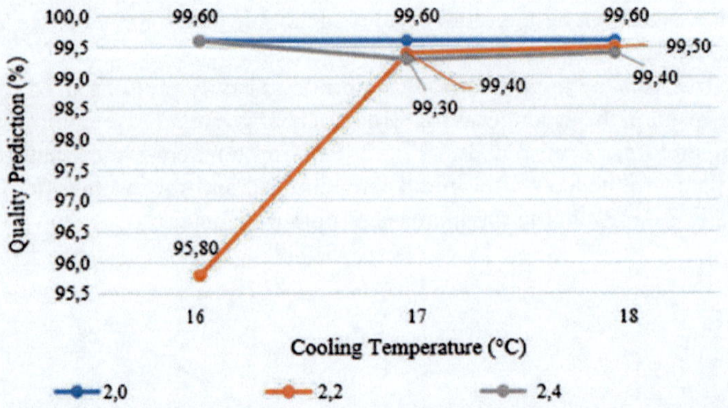

Fig. 3. Relationship of quality prediction to cooling temperature and injection time

3.3 Weld Line

Product quality simulation results in Autodesk Moldflow Adviser software cause weld line defects that are shown in Fig. 4. While the form of the weld line is shown in Fig. 5. The toothbrush head product that has been analyzed has a minimum weld line dimension value of 0.0277°.

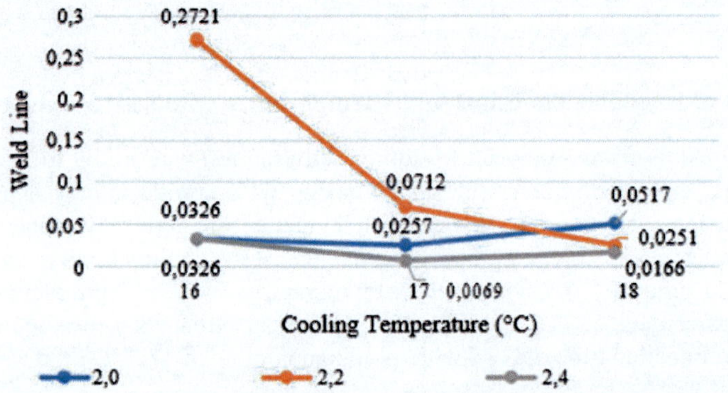

Fig. 4. Graph of weld line's relationship to cooling temperature and injection time

The graph in Fig. 4. shows that the smallest weld line occurs at a cooling temperature of 17 °C and an injection time of 2.4 s with a dimension of 0.0069°. The melted plastic hardens too quickly while the melting temperature is not high enough. When the melting flow of one material crosses the melt flow of another material, the two ends of the plastic cannot be properly fused because the melted plastic slowly hardens [10]. Weld line is caused by the presence of two flows that meet, this cannot be avoided when the front flow is split and fused around the mold cavity or if the part has many gates location. The presence of a weld line can indicate structural weakness or surface [11].

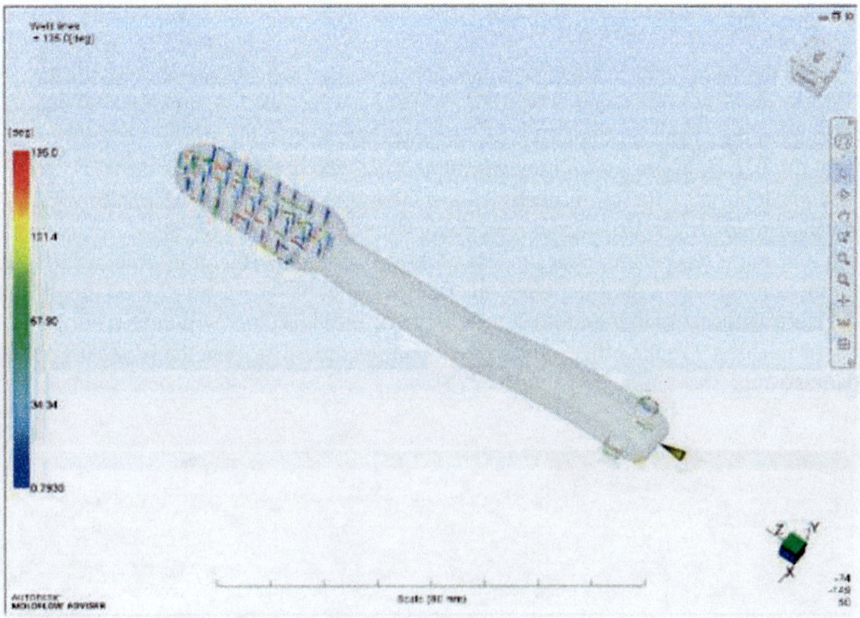

Fig. 5. Weld line product

3.4 Product Defect

Types of defects that occur in the product are short shot and shown in Table 2. Defects that occur affect the volume and weight of the product.

Table 2. Product defects data

No.	Mold temperature (°C)	Injection time (s)	Disability product	Weight (g)	Depreciation (%)
1	41.07	2.0	*Short shot*	5.760	24,0
2	41.98	2.2	*Short shot*	5.762	24.1
3	42.83	2.4	*Short shot*	5.767	23.3
4	42.83	2.0	*Short shot*	5.759	23.8
5	41.07	2.2	*Short shot*	5.785	21.5
6	41.98	2.4	*Short shot*	5.767	23.3
7	41.98	2.0	*Short shot*	5.759	23.8
8	42.83	2.2	*Short shot*	5.763	23.7
9	41.07	2.4	*Short shot*	5.789	21.1

3.5 Short Shot

Defect in the form of short shots is shown in Fig. 6. Short shots that occur in the analysis of this product are good because the defects that occur are very small and do not damage the dimensions of the product. The location of the short shot is found in the middle of the toothbrush head product. The factor that affects the occurrence of short shot defects in the production of the toothbrush head is the improper melting temperature value of the material due to the use of raw materials [12]. Short shots occur in this analysis with a setup of mold temperature parameters of 41.07 °C and an injection time of 2.0 s. The cooling temperature will affect when the liquid plastic has the formation of the product. If the cooling process in the mold is very slow, then the product will have damage in the form of flashing. On the other hand, if the cooling is too fast, then the product will have a damage-type short shot [13].

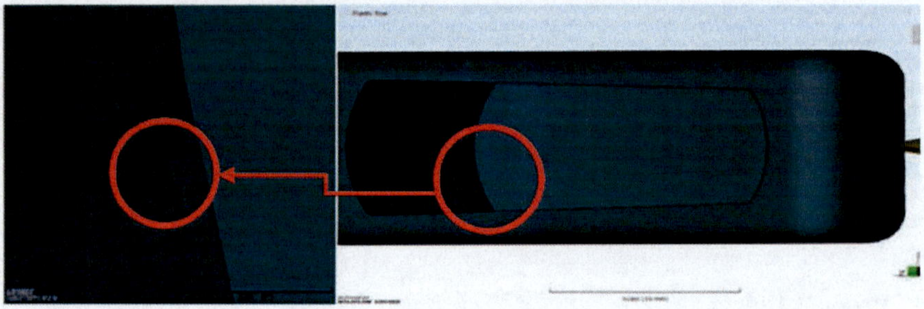

Fig. 6. Short shot

4 Conclusion

This simulation aims to analyze the cooling temperature and injection time on the quality of the toothbrush head product. The cooling temperature affects the filling process of the product. The rapid cooling process will result in bubble defects and shrinkage of the product. The injection time will affect the process of liquid plastic entering the cavity. Too short time before filling the cavity will cause a short shot and shrinkage defects. The results of the simulation and optimization of the toothbrush head product obtained the optimal setup parameters to produce good product quality, namely, the cooling temperature parameter of 16 °C and injection time of 2.0 s resulting in a mold temperature of 41.98 °C which has a product quality percentage of 99.6%. The response of cooling temperature and the melting temperature has a significant effect on product quality and shrinkage defects that occur, while the process variable that most affects shrinkage defects is the cooling temperature.

References

1. Purnwaningrum P (2016) Efforts to reduce the generation of plastic waste in the environment. J Urban Environ Technol
2. Mohan M, Ansari MNM, Shanks RA (2017) Review on the effects of process parameters on strength, shrinkage, and warpage of injection molding plastic component. Polym-Plast Technol Eng
3. Gusniar IN (2018) Method of making hexagon paving blocks made of plastic waste with injection molding machine. J Barom
4. Amran MAM, Idayu N, Faizal KM, Sanusi M, Izamshah R, Shahir M (2016) Part weight verification between simulation and experiment of plastic part in injection moulding process. IOP Conf Ser Mater Sci Eng
5. Wildan A, Andoko (2021) Effect of coolant Suhue on product quality on injection molding machines using Autodesk Moldflow adviser software. IOP Conf Ser Mater Sci Eng
6. Rao S (2010) The finite element method in engineering, 5th edv. Butterworth-Heinemann
7. Kovacs J, Siklo B (2011) Investigation of cooling effect at corners in injection molding. Int Commun Heat Mass Transf
8. Andoko, Saputro NE (2018) Strength analysis of connecting rods with pistons using finite element method. MATEC Web Conf
9. Hermawan Y (2009) Optimization of shrinkage defects for 120 Ml Chamomile products in the injection molding process with response surface method. Pros Semin Nas Manaj Teknol X
10. Anggono AD (2015) Prediksi Shrinkage Untuk Menghindari Cacat Produk Pada Plastic Injection. Media Mesin Maj Tek Mesin 6(2)
11. Dzulkipli AA, Azuddin M (2017) Study of the effects of injection molding parameter on weld line formation. Proc Eng 184:663–672
12. Devalia PT, Arief TM (2019) Analisis dan Optimasi Parameter Proses Injeksi Plastik Multi Cavity untuk Meminimalkan Cacat Short Mold. Ind Res Work Natl Semin 10(1):553–560
13. Purba SA, Albana MH (2017) Effect of cooling media temperature and circle time on defect crack line on SP 04 haemonetics products. J Integr

Modularity Design for Archimedes Screw Turbine Shaft

Herman Budi Harja(✉), Fajar Halim Mulyadi, Haris Setiawan, and Suyono

Politeknik Manufaktur Bandung, Bandung, Indonesia
herman@polman-bandung.ac.id

Abstract. This paper focuses on the modularity of Archimedes screw turbine (AST) shaft for obtaining features that have easy assembly and handling, especially in remote areas. Modularity enables the AST shaft to arrange into several component modules. Modularity requirements should be considering optimum dimensions of screw turbine blades for efficient energy conversion, shaft deflection, and joining strength between modules. The research uses three methods: Rorres formulation for calculation optimization dimensions of the screw turbine, the VDI 2222 method to guide its design meets the design requirements, and simulation analysis solid work to verify shaft deflection on each screw blade modularity. Modularity for AST shaft has been designed and resulted in a design draft of AST shaft modular. Finite element deflection analysis has been tested for the AST shaft with variations module simulation in the pitches number for 1, 2, 4, and 6 of screw blade per module. The greater the pitch number of the screw blade per module, the smaller the maximum deflection value of the shaft. Optimum dimensions of AST on research object have found that the volume maximum of each bucket is 0.000855 m^3, with the number of pitches as much as 24.8, then the maximum total volume of the bucket is 0.02124 m^3.

Keywords: Modularity · Easy assembly and handling · Pitch number per module · Optimum dimension · Shaft deflection

1 Introduction

The potential fluid energy at the low head is one of the potential energy sources to be exploited in Indonesia. This energy source can acquire through rivers located in various places in Indonesia. Based on the Central Statistics Agency (BPS) data in 2015 regarding "Daily Average River Flow, Flow Height, and Water Volume in Several Rivers with More Than 100 km^2 of Watershed", rivers in Indonesia have the potential for low head water energy with an average head of 1.6 m [1]. Many of these rivers are in remote areas, which are access difficult for transportation.

Figure 1 displays a graph of the relationship range between head and flow rate for various water turbines such as Pelton, Francis, Crossflow, and Kaplan, as well as Water wheel and Archimedes Screw. Archimedes screw turbine is a suitable type of water turbine utilized to operate at a potential source of water energy with a very low head from 0.1 to 10 m and a low water flow rate of 0.1–10 m^3/s.

I. Tolj et al. (eds.), *Recent Advances in Mechanical Engineering*, Lecture Notes in Mechanical Engineering, https://doi.org/10.1007/978-981-19-0867-5_24

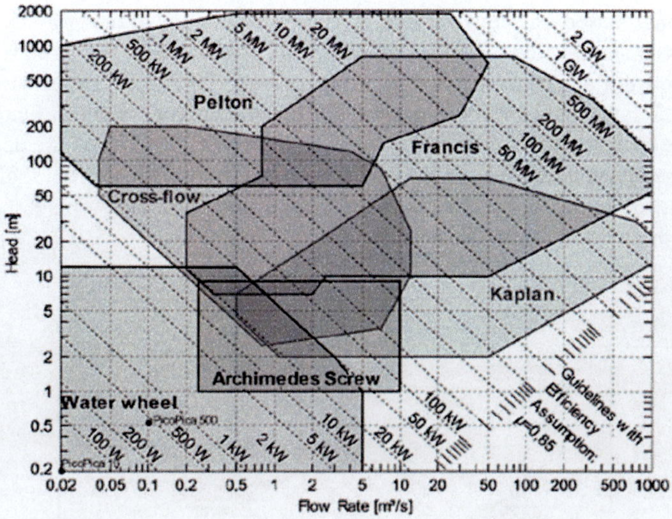

Fig. 1. The range of head and flow rate corresponding to the generated power of different turbines [2]

One of the main problems in applying screw turbines in remote areas is the transport and installation handling of screw turbines in those remote areas. Therefore, this research was conducted on the modularity implementation on the Archimedes screw turbine shaft to obtain a verified design of the screw turbine shaft which is arranged in several segment modules of the screw turbine blade components. Hence, its modularity feature could be a solution when handling and assembling the Archimedes Screw Turbine in a remote area.

2 Material and Method

The research has three stages that consist of (i) calculation of the optimum dimension of Archimedes Screw Turbine, (ii) design of the modularity construction of Archimedes Screw Turbine shaft, and (iii) Verification of the joining strength between part segments and the shaft deflection on each modularity variation on a single screw blade.

The case study of Archimedes Screw Turbine is determined based on the assumption of the potential energy source of water with a head of 2 m and flowrate 0.01 m^3/s. First stage research using the Rorres formulation for obtaining the optimum AST dimensions, especially in determining the maximum bucket volume of screw blade dimensions. The bucket volume value has a significant contribution to the energy conversion efficiency of the AST [3].

Figure 2 show the Profile parameters of Archimedes Screw Turbine. Figure 3show the relation between (H) head, (θ) slope of the screw, (L) screw turbine length, and (Lh) screw turbine horizontal length. The Archimedes Screw Turbine geometry has external dimensions and internal dimensions. External dimensions are (R_o) outer radius of screw blade, (Λ) pitch of screw blade, and (θ) slope of the screw. Its external dimensions are determined based on screw turbine location, availability of material, and flow rate. Generally, the slope value of the screw parameter is between 30° and 60°, Harja et al. (2014) stated the optimum slope value of a screw to be 32° [4]. Internal dimensions consist of the inner radius of the screw blade (R_i), the number of screw blades (N), and the distance between screw blades. Its internal dimension parameters are determined to obtain an optimum dimension of the screw turbine. The length of AST shaft (L), outer radius (R_o), the maximum bucket volume (V^t), the pitch of screw blade (Λ), and the number of pitch (m) should be determined using Eqs. (1)–(5) [3, 5, 6].

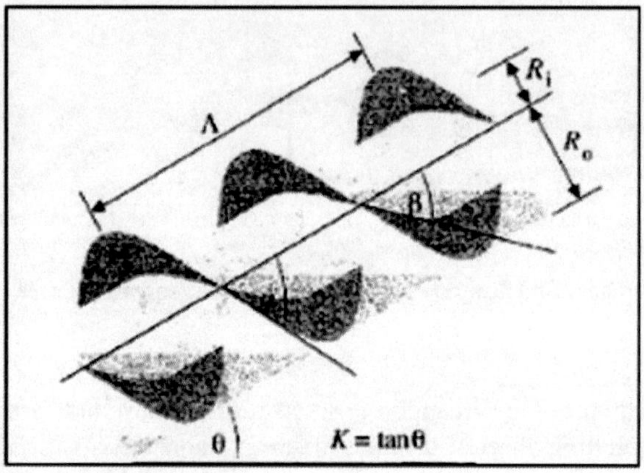

Fig. 2. Profile parameters of Archimedes Screw Turbine [3]

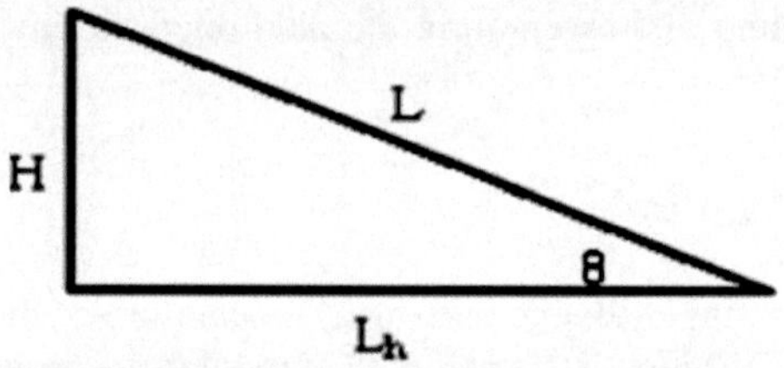

Fig. 3. Relation between (H) Head, (θ) slope of the screw, (L) Screw Turbine Length, and (L_h) Screw turbine horizontal length

$$L = \frac{H}{K} \tag{1}$$

$$Ro = \frac{Ri}{\rho^*} \tag{2}$$

$$V_T^* = \pi R_o^2 \Lambda^* \nu^* \tag{3}$$

$$\Lambda^* = \frac{\pi R_o \lambda^*}{K} \tag{4}$$

$$m = \frac{L}{\Lambda^*} \tag{5}$$

In the second stage, VDI 2222 method is applied to design the best AST shaft modularity construction. Modularity construction should be considering several things, such as the accuracy of dimensional and geometrical of each part segment, alignment of AST shaft, and joining forces on each joining of the screw turbine shaft. The third stage is doing calculation analysis and simulation using Computer-Aided Engineering (CAE) Solid work software. It is to verify the joining strength between part segments and the shaft deflection on each modularity variation on a single screw blade.

3 Result and Discussion

3.1 Calculation of Optimum Dimensions Archimedes Screw Turbine

The potential energy of water is 2 m head and 0.01 m^3/s or 10 L/s of flowrate. Determine the maximum bucket volume is the way for obtaining the optimum energy conversion of the screw turbine. The maximum bucket volume of the screw blade is conducted using the Rorres formula to calculate the optimum dimensions of Archimedes screw turbine. Its calculation uses 32° of the slope value of the screw parameter [4]. Table 1 shows the calculation result of optimum dimensions Archimedes screw turbine. Those are calculated using Eqs. (1)–(5) as Rorres formulation. The value of optimal radius ratio (ρ) and optimal pitch ratio (λ) are 0.5369 and 0.1863, its values are taken from Rorres' recommendation [3].

Table 1. The dimension value of Archimedes screw turbine parameters

Parameter	Value	Unit
Flowrate	0.1	m^3/s
Head	2	m
(R_o) Outer radius	0.083	m
(Λ) Pitch of screw blade	0.143	m
(θ) Slope of screw	32	°
(R_i) Inner radius	0.0446	m
(N) The number of screw blade	2	pcs
(L) Screw turbine length	2.94	m
(V^*_t) Max buckets volume	0.000855	m^3
(m) Pitch number	24.8	Pcs
(V_t) Total buckets volume		m^3

3.2 Designing the Construction of Modularity Shaft Archimedes Screw Turbine

The principle of modularity in the Archimedes screw turbine is applied by splitting the screw turbine shaft into several segments or part modules. This modularity should consider the optimum dimensions of screw turbine blades for efficient energy conversion, AST shaft alignment, joining strength, and assembly–disassembly ability of its module.

VDI 2222 method is used to design the best AST shaft modularity construction. The VDI 2222 encompasses the four main design phases: task clarification, conceptual design, embodiment design, and detail design [7]. The phase of the task clarification is conducted by collecting screw turbine data, literature study, parameter identification of screw turbine shaft modularity, and making product requirements as the primary and secondary requirements. Observation and literature have identified several parameters. Based on observation and literature study, several parameters have identified as (i) the accuracy of dimensional and geometrical of each part segments; (ii) alignment of AST shaft; and (iii) joining force on each joining segment of the screw turbine shaft. Tables 2 and 3 show the primary and secondary requirements of AST shaft modularity.

Table 2. Primary requirement list of AST shaft modularity

Requirement	Quantification/qualification
Alignment of AST shaft	<Misalignment tolerance of bearing
Joining force of part module	Able to accept the force that occurs in the AST shaft
Assembly and disassembly of module joining	Time assembly standard

Table 3. Secondary requirement list of AST shaft modularity

Requirement	Quantification/qualification
Pitch number of modules	1–6
Rotation shaft	90–300 rpm

The conceptual design phase determines function unit input, process, and output. After identifying needs by producing a requirement list, the screw turbine concept is classified based on the processes sequence described in the black box as shown in Fig. 4.

The working principle of the Archimedes screw turbine is that water from the top end flows into the space between the screw blades (bucket) and out of the lower end. The gravity and hydrostatic forces of the water in the bucket along the rotor cause the screw blades to generate torque and rotate the rotor on its axis [4, 8]. The difference in hydrostatic force is due to the difference in the water head between the water head above the blade (head upper) and the water head below the blade (head down) and the level of water immersion [9]. Based on the black box, several part functions are produced and consisted of three-part functions. Those part functions are joining system between

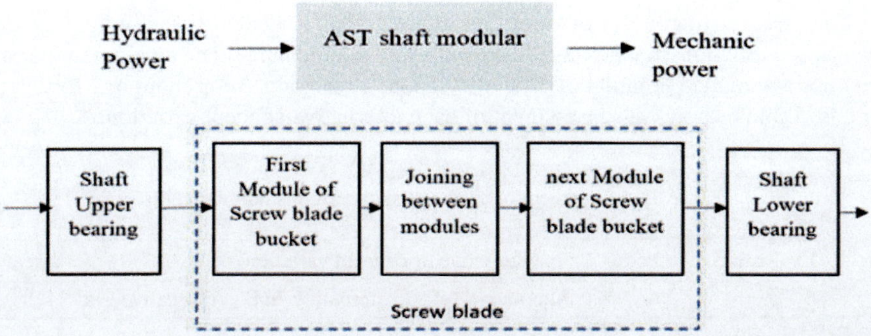

Fig. 4. Black box concept

module, support system for shaft upper bearing, and support system for shaft lower bearing. The next stage is defining some alternatives of part function, evaluating, and determining concept variation alternative.

Tables 4 and 5 show the alternative part function and concept variation alternative. Each variation alternative is analyzed to produce a product with the highest sub-function principal efficiency value [10].

Table 4. The alternative of part function

	Function system	Alternative 1st	Alternative 2nd	Alternative 3rd
A	Joining system between module	Pen	Thread	Bolt and nut
B	Support system for the upper bearing	Self-aligning ball bearing dan thrust ball bearing	Tandem taper roller bearing	Spherical roller bearing and spherical thrust roller bearing
C	Support system for the lower bearing	Solid bronze bushing	Sintered bronze bushing	Wrapped bronze bushing

Table 5. The alternative of concept variation

	Function system	Alternative 1st	Alternative 2nd	Alternative 3rd
A	Joining system between module	A1	A2	A3
B	Support system for the upper bearing	B1	B2	B3
C	Support system for the lower bearing	C1	C2	C3
Alternative overall function		AFK1	AFK2	AFK3

The assessment concept of variations alternative is evaluated based on four aspects such as construction, manufacture, assembly, and maintenance. The assessment parameters use a weighted nominal of 5 for very good, 4 for good, 3 for enough, 2 for less, 1 for bad. Table 6 shows the assessment of each alternative concept variation.

Table 6. Assessment of each alternative of concept variation

No.	Evaluation aspect	Point	An alternative of concept variation						Ideal point
			Alternative 1st		Alternative 2nd		Alternative 3rd		
1	Construction	5	2	10	2	10	3	15	25
2	Production	5	3	15	5	25	5	25	25
3	Assembly	5	4	20	4	20	4	20	25
4	Maintenance	5	3	15	3	15	4	20	25
	Total point			60		70		80	100
	Percentage			60		70		80	100

Based on these aspects, the third alternative was the best alternative chosen. Figure 5 shows the design draft of AST shaft modularity. Its design draft consists of spherical roller bearing and spherical thrust roller bearing as a support system for upper bearing, joining system between modules using bolt and nut, and wrapped bronze bushing as a support system for lower bearing.

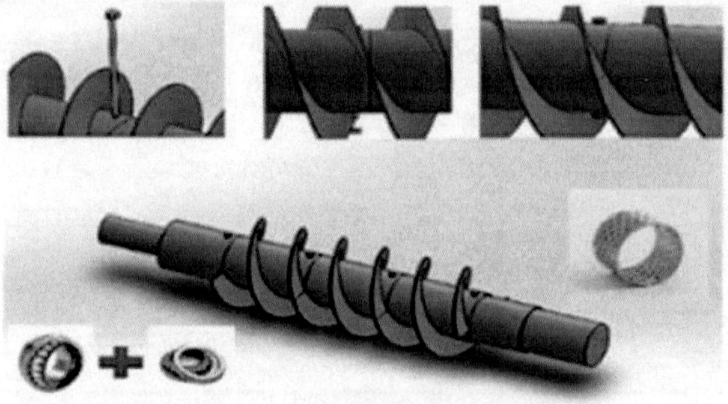

Fig. 5. Function diagram

3.3 Verification of the Shaft Deflection on Each Modularity Variation

Simulation analysis of AST shaft deflection uses (Computer Analysis Engineering) CAE Solidwork software. AST shaft object was for a 2 m head and four variations of screw blade modularity. Its variation describes the pitches number for each module that consists of 1, 2, 4, and 6 pitches per module. It shows in Table 7. The simulation determines the maximum deflection value along the shaft of Archimedes Screw Turbine and at the upper and lower bearing area.

Table 7. Variation of the pitch number for each module

	One pitch	Two pitches	Four pitches	Six pitches
Plane view				
ISO view				

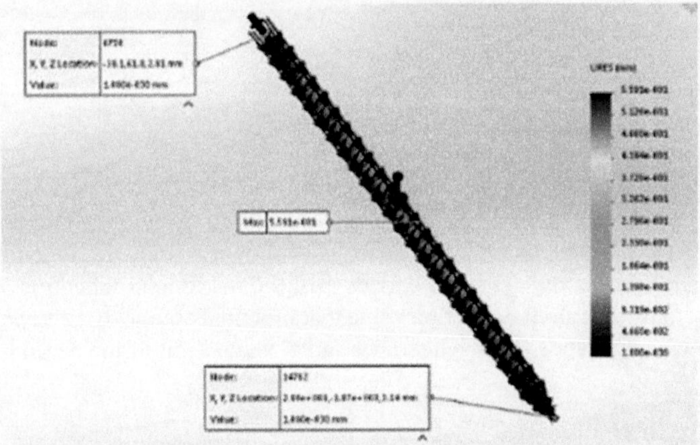

Fig. 6. Deflection plot for Archimedes Screw Turbine shaft with one pitch screw blade per module

Figure 6 shows the finite element deflection analysis result for the Archimedes Screw Turbine shaft with one pitch of screw blade per module. It has a maximum deflection of about 0.5591 mm. The deflection value at upper and lower bearing is about 1×10^{-30} mm. Figure 7 displays the deflection analysis for the Archimedes Screw Turbine shaft with six pitches screw blade per module. It has a maximum deflection of about 0.4018 mm, and the deflection value at upper and lower bearing is about 1×10^{-30} mm.

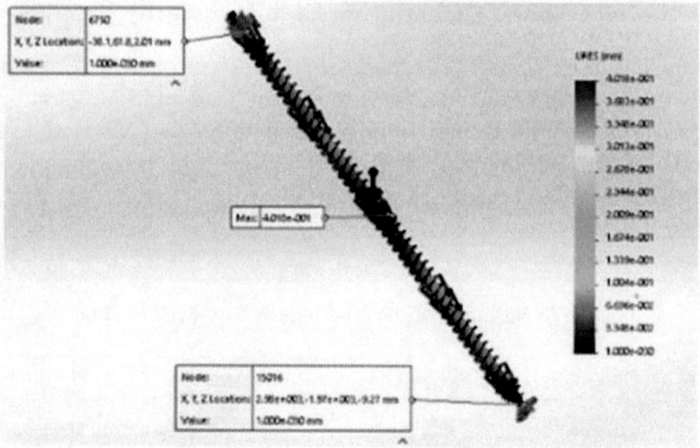

Fig. 7. Deflection plot for Archimedes screw turbine shaft with six pitch screw blade per module

Table 8. The result of Finite element deflection analysis for AST shaft with variation number of pitches per module

Number of pitches per module	Maximum deflection (mm)	Deflection at upper and lower bearing area (mm)
1	0.5591	1×10^{-30}
2	0.4602	1×10^{-30}
4	0.4150	1×10^{-30}
6	0.4018	1×10^{-30}

Based on finite element deflection analysis as shown in Table 8, if the pitch number of screw blades per module is more added, then the maximum deflection value of the shaft is less small. The shaft deflection value that mounted area with the upper and lower bearing can be a reference for the clearance radial value limit of the bearing usage.

4 Conclusion

Modularity for AST shaft has been designed and resulted in a design draft of AST shaft modular. Finite element deflection analysis has been tested for the AST shaft with variations module simulation in the pitches number for 1, 2, 4, and 6 of screw blade per module. The greater the number of the pitch of the screw blade per module, the smaller the maximum deflection value of the shaft. Optimum dimensions of AST for the potential energy of water with 2 m head and 0.01 m^3/s have found that 0.0446 m of inner radius, 0.083 m of outer radius, 0.143 m of screw blade pitch, 2.94 m of screw turbine length, 24.8 of pitch number, and 0.02124 m^3 of maximum bucket volume.

References

1. Rata-rata harian aliran sungai tinggi aliran dan volume air di beberapa sungai yang daerah pengalirannya lebih dari 100 km^2, https://www.bps.go.id/statictable. Accessed 03 Mar 2020
2. Map of Hydro Turbines. https://www.commons.wikimedia.org/wiki/File:Kennfeld_Wasserturbinen.svg. Accessed 13 Mar 2021
3. Rorres C (2000) The turn of the screw: optimal design of an Archimedes. J Hydraul Eng 126(1):72–80
4. Harja HB, Abdurrahim H, Yoewono S, Riyanto H (2014) Penentuan dimensi sudu turbin dan sudut kemiringan poros turbin ulir Archimedes. J Metal 36(1):26–33
5. Nagel G (1968) Archimedes screw pump handbook. Prepared for Ritz-Atro Pumpwerksbau GMBH Roding, Nu¨rnberg, Germany
6. Nagel G, Radlik K (1988) Wasserförderschnecken [Water lifting screws] Bauverlag. Wiesbaden/Berlin, German
7. Pahl G, Beitz W, Feldhusen J, Grote KH (2007) Engineering design, 3rd edn. Springer-Verlag, London
8. Aries DH, Akhmad N (2018) Kajian Teoritis Kerja Turbin Archimedes Screw Pada Head Rendah. Jurnal SIMETRIS 9
9. Dellinger G, Terfous A, Garambois P (2016) Experimental investigation and performance analysis of Archimedes screw generator. J Hydraul Res 54(2):197–209
10. Wittel H, Jannasch D, Voßiek J, Spura C (2019) Roloff/Matek Maschinenelemente. In Roloff / Matek Maschinenelemente, 24th edn. Springer-Verlag, Wiesbaden

Determination of Horizontal Turbine Airfoil Type for Power Plant in Tulungagung—East Java

Akhmad Safi'i, Triyogi Yuwono(✉), and Bambang Arip Dwiyantoro

Departement of Mechanical Engineering, Institut Teknologi Sepuluh Nopember, Surabaya, Indonesia
triyogi@me.its.ac.id

Abstract. The wind turbine is owned by PT Pembangkit Jawa Bali (PT.PJB); the wind turbine is a research project conducted by PT. PJB. This study aims to determine the appropriate airfoil for a horizontal type wind turbine installed at the Wind Power Plant—Tulungagung, East Java. The installed turbine airfoil cannot produce the expected power at wind speeds, at the site, in the range of 3–5 m/s. Three types of airfoils, namely SD 7080, SG 6043, and NACA 4412, are proposed to be selected as the best compared to each other and with the installed turbine over the wind speed range at the site. A QBlade software was used to assess the performance of the four types of airfoils. The simulation results show that the SD 7080 type airfoil can produce the highest power compared to the other three airfoils; it gives 49.77 Watts at a wind speed of 5 m/s. This power is 4.25% higher than the airfoil of the installed turbine.

Keywords: Horizontal turbine · Simulation · QBlade · Airfoil SD7080 · SG6043 · And NACA 4412

Nomenclature

Cp	Coefficient of Power
BEM	Blade Element Momentum
TRS	Tip Speed Ratio
HAWT	Horizontal Axis Wind Turbine
C_L	Coefficient lift
PT. PJB	PT. Pembangkit Jawa Bali
C_D	Coefficient drag
CFD	Computational Fluid Dynamics

I. Tolj et al. (eds.), *Recent Advances in Mechanical Engineering*, Lecture Notes in Mechanical Engineering, https://doi.org/10.1007/978-981-19-0867-5_25

1 Introduction

The wind turbine owned by PT Pembangkit Jawa Bali (PT.PJB) is located in the Dukuh Sindem area, Popoh, Besole Village, Kec. Besuki, Tumpak Tulungagung, East Java, as shown in Fig. 1a. This wind turbine is a research project conducted by PT. PJB. There are 5(five) wind turbines with Horizontal Axis Wind Turbine (HAWT) type which has 3 (three) blades installed at the location as indicated in Fig. 1b, where each turbine is designed to produce 500 Watts at a wind speed of 12 m/s [1]. Unfortunately, the available wind speed in location is only in the arrange of 3–5 m/s, so the wind turbine cannot produce the designed power, which is 500 Watts.

Fig. 1. **a** Wind turbine location, **b** 5 attached wind turbines, and **c** turbine No.5

For that reason, this research is expected to optimize the performance of wind turbines at speeds of 3–5 m/s, and this is done by examining several types of the airfoil proposed and determining which airfoil is suitable and provides the highest performance for the wind speed range at the wind turbine installation site. In this case, the research object will be turbine No. 5, as shown in Fig. 1c. Several studies have been done for academic purposes and various other technical applications to investigate airfoil characteristics at low speeds. For example, the author [2] examines Airfoil NACA 4412 and 4415 using Qblade software, and the study results show that NACA 4412 has a better performance than NACA 4415. Authors [3], with the help of Qblade software, analyze the characteristics of the aerodynamics of 10 (ten) airfoils, namely Aquila, BW-3, E387, FX63-137, NACA0012, NASA LS-0413, RG-15, S1223, SD7080, and SG6043. The authors selected SD7080, an airfoil that is suitable for producing maximum power for low wind speed. Authors [4] have studied 4 (four) airfoils operated at low wind speed, namely airfoil SG6040, SG6041, SG6042, and SG6043. They have concluded that the airfoil SG6043 is the best airfoil. Authors [4] have experimentally and numerically analyzed the performance of the HAWT with 3 (three) blades. The authors have found that the differences in experimental and numerical testing errors are a maximum of 5%. Authors [5] have examined the effect of pitch angle on the performance characteristic of a turbine with the help of turbine CFD software. They have confirmed that the power maximum generated by the turbine has occurred at an optimum pitch angle. The purpose of this study is to determine the appropriate airfoil for a horizontal type wind turbine installed at the PLTB-Tulunggung, East Java, with the wind speed range at the site, which is 3 to 5 m/s. QBlade software was used to examine the three proposed airfoil types, namely SD 7080, SG 6043, and NACA 4412, and select which airfoil had the best performance.

2 Methodology

2.1 Retrieval Method Data

Based on these previous studies, to achieve the objectives, this study examines 3(three) types of airfoils, namely SD7080, SG6043, and NACA4412, as shown in Fig. 2.

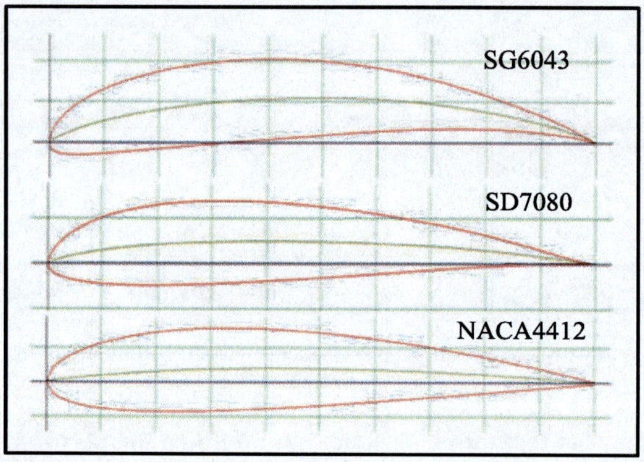

Fig. 2. Form of airfoil SG6043, SD7080, and NACA4412

Research begins with measuring the size of the installed blade turbine using the 3D scanner model CREAFORM handy scan 700 portable as shown in Fig. 3. It is to get the size and contour of the blade installed in the field, then proceed with a detailed description of the blade size.

SolidWorks 2018 software was used to draw the blades, as shown in Fig. 4 for the installed blade, then using the QBlade software [6] to simulate the airfoil as shown in Fig. 5.

After obtaining the shape and size of the four airfoils (in this case, the SD 7080, SG6043, and NACA4412 airfoils were obtained from the existing literature), then a simulation was carried out using QBlade software through the Blade Element Momentum (BEM) method to obtain the performance of each airfoil. The performance of each airfoil from the simulation is then compared with each other to select the best performance of the airfoil. The performances include output power, power coefficient (Cp) lift, and drag coefficient ratio (C_L/C_D).

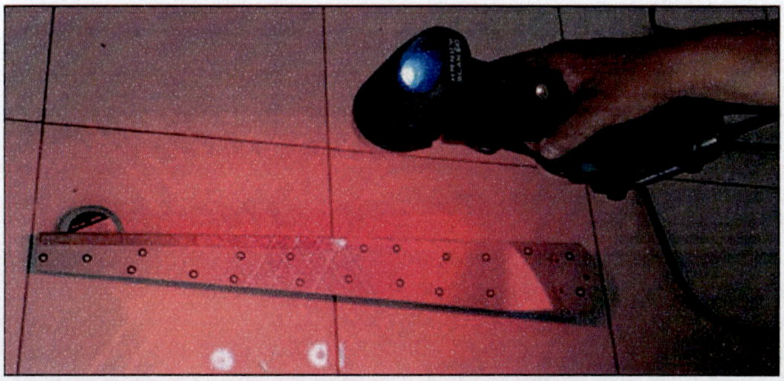

Fig. 3. Process 3D scanner

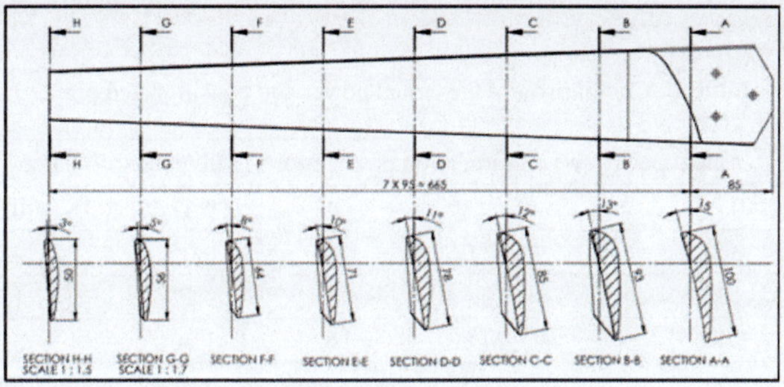

Fig. 4. Sizes of airfoil wind turbines installed use SolidWorks 2018 software

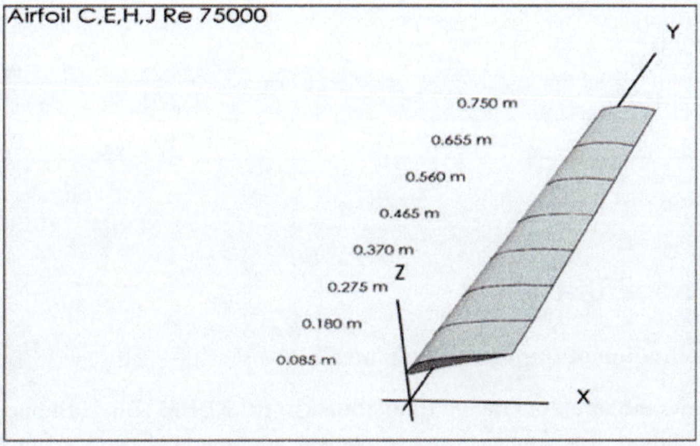

Fig. 5. Installed wind turbine blade shape formed using QBlade software

2.2 Airfoils Simulation Parameters

The simulation was carried out using the BEM method with a maximum wind speed limit of 12 m/s and a TSR of 1–10 with a delta of 0.5. Then proceed with the simulation of the BEM rotor with the wind speed range 1–12 m/s delta 1 m/s, the rotation speed range 68–680 rpm, with a pitch range of 0–6°.

2.3 Validation of Simulation Results

The proposed simulation method is validated by comparing the simulated power using QBlade software with the actual power obtained by directly measuring the installed turbine. The comparison of the two powers strongly agrees with each other with an average difference of about 3.56%, as shown in Table 1. This error is smaller than that obtained by previous studies [2], 7. So it can be concluded that using the BEM method of the QBlade software for simulating airfoil performance is considered valid.

Table 1. Comparison of the actual power and the simulated power

Speed (m/s)	Actual power (w)	Simulation power (w)	Difference (w)	Percentage (%)
1	0	0.37	0.37	0.37
2	3	3.06	0.06	2.0
3	10	10.42	0.12	1.2
4	24	24.80	1.20	5.1
5	46	48.57	2.67	5.8
6	80	84.01	3.81	4.8
7	128	133.52	5.12	4.0
8	196	199.43	3.63	1.9
9	275	284.15	9.64	3.5
10	370	359.14	−10.86	−2.9
11	480	427.36	−52.65	−11.0
12	500	500.67	0.67	0.1

3 Results and Discussion

3.1 Determination of the Value of Cp on TSR

Figure 6 shows the result of Qblade simulation using the BEM (Blade Element Momentum) rotor method. The figure shows the power coefficient (Cp) as a function of Tip Speed Ratio (TSR), indicating that the SD7080 airfoil has the highest Cp among all airfoils tested in this study [8]. It has a Cp of 0.39 at TSR 5.5. While the installed airfoils,

SG6043 and NACA 4412, have Cp values of 0.293 at TSR 4.5, 0.321 at TSR 4, and 0.271 at TSR 5.5, respectively. It shows that the value of Cp will continue to increase until it reaches a maximum point with the increase in TSR at the Optimum point [9]. After reaching the maximum point, the Cp value will decrease even though the TSR continues to increase. The best performance of the SD7080 airfoil is also caused by a wider range of TSR values which result in a larger Cp power coefficient [3]. In addition, the SD7080 airfoil has a wider TSR range, successively with a TSR value of 1–9.25 with a Cp of 0.355, while the installed airfoil has a TSR area of 1–8 with a Cp of 0.293, while the SG6043 airfoil has a TSR area of 1–8, 25 with a Cp of 0.321 and a NACA4412 airfoil, the TSR area is 1–9 with a Cp of 0.271 so that the SD7080 airfoil has the widest operating range among the airfoils tested in this study and also the highest of the maximum Cp.

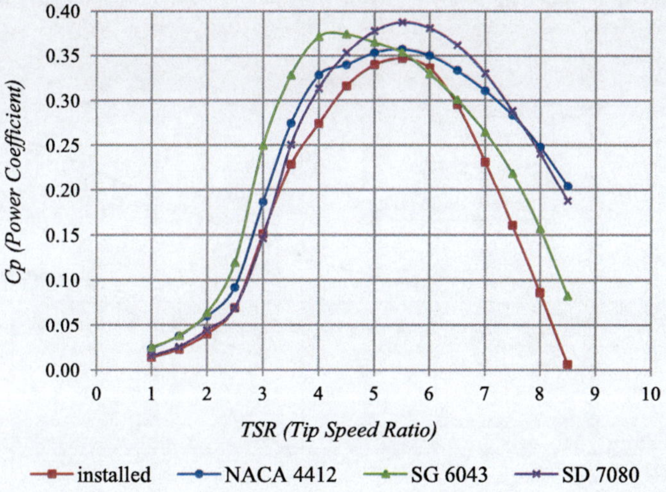

Fig. 6. Comparison of the coefficient of power as a function of TSR for all airfoils tested

3.2 Airfoil Selection

Further discussion is important to explore the power data in the wind speed range available at the site, which is 3–5 m/s, as shown in Table 2. It proves that the SD 7080 airfoil has the highest power of all airfoils tested. The highest power is obtained at a wind speed of 5 m/s, where the SD 7080 airfoil can reach a power of 49.77 Watts.

This power provides a 4.25% higher increase than currently installed airfoils. Meanwhile, at a wind speed of 5 m/s, the SG 6043 and NACA 4412 airfoils produce about 49.02 Watts and 28.61 Watts, respectively. As for the thrust value, it can be seen that the SD 7080 airfoil has a lower thrust than the SD 6043 and NACA 4412 airfoils, although it is still higher than the installed airfoils. The characteristics of the SD 7080 airfoil, indicated by the C_L/C_D ratio in Fig. 7, are the best among the tested airfoils. It can be

Table 2. Qblade simulation results, power and thrust for each airfoil

Wind speed (m/s)	Installed		SG 6043		SD 7080		NACA 4412	
	Power (W)	Thrush (N)	Power (W)	Thrush (N)	Power (W)	Thrush (N)	Power (W)	Thrush (N)
3	10.42	10.55	8.13	13.51	10.75	12.32	7.75	5.56
4	24.80	17.75	24.37	21.42	25.49	20.35	18.26	10.16
5	48.57	26.83	49.02	31.52	49.77	30.63	28.61	16.11

understood that high C_L/C_D indicates high C_L and/or low C_D. The high lift on the HWAT type turbine will produce high power, while the low drag will contribute to the low thrust value.

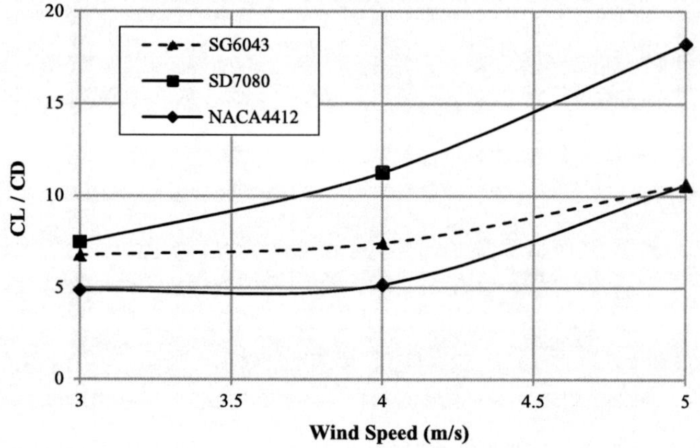

Fig. 7. Evolution of $(C_L/C_D)_{max}$ as a function of wind velocity

4 Conclusion

Simulations carried out with Qblade software between the actual and the simulation results in an average validation error of 3.53%. By the BEM rotor simulation analysis at wind speeds in the field ranging from 3, 4, and 5 m/s, the SD7080 airfoil has the best performance with a Cp value of 0.35 at TSR = 5. From the multi-parameter simulation analysis of BEM airfoil, the airfoil SD7080 has the highest power, where its power output is about 49.77 Watts. The power is about 4.25% higher than the installed airfoil at a wind velocity of 5 m/s. From the analysis of thrust selection, the SD7080 and SG6043 type airfoils almost have the same thrust value, while the thrust value of NACA

4412 has the smallest value. However, the SD7080 airfoil still has a higher thrust value than the validated installed airfoil of 15.23%. Analysis based on the maximum C_L/C_D value versus wind speed shows that the maximum C_L/C_D value of SD 7080 has the highest value among other airfoils. Author [10] conducted an experimental and numerical analysis of the performance of a 3-blades horizontal axis wind turbine. The author states that for a wind speed of 2.6 m/s, the turbine power coefficient is 0.24 at a tip speed ratio of 5.39. The difference between experimental and numerical test errors on the tip speed ratio is a maximum of 5%.

References

1. http://www.awing-i.com/english/500W_wind_turbine.html. Accessed 19 June 2020
2. Burhannudin D (2016) Design of wind turbine based on the Naca 4412 and 4415 using mahogany (Swietenia macrophylla) and pines wood (Pinus merkusii). Thesis, ITS, Indonesia
3. Suresha A, Rajakumar A (2020) Design of small horizontal axis wind turbine for low wind speed rural applications. Mater Today Proc 23(1):16–22
4. Lyon CA, Broeren AP, Giguere P, Gopalarathnam A, Selig MS (1997) Summary of low-speed airfoil data, vol 3. Department of Aeronautical and Astronautical Engineering University of Illinois at Urbana-Champaign
5. Sudhamshu AR, Pandey MC, Sunil N, Satish NS, Mugundhan V, Velamati RK (2016) Numerical study of the effect of pitch angle on performance characteristics of a HAWT. Eng Sci Technol Int J 19(1):632–641
6. Marten D, Wendler J (2013) QBlade guidelines V06. TU Berlin
7. Marten D, Wendler J, Pechlivanoglou G, Nayeri CN, Paschereit CO (2013) Qblade an open source tool for design and simulation of horizontal and vertical axis wind turbines. Technical University Berlin, Germany
8. Mathew S (2006) Wind energy fundamentals, resource analysis and economics. Springer, Berlin, Germany
9. Hau E (2016) Wind turbines fundamentals, technologies, application economics. Springer, Germany
10. Moussa MO (2020) Experimental and numerical performances analysis of a small three blades wind turbine. Energy (23):117807

An Experimental Study on the Effect of Three-Stage Air Intake Using Temperature Control System on the Performance of Refuse Derived Fuel (RDF) Gasification

Harsono[1(✉)], Bambang Sudarmanta[1], and Sigit Mujiarto[2]

[1] Department of Mechanical Engineering, Institut Teknologi Sepuluh Nopember (ITS), Surabaya 60111, Indonesia
harsono.rusmina@gmail.com
[2] Department of Mechanical Engineering, Faculty of Engineering, Universitas Tidar, Magelang, Indonesia

Abstract. Refuse derived fuel (RDF) is waste that can't be reprocessed. One way to use RDF is to convert it to alternative fuels. A gasification is a tool that can be used to convert RDF into syngas (gas fuel) so that it can be used for internal combustion engines and power plans. This research was conducted using a downdraft gasifier with a three-stage gasifying agent with variations in the air ratio (AR) at the air intake in the pyrolysis, oxidation, and reduction zones (0:10:0, 2:6:2, 1:6:3, 3:6:1, 2:5:3, 3:5:2, 1:8:1, 2:7:1, and 1:7:2), besides that the temperature control system in the oxidation zone is varied from 600 °C, 700 °C, 800 °C, 900 °C, to 1000 °C. Parameters in the gasification process were measured such as combustible syngas composition, LHV syngas, and tar content. The results obtained using AR 1:7:2 and setting 1000 °C are the most optimal results, where the highest combustible syngas such as CO is 2.67 %vol and H2 is 11.54 %vol while the LHV was obtained at 4201.02 kJ/kg and the tar content was obtained at 61.89 mg/Nm^3. The presence of the air is limited to the pyrolysis and reduction of the gasification process.

Keywords: Refuse derived fuel (RDF) · Multi-stage downdraft · Air ratio (AR) · Syngas · Gasification temperature control

1 Introduction

The waste generated by urban communities better known as municipal solid waste (MSW) is the most generated waste. MSW has a composition consisting of paper = 12%, plastic = 13%, textile = 3%, glass = 1%, metals = 1%, miscellaneous = 2%, yard waste = 43%, and kitchen waste = 25% [1]. MSW there is still waste that can be reused or recycled, so a reprocessing process needs to be carried out so that waste that cannot be used anymore is called refuse derived fuel (RDF). RDF has a composition consisting of paper = 63%, plastics = 19%, textile = 1%, and foil = 15% [2]. One way that RDF can be useful is by converting it into alternative fuels [3].

I. Tolj et al. (eds.), *Recent Advances in Mechanical Engineering*, Lecture Notes in Mechanical Engineering, https://doi.org/10.1007/978-981-19-0867-5_26

Energy conversion that can be carried out on RDF materials is to use gasification process technology, where the process can reduce pollution and the gas produced can be used directly [4]. The gasification process can be carried out using the downdraft gasifier, which is one of the widely used gasifiers, because it is very suitable for direct application by engines and other heat uses, besides that the gas produced is clean enough to be used as fuel [5].

RDF waste before it is used needs to be used, where one process of analyzing the feasibility of waste RDF can be done by testing the composition and calorific value generated notice [3], analyze the properties of RDF produced by organic waste [6], evaluation of syngas content and carbon monoxide production rate [7], analysis of gas production produced by using air, oxygen, a mixture of oxygen and water vapor as a gas reaction medium for the gasification process [2], temperature analysis of the gas production produced by the gasification process [8], performance experiments or performance in the gasification process [9]. From some of the discussions above, all of the RDF that has been used for the gasification process were transformed into pellets and briquettes, and the results obtained in the gas content were dominated by CO and H_2 content. So that RDF material can be converted into syngas and can be used as fuel.

The effort to optimize the gasification process began to be developed through several efforts carried out. One of them is by developing a control system to be applied to the gasification process. The process aims to improve the gasification performance expressed with combustible syngas composition, LHV syngas, and tar content in the syngas. The one addition is the control system to adjust the rotation speed of the blower so that the resulting temperature remains stable and gasification processes that do remain are continuous [10]. The purpose of this study was to determine the effect of three levels of air input, namely in the pyrolysis, oxidation, and reduction zones with variation in AR (0:10:0, 2:6:2, 1:6:3, 3:6:1, 2:5:3, 3:5:2, 1:8:1, 2:7:1, and 1:7:2) with the temperature setting (600 °C, 700 °C, 800 °C, 900 °C, and 1000 °C) is used, when air enters the pyrolysis zone and reduction can change the endothermic reaction into exothermic in the zone, therefore the devolatilization process that occurs will increase due to changes in reactions that occur in the pyrolysis zone.

2 Methods

2.1 Materials

The raw materials used are RDF briquettes with a composition of 60% organic and 40% inorganic. RDF briquettes have a diameter of 3 cm and a height of 3 cm, the materials are obtained from PT. Bakti Bumi Sidoarjo. The production of RDF briquettes is carried out through several stages consisting of (a) drying aims to reduce the water content when RDF is obtained from landfills, besides that RDF is easier to sort when the material is dry, (b) separation is carried out to separate non-combustible wastes such as metal and glass, (c) enumeration is carried out to facilitate further confectionery such as changing the shape of RDF, (d) RDF briquettes are carried out to increase the density and reduce the humidity of the RDF. Before being used, a proximate analysis test was carried out on RDF, the full results of the proximate analysis of RDF briquettes are in Table 1.

Table 1. Characteristics of RDF proximate analysis (Energy Laboratory ITS, 2021)

Sample	Test type	Results	Unit	Method test
Refuse derived fuel	Moisture in sample	3.68	%, adb	ASTM D 317/D 3173 M-17a
	Volatile matter	76.70	%, adb	ISO 562-2010
	Ash content	14.71	%, adb	ASTM D 3174-12(2018)
	Fixed carbon	4.1	%, adb	ASTM D 3172-13
	Calorific value	5.378	Kal/gr, adb	Bomb calorimeter

2.2 Tool Schematic

The equipment of the gasification reactor used is as shown in Fig. 1. The dimensions of the reactor are 1.26 m and 0.5 m in diameter, 4 thermocouples are installed along the height of the reactor and 1 thermocouple is used to control the speed of the blower which is controlled by 1 Arduino Uno which functions as a temperature control system in the oxidation zone.

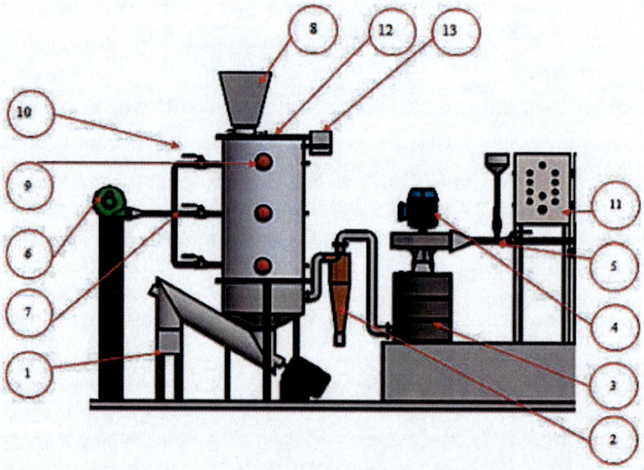

1. Ash box
2. Cyclone
3. Dry filter
4. Exhaust fan
5. Syngas flow pipe
6. Blower
7. Intake valve
8. Feeding Inlet
9. Thermocouple
10. Valve inlet
11. Panel box
12. Mixer
13. Mixer motor

Fig. 1. Schematic of gasification equipment

2.3 Operating Conditions

This research was conducted by applying nine variations of the air ratio (AR) and using temperature variations of 600 °C, 700 °C, 800 °C, 900 °C, and 1000 °C in the oxidation zone. While the AR variation used in the intake air is regulated using a valve located on the inlet pipe in the reactor, the air flow of each pipe is measured using a manometer. The application of AR variation is: 0:10:0, 1:8:1, 2:7:1, 1:7:2, 1:6:3, 2:6:2, 3:6:1, 2:5:3, and 3:5:2 air entering the pyrolysis, oxidation, and reduction zones. Oxidation temperature value input into the control system so that the rotation speed of the blower can be changed. Where the temperature control system works if there is a change in temperature that occurs in the oxidation zone in the reactor when the temperature of the oxidation zone changes the blower speed changes. The purpose of adding a control system is to adjust the change in blower rpm so that it can maintain the temperature variation specified in the oxidation zone. Data retrieval is carried out at each temperature setting used and the use of AR variations used, for example, at 600 °C settings, nine variations of AR are applied and the data collection is carried out.

3 Results and Discussion

3.1 Temperature Distribution

The temperature distribution using a temperature setting of 600–1000 °C in the drying zone experienced an increase in temperature of 120 °C. The increase in temperature in the drying zone in each temperature setting was caused by the large temperature produced in the oxidation and pyrolysis zones. The pyrolysis zones of 600–1000 °C increased by 477 °C.

The oxidation zone uses temperature settings of 600 °C, 700 °C, 800 °C, 900 °C, and 1000 °C all temperature settings are reached, but by using AR variations of 2:5:3 and 3:5:2 with a temperature setting of 900 °C and 1000 °C are not reached. The cause of the two variations is due to the lack of air entering the oxidation zone because the oxidation zone is the source of the combustion process and is the main source of exothermic reactions. So the exothermic reaction requires sufficient air to increase the temperature in the oxidation zone [9].

Drying zone using AR 1:7:2 variations with setting 1000 °C obtained the highest temperature of 250 °C. The pyrolysis zone with AR 2:7:1 variation and setting 1000 °C obtained the highest temperature of 625 °C. Oxidation zone using the setting of 1000 °C and all the AR used was reached, but there were variations in the AR that the temperature was not reached, namely AR 2:5:3 and 3:5:2. The reduction zone using the AR 1:7:2 variations with a setting of 1000 °C obtained the highest temperature of 545 °C. The temperature results are displayed only at the 1000 °C setting, because the highest temperature is using the 1000 °C. The resulting temperature distribution results are shown in Fig. 2.

3.2 Syngas Composition

Combustible syngas composition is shown only using a temperature setting of 1000 °C because it is the maximum temperature setting for the oxidation zone compared to other temperature settings. The results using all variations of AR increase with increasing temperature settings used and the most optimal AR variation uses 1:7:2 variations. The results of the highest combustible syngas composition of CO gas were 22.67 %vol and the highest H_2 was 11.54 %vol [11]. Where the increasing temperature setting in the partial

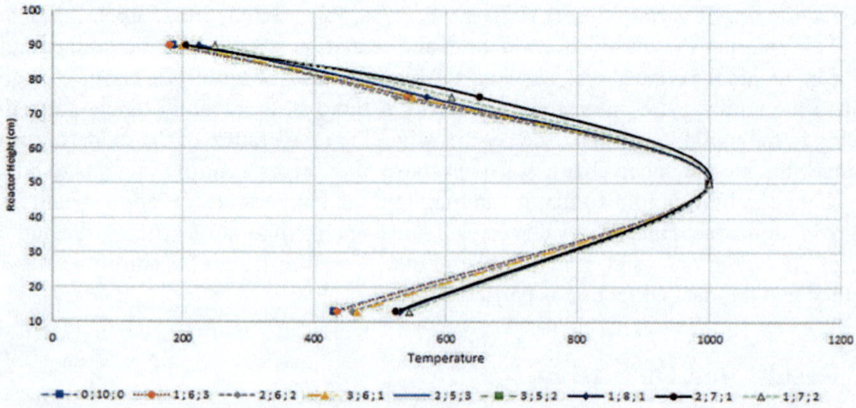

Fig. 2. Temperature distribution

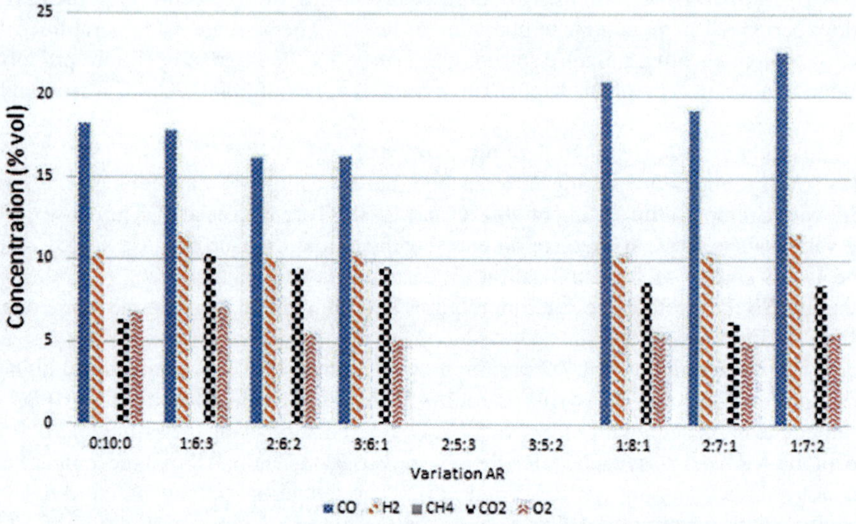

Fig. 3. Syngas composition

oxidation zone, the composition of the combustible gas produced will also increase. The results of the syngas composition are shown in Fig. 3.

The ratio of incoming air to the pyrolysis zone and the maximum reduction is between 10% of the total incoming air because the pyrolysis process basically does not need air. However, to change an endothermic reaction, it takes a little air to enter so that it becomes an exothermic reaction.

3.3 Lower Heating Value (LHV) Syngas

The increase in LHV produced tends to be caused by CO and H_2 gases because these gases are the most flammable gases produced in the gasification process. By using 10% air input in the pyrolysis zone and 20% in the reduction zone, it can increase the LHV produced, therefore it is necessary to limit the intake of air to the pyrolysis and reduction zones, in order to produce more combustible gas. In accordance with research [12] by limiting air intake in the pyrolysis and reduction zones, it can increase the LHV produced. LHV continues to increase along with the increase in the temperature setting used, the most optimal AR variation of 1:7:2 and the set temperatures of 1000 °C is obtained at 4201.025 kJ/kg, as shown in Fig. 4.

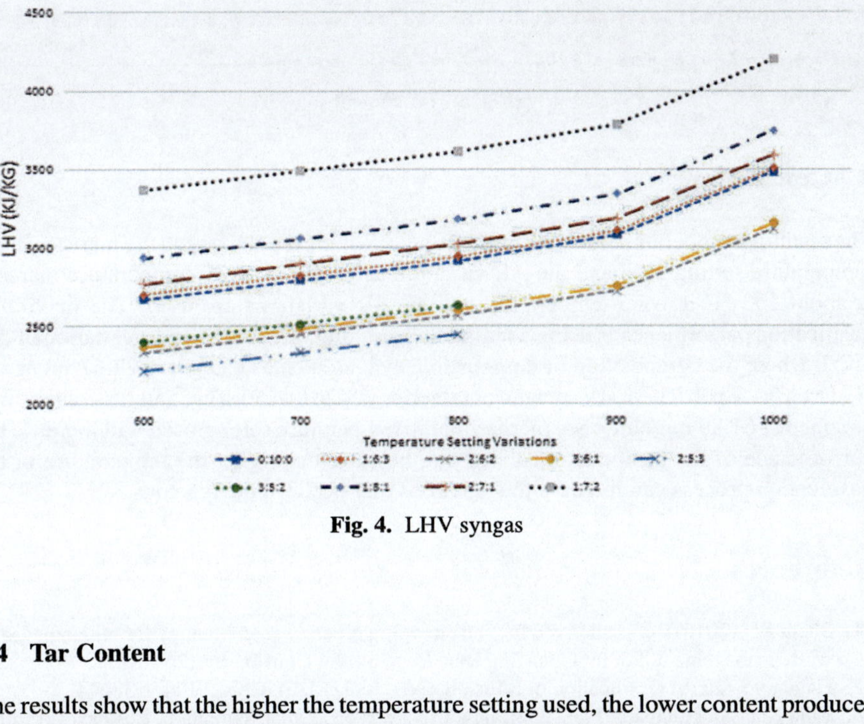

Fig. 4. LHV syngas

3.4 Tar Content

The results show that the higher the temperature setting used, the lower content produced for all variations used. The maximum AR variation is AR 1:8:1 at 600 °C settings and 1:7:2 at 800–1000 °C settings. The resulting LHV for all temperature variations and AR is shown in Fig. 5. The maximum yield was obtained at 61.89 mg/Nm^3 using a temperature setting of 1000 °C and AR variations of 1:7:2. Tar formation occurs initially in the pyrolysis zone and continues to decrease as it passes through the oxidation and reduction zones. Primary tar begins to decrease starting at 500 °C, so the primary tar will continue to decrease and will run out when the temperature reaches 800 °C. The decrease in tar content at a temperature setting of 800–1000 °C is the dominant type of secondary and tensioner tar, due to the occurrence of secondary and tertiary tars starting above 800 °C.

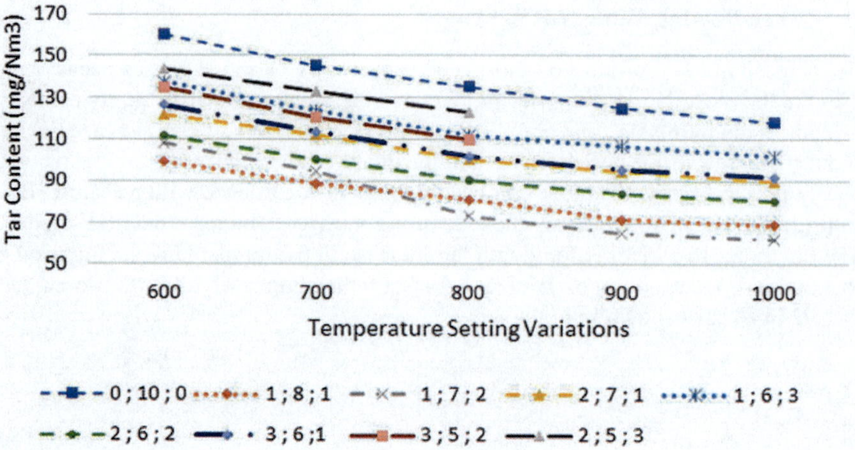

Fig. 5. Tar content

4 Conclusion

The resulting temperature distribution continues to increase along with the increase in the temperature setting used and the AR variation used. The average temperature increase is about 100 °C at each temperature and all AR variations are used. The maximum gasification performance is using a temperature setting of 1000 °C and a variation of AR 1:7:2, where the composition of combustible syngas such as CO gas is 22.67 %vol and H_2 is 11.54 %vol. The LHV syngas obtained was 4201.025 kJ/kg, and tar content was obtained at 61.89 mg/Nm^3. So, increasing the temperature setting used will increase the performance of the gasification carried out, because the higher the temperature in the gasification process can increase the reactions that occur in the reactor.

References

1. Sheth KN (2016) Refuse derived fuel-an emerging processing technology for municipal solid waste management. 9th Int Conf Eng Bus. Educ 6th Int Conf Innov 126–182
2. Haydary J (2016) Gasification of refuse-derived fuel (RDF). Geosci Eng 1:37–44
3. Irma Natasya Hutabarat MH, Ika Bagus Priyambada, Ganjar Samudaro, Baskoro Lokahita, Sarifudin, Irawan Wisnu Wardhana (2018) Potensi material sampah combustible pada zona pasif tpa jatibarang semarang sebagai bahan bakar baku RDF (Refuse Derived Fuel). J Tek Mesin 7:24–28
4. Násner AML et al (2017) Refuse derived fuel (RDF) production and gasification in a pilot plant integrated with an Otto cycle ICE through Aspen Plus™ modelling thermodinamic and economic variability. Waste Manag 69:187–201
5. Reed TB, Das A (1988) Handbook of biomass downdraft gasifier engine system. SERI US Dep Energy
6. Dianda P, Mahidin M, Munawar E (2018) Production and characterization refuse derived fuel (RDF) from high organic and moisture contents of municipal solid waste (MSW). IOP Conf Ser Mater Sci Eng 334(1)

7. Kunkajit C, Prateepchaikul G, Kaosol T (2015) Influence of plastic waste for refuse-derived fuel on downdraft gasification. Elsevier B.V. 79
8. Ajay GJS, Dalai K, Batta N, Eswaramoorthi I (2009) Gasification of refuse derived fuel in a fixed bed reactor for syngas production. Waste Manag 29:252–258
9. Arifin IR, Sudarmanta B (2020) Refuse-derived fuel pada downdraft gasifier menggunakan masukan udara tiga tingkat. Julnal Teknik Mesin ITS 2337–3539
10. Striugas N, Zakaraukas K, Džiugys A, Navakas R, Paulauskas R (2014) An evaluation of performance of automatically operated multi-fuel downdraft gasifier for energy production. Appl Therm Eng 73(1):1151–1159
11. FIcer B, Sudarmanta B (2020) Pengaruh Variasi Settingan Suhu Menggunakan Sistem Pengendali Suhu Pada Zona Oksidasi Parsial Terhadap Performa Gasifier Tipe Downdraft. Julnal Teknik Mesin ITS 2337–3539

Aerodynamic Characteristics of the Wing Airfoil NACA 43018 in Take Off Conditions with Slat Clearance and Flap Deflection

S. P. Setyo Hariyadi[1(✉)], Nyaris Pambudiyatno[2], Sutardi[3], and P. Fungky Dyan[3]

[1] Department of Aircraft Engineering, Politeknik Penerbangan Surabaya, Jemur Andayani I No 73 Wonocolo, Surabaya, Indonesia
setyo_hariyadi@poltekbangsby.ac.id

[2] Department of Air Navigation, Politeknik Penerbangan Surabaya, Jemur Andayani I No 73 Wonocolo, Surabaya, Indonesia

[3] Department of Mechanical Engineering, Institut Teknologi Sepuluh Nopember, Surabaya 60111, Indonesia

Abstract. The use of slat slots or what is often called slat clearance and flaps is mandatory on aircraft today. Its use is very useful during take off, cruising, and landing. However, these conditions are very different for each wing depending on the configuration and environmental conditions. This study examines the use of wings with the NACA 43018 airfoil during take off. Numerical simulation uses Ansys 19.1 with the turbulent model k-ε Realizable. Free stream flow rate used is 120 m/s with angle of attack (α) = 0°, 2°, 4°, 6°, 8°, 10°,12°,15°, 16°,17°,19°, and 20°. The conditions being compared are plain wing, flap deflection 0°, and 15° with fixed slat clearance during take off conditions. Using a fixed slat clearance, flap deflection $\delta_T = 15°$ produces the highest lift coefficient at angle of attack $\alpha = 12°$ compared to other configurations. However, the drag coefficient produced by the flap deflection $\delta_T = 15°$ is also the highest. With the highest lift coefficient, it is possible to use the maximum take off angle of attack $\alpha = 12°$.

Keywords: Slat clearance · Take off · NACA 43018 · Flap deflection · Lift

1 Introduction

High lift devices have become common in aircraft use. Usually, the equipment is mounted on the wing because most of the lift is formed in that section. High lift devices can also be used as a means of controlling aircraft movements, so they are often called primary control surfaces. High lift devices used in aircraft include flaps, slats, slots, elevators, ailerons, and others. Each aircraft manufacturer has a different combination of high lift devices. For example, the combination of high lift devices between Boeing, Airbus, McDonnell Douglas has different slat clearance settings and flap angles [1]. Each set has a different effect depending on the use of the aircraft. The use of high lift device settings, of course, has very different effects during take-off, cruise, and landing. Therefore, each manufacturer has different settings depending on the movement of the

I. Tolj et al. (eds.), *Recent Advances in Mechanical Engineering*, Lecture Notes in Mechanical Engineering, https://doi.org/10.1007/978-981-19-0867-5_27

aircraft it experiences. Likewise, with the operating area of the aircraft, the aircraft will use different speeds depending on the altitude and engine of the aircraft used. Airplanes with turboprops usually use trailing edge flap slots and do not use slat clearance on the leading edge. This is of course different from the design of Boeing, Airbus, and McDonnell Douglas aircraft. With different types of engines and operating heights, the resulting aerodynamic performance is also different. And planes with turboprops are more suitable for use in archipelagic areas such as Indonesia because they require a shorter runway length. On the wings, one of the airfoils used on turboprop aircraft is the NACA 43018 which is equipped with a trailing edge slotted flap. It will be interesting to be used as research material by providing some aerodynamic modifications to the wings.

Traub et al. [2] investigated the effect of using leading-edge slats on UAVs with low Reynolds numbers. The study was conducted at Re = 250,000 with a freestream velocity of 35 m/s on the wing airfoil A S8036 with an aspect ratio (AR) = 3. At the angle of attack $\alpha = 0°$, it looks the same between the plain wing and the G5R0 model. At the angle of attack $\alpha = 10°$, slat causes a wide trailing edge separation when compared to the plain wing. This explains the large lift loss and the sharp increase in drag with the addition of slat. However, the G5R0 model shows a potential reattach on the backside of the trailing edge so that if there is an additional extension the flow will return to the model surface.

Mondher Yahyaoui [3] used the vortex lattice method to formulate the lifting surface of the flap and aileron deflection. This method takes into account the variations used on wings equipped with flaps and ailerons, including camber, taper, sweep, washout, and dihedral. MATLAB code was used to formulate and validate the results of Bertin and Smith's simulations and experiments, experiments on NACA NACA 4415, and NASA TN D-6142. From this study, it was found that a fairly large error of 30% was obtained in the deflection of the flap and aileron at a high angle of attack. The low deflection will produce a low error.

Abbas et al. [4] conducted experiments on the Clark Y-14 wing airfoil with a Reynolds number of 2.1×10^5 and a freestream velocity of 35 m/s. The experimental results on the wind tunnel are compared with the results of the two-dimensional simulation on the Javafoil application. The lift coefficient results show that the open slot provides a better value compared to other model configurations. Rotating slat with an angle—10° gives the best performance in this study.

This study explores further the use of the NACA 43018 wing airfoil which is equipped with slat clearance (slat slot) and slotted flap deflection. The selected condition is when the plane takes off where the highest lift coefficient is sought at that time. The highest lift coefficient is expected to soon reach the desired height.

2 Methods

2.1 Simulation Domain and Boundary Conditions

This numerical simulation study uses Ansys 19.1 with a turbulent model of k-ε Realizable based on research by Mulvany [5] which shows that k-ε Realizable produces outputs that are closer to experimental research than other turbulent models. The geometry of the study uses airfoil on the NACA 43018 wing in the form of a plain wing which is compared with a combination of slat clearance and flap deflection. The simulation domain of this research is based on research by Mulvany [5] which was modified with a section behind the trailing edge 10 times the chord line to show the effect of the wing on the airflow behind the wing [6–8]. The angles of attack observed were 0°, 2°, 4°, 6°, 8°, 10°, 12°, 15°, 16°,17°,19°, and 20°. The simulation domain is made as the condition of the NACA 43018 wing airfoil in a wind tunnel with a freestream velocity of 120 m/s (Re = 4,538 × 10^6) with a pressure of 1 atm on the inlet side and 0 atm on the outlet pressure. The simulation domain can be seen in Fig. 1, and the test object is shown in Fig. 2. The terms and nomenclature on flap angle and slat clearance refer to Gunther's research [9].

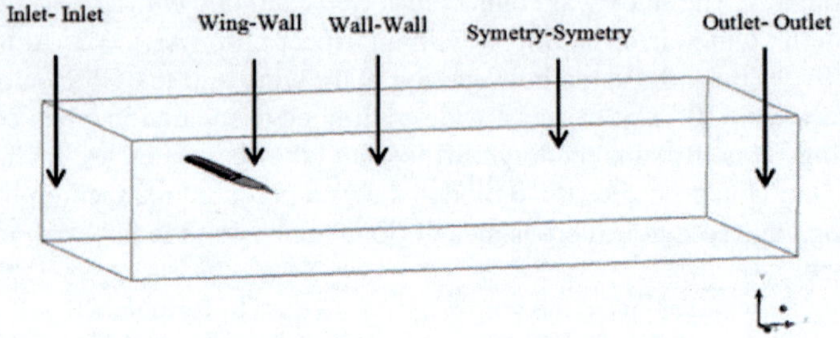

Fig. 1. Domain simulation

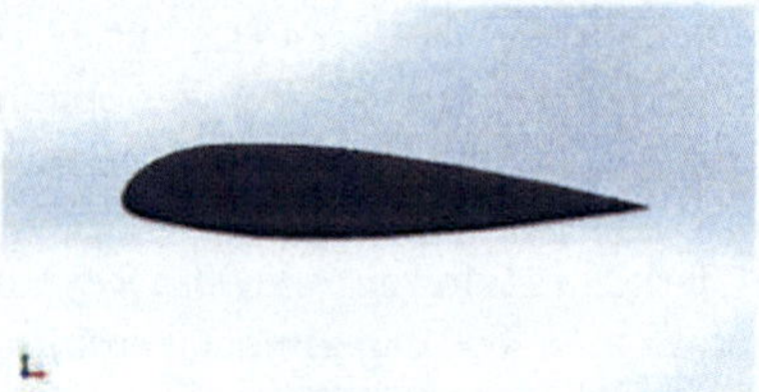

a. Plain Wing

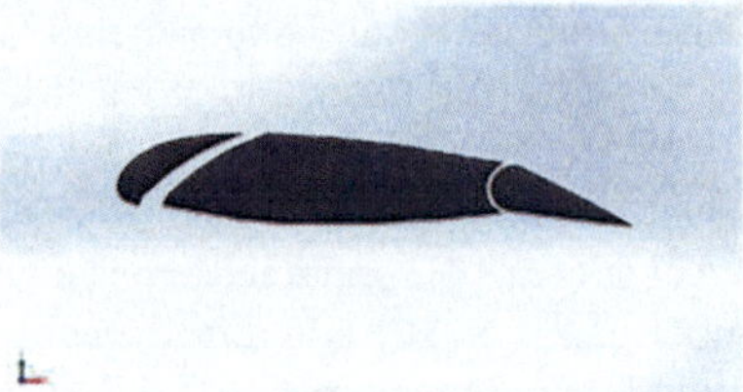

b. Flap $\delta_T = 0^\circ$, Slat $\delta_L = 15^\circ$

Fig. 2. Test object

2.2 Meshing and Grid Independency

In this numerical simulation, we use the form of meshing cut cells in all parts, including the extension behind the trailing edge. Grid independence in this study was compared with that of Eastman and Abbot [10]. Table 2 shows the variation of the meshing of the 3-dimensional test model at the Reynolds number 4,538 × 10^6. Based on Table 2, the C_D value that tends to approach Eastman and Abbot's research [10] occurs in Meshing A and Meshing B. The smallest error value according to the results of Eastman and Abbot's research [10] is Meshing A. In addition, with meshing cut cells, y+ formed on all meshing meet the criteria in Kontogiannis' research [11]. One of the considerations in performing numerical simulations is the time and memory used, so the meshing used for the next simulation is Meshing A.

Table 1. Details of the geometry between main airfoil, slat clearance, and flap

	$\delta_L = 0°$	$\delta_T = 15°$
Slat Clearance	5%	–
Flap	–	1%

Table 2. Grid independency

Meshing type	Number of nodes	C_D	C_D [10]	Error
Meshing A	3.504.086	0.08	0.04	0.04
Meshing B	3.639.000	0.15	0.04	0.11
Meshing C	3.832.000	0.216	0.04	0.176
Meshing D	3.862.642	0.297	0.04	0.257

3 Results and Discussion

3.1 Aerodynamic Performance

Aerodynamic performance is highly dependent on the type of airfoil and the equipment attached to the wing. The effect of the equipment, of course, affects the formation of the lift coefficient, drag coefficient, and lift to drag ratio. Figure 3 shows the ratio of drag coefficient of plain wing, flap $\delta_T = 0°$, and flap $\delta_T = 15°$. At the angle of attack $\alpha = 0–12°$, the flap $\delta_T = 15°$ produces the highest drag coefficient compared to other configurations, while at the angle of attack $\alpha = 15–19°$, the plain wing shows the highest drag coefficient. Flap $\delta_T = 0°$ shows the lowest drag coefficient compared to other configurations at all angles of attack.

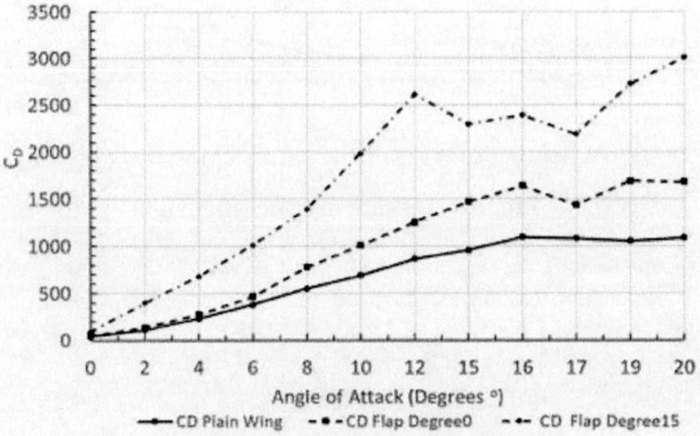

Fig. 3. Total drag coefficient

Figure 4 shows that the effect of increasing the wing surface with the addition of flaps and slat slots only gives the effect of adding a little friction drag. The addition of friction drag with the presence of flaps and slat slots only causes a small value. Figure 5 shows that the flap $\delta_T = 15°$ causes a significant increase in pressure drag. This is possible because the frontal flap area $\delta_T = 15°$ with respect to the flow direction contributes greatly to the pressure drag coefficient. Therefore, the value of the pressure drag coefficient produced by the flap $\delta_T = 15°$ has a large enough difference compared to other configurations.

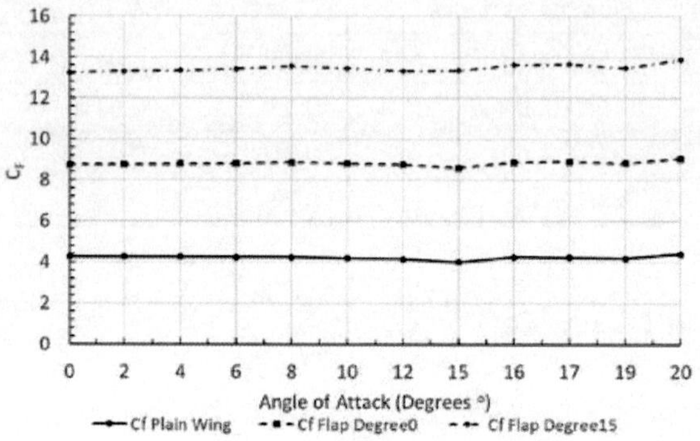

Fig. 4. Friction drag coefficient

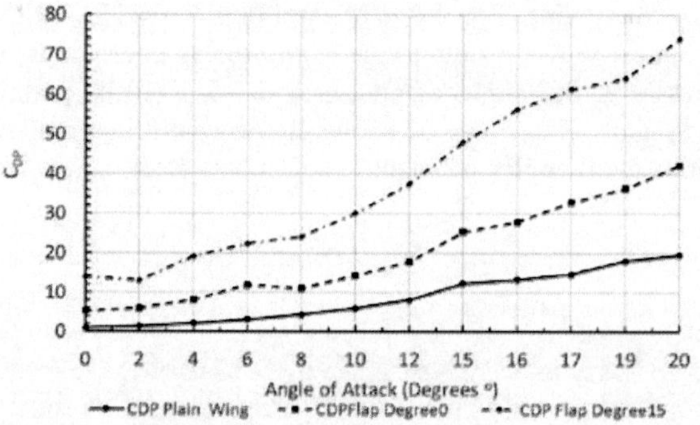

Fig. 5. Pressure drag coefficient

Figure 6 is closely related to Fig. 7 where a high lift coefficient will result in a high induced drag coefficient as well. Figure 7 shows that the flap $\delta_T = 15°$ produces the highest lift coefficient compared to other configurations. Therefore, the flap $\delta_T = 15°$ produces the highest induced drag coefficient in Fig. 6. Figure 7 shows that the flap $\delta_T = 0°$ produces the highest lift to drag ratio compared to other configurations, followed by the flap $\delta_T = 15°$ and the plain wing. By paying attention to Figs. 7 and 8, it is found that the flap $\delta_T = 15°$ is the most appropriate for the takeoff process where this configuration is used to get the highest lift so that it reaches the desired height as fast as possible. In addition, the flap $\delta_T = 0°$ is most appropriate for the cruising process where it uses the best aerodynamic performance compared to other configurations. Overall, it was found that the configuration using flaps with the addition of slat slots in this study was better than plain wing.

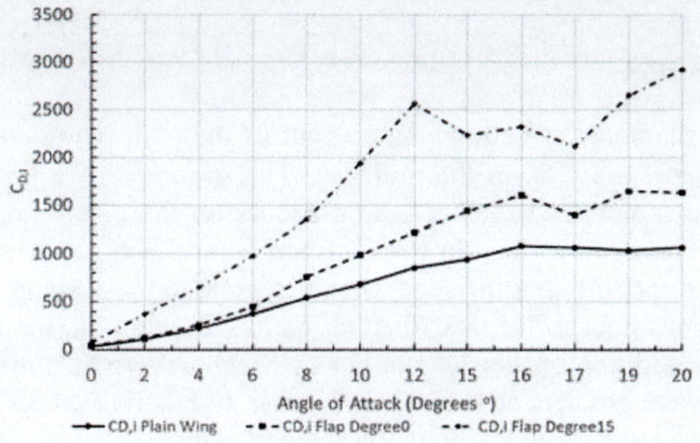

Fig. 6. Induced drag coefficient

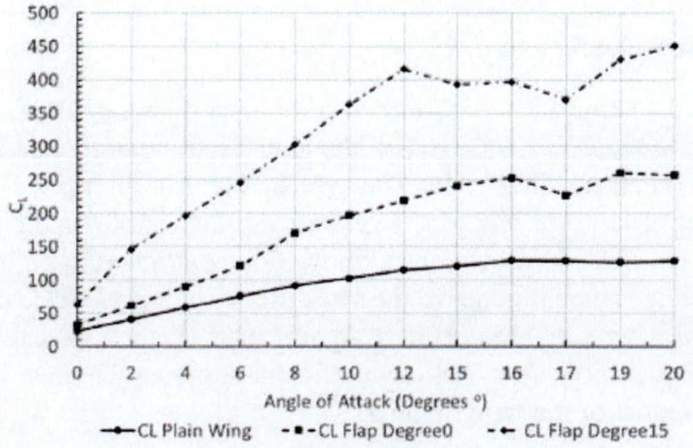

Fig. 7. Lift coefficient

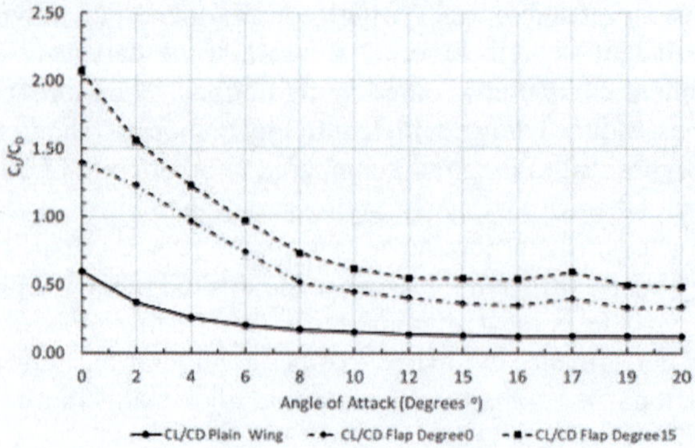

Fig. 8. Lift to drag ratio

3.2 Velocity Pathline

Figure 9 shows the velocity pathline as a result of the plain wing configuration, flap $\delta_T = 0°$, and flap $\delta_T = 15°$ in the midspan. The presence of a flow vortex on the trailing edge shows how fast the separation occurs on the upper surface. Flap $\delta_T = 15°$ shows the separation formed in the midspan from the angle of attack $\alpha = 0°$. It is therefore not surprising that the separation has gone to the leading-edge faster than other configurations because of its earlier formation. In addition, the flow momentum generated by the addition of the slat slot has not been able to face the frictional force and adverse pressure gradient so that the flow separation is formed earlier. To see more deeply the effect of slat slots and flaps on the entire upper surface, it is necessary to add three-dimensional visualization so that it can be seen how wide the part has been separated.

3.3 Skin Friction Coefficient

Figure 10 shows the skin friction coefficient contour of the research results on the upper surface. At plain wing $\alpha = 15°$ (Fig. 10), the skin friction coefficient has a high value and a large area on the leading edge. The area that shows the tip vortex effect is also visible in the area around the wingtip. The separation is visible on the upper surface of the flap $\delta_T = 0°$ $\alpha = 15°$. However, the skin friction coefficient has a low value around the leading edge due to the influence of the slat slot. In the flap $\delta_T = 15°$ $\alpha = 15°$, the flow separation on the wing root side extends significantly enough to reach the mid-wing. The fairly high value of the skin friction coefficient has been flattened in all parts of the leading edge and most of the wingtip area.

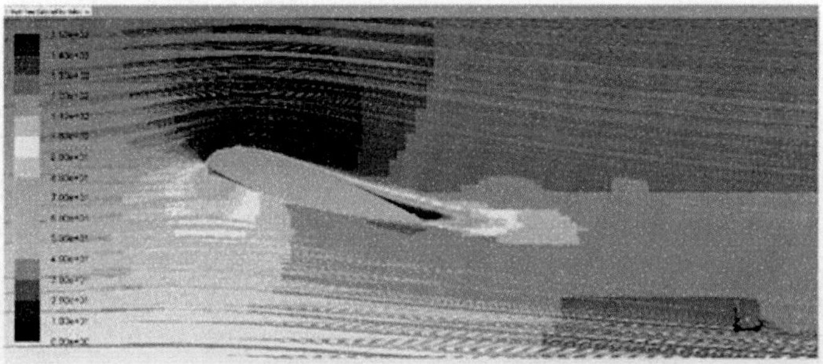

a. Velocity Pathline Plain Wing $\alpha = 15^{o}$

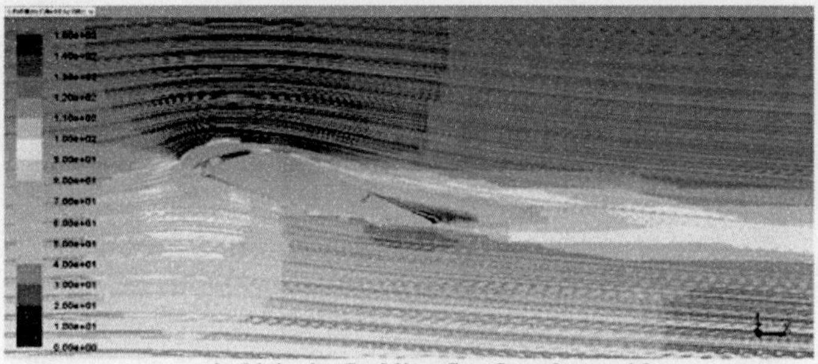

b. Velocity Pathline flap $\delta_T = 0^{o}$ $\alpha = 15^{o}$

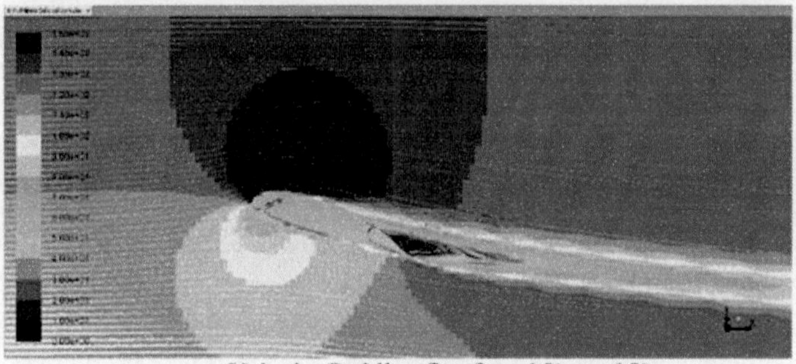

c. Velocity Pathline flap $\delta_T = 15^{o}$ $\alpha = 15^{o}$

Fig. 9. Velocity pathline results in the midspan

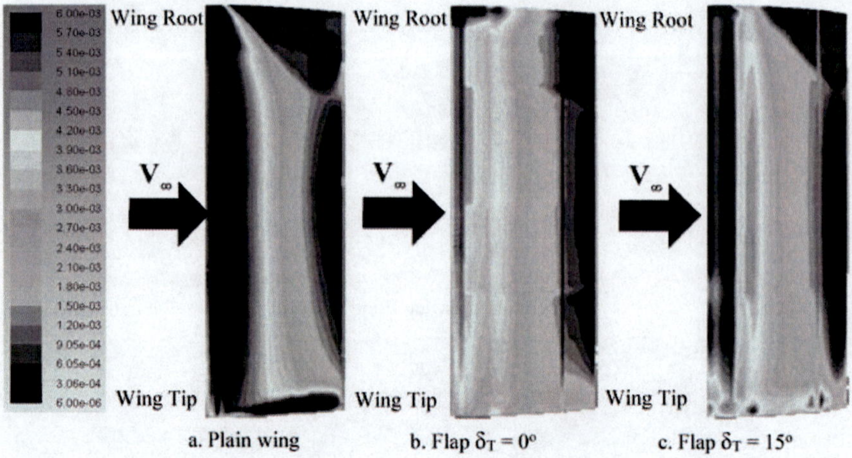

Fig. 10. Skin friction coefficient contour results on the wing upper surface $\alpha = 15°$

4 Conclusions

A numerical simulation study comparing the use of plain wing and combination of flap $\delta_T = 0°$ and flap $\delta_T = 15°$ has been carried out. From the numerical simulations carried out, several things can that can be noted, including

- The combination of flap $\delta_T = 15°$ and slat $\delta_L = 0°$ has a higher total drag coefficient than plain wing due to the dominant role of the induced drag coefficient on the wing.
- Lift to drag ratio produced by flap $\delta_T = 0°$ is better than the combination of flap $\delta_T = 15°$ and plain wing
- The use of flaps $\delta_T = 15°$ accelerates flow vortex and wing separation. However, flap $\delta_T = 15°$ produced the highest C_L compared to other configurations.
- The effect of using slat slots can be seen by paying attention to the value of the skin friction coefficient in the leading-edge area which is lower than the plain wing because the flow gains additional momentum so that it can face the surface friction force of the wing.
- The pattern of flap separation area $\delta_T = 0°$ and $\delta_T = 15°$ began to form and was concentrated in the upper surface flap area which then expanded around the wing root area.

References

1. Neigapula SNV, Maddula SP, Nukala VB (2020) A study of high lift aerodynamic devices on commercial aircrafts. Aviation 24:123–136. https://doi.org/10.3846/aviation.2020.12815
2. Traub LW, Kaula MP (2016) Effect of leading-edge slats at low Reynolds numbers. Aerospace 3. https://doi.org/10.3390/aerospace3040039

3. Yahyaoui M (2014) Generalized vortex lattice method for predicting characteristics of wings with flap and Aileron deflection. 8:1660–1668
4. Askar M, Eng Mohammed Kheiraldeen Abbas A, Hasan Khudhur Abbas E (2016) Aerodynamic study of Slot Effect on ClarkY-14 in low speed wind tunnel engineering analysis with ANSYS software view project aerodynamic study of slot effect on ClarkY-14 in low speed wind tunnel. Int J Art Human Sci (IJAHS) www.ijahs.com 3:21–29
5. Mulvany N, Chen L, Tu J, Anderson B (2004) Steady-state evaluation of two-equation RANS (Reynolds-Averaged Navier-Stokes) turbulence models for high-reynolds number hydrodynamic flow simulations. Department of Defence, Australian Government, pp 1–54
6. Setyo Hariyadi SP, Sutardi, Widodo WA, Sonhaji I (2021) Numerical study of secondary flow characteristics on the use of the winglets. J Phys: Conf Series 1726. https://doi.org/10.1088/1742-6596/1726/1/012012
7. Putro SHS, Sutardi, Widodo WA (2019) Numerical study of three-dimensional flow characteristics around the wing airfoil E562 with forward and rearward wingtip fence. Presented at the (2019). https://doi.org/10.1063/1.5138272
8. Setyo Hariyadi SP, Sutardi, Widodo WA, Mustaghfirin MA (2018) Aerodynamics analysis of the wingtip fence effect on UAV wing. Int Rev Mech Eng 12. https://doi.org/10.15866/ireme.v12i10.15517
9. Günther B, Thiele F, Petz R, Nitsche W, Sahner J, Weinkauf T, Hege HC (2007) Control of separation on the flap of a three-element high-lift configuration. Collection of Technical Papers—45th AIAA Aerospace Sciences Meeting 5:3259–3273 https://doi.org/10.2514/6.2007-265
10. Aeronautics FOR (1939) ADVISORY
11. Kontogiannis SG, Mazarakos DE, Kostopoulos V (2016) ATLAS IV wing aerodynamic design: from conceptual approach to detailed optimization. Aerosp Sci Technol 56:135–147. https://doi.org/10.1016/j.ast.2016.07.002

The Performance Augmentation of Savonius Wind Turbine Using Staggered Cylinder Circular as Passive Flow Control

Gunawan Sakti, Triyogi Yuwono(✉), and Wawan Aries Widodo

Department of Mechanical Engineering, Institut Teknologi Sepuluh Nopember, Surabaya 60111, Indonesia
triyogi@me.its.ac.id

Abstract. There are two flow control methods for reducing drag force on the cylinder: active and passive control. The passive control method regulated the airflow by varying the body shape or installing additional rods or surface roughness. This airflow control procedure is applied to the returning blade of the Savonius wind turbine by installing a circular cylinder to reduce pressure drag and improve its performance. A circular cylinder varies with a stagger angle $-10° \leq \alpha \leq 90°$ and within the center to a center distance fixed at *S/D* = 1.4 were investigated numerically. The transient numerical calculation was performed with Ansys Fluent 19.1 to identify the dynamic nature of the 2D modeled turbine. The numerical calculation set with constant freestream velocity (U_∞) 5 m/s at Reynolds number *Re* = 105.000. The two-equation URANS and *k-ε enhance wall function* turbulence models are used in this simulation. The simulation was also verified with experimental data and showed healthy agreement so that the boundary conditions, grid, turbulent model, and solver setup settings can be maintained for further consideration. The analysis is carried out on the power coefficient and moment coefficient compared between conventional cylinders and cylinders with staggered cylinders. The result shows that staggering cylinder circular enhances the conventional turbine performance by 9% higher. The optimum staggered angle was founded at α = 70° within TSR = 0.8.

Keywords: Savonius wind turbine · Cylinder circular · CFD · Pressure drag · Coefficient of power

1 Introduction

Savonius wind turbine is one type of VAWT turbine which is still a prevalent research object [1]. This turbine has many advantages compared to other turbine types, including simple construction, high starting capability, not being affected by wind direction, and still being able to rotate at low wind speed. Ironically, this turbine performance is still low, so many mysteries must be revealed scientifically [2]. The rotation of the Savonius turbine is the result of the difference in drag force between the advancing and returning blades. The difference in the drag force experienced by the two turbine blades will produce a moment force so that the turbine rotates to absorb and convert energy [3].

I. Tolj et al. (eds.), *Recent Advances in Mechanical Engineering*, Lecture Notes in Mechanical Engineering, https://doi.org/10.1007/978-981-19-0867-5_28

In this scientific report, an attempt will be made to improve the performance of the Savonius turbine by placing the cylinder in the direction of the incoming airflow of the returning blade as passive flow control. Passive flow control is one method that regulates airflow by varying the body shape or installing additional rods or roughness [4]. The cylinder is also varied in its staggered angle $-10° \leq \alpha \leq 90°$ in order to analyze the optimum position relative to the returning blade center point. The installation of this cylinder is expected to affect the pressure of the upstream and downstream of the returning blade so that the pressure drags decrease, turbine torque and power increase.

2 Literature Reviews

The previous report of Savonius wind turbine has been reviewed for design innovation idea of advance turbine arrangement. The author [5, 6] reported the comparison of the power coefficient on the rotor with 2, 3, and 4 blades with the results that the power coefficient for the two blades rotor was higher than the power coefficient obtained by the Savonius turbine with 3 and 4 blades. The numerical result reported by [2] with twisted blade design can improve the operating range of the Savonius rotor where the maximum power coefficient is 0.3 at TSR 1.4 with a fluid velocity of 2 m/s. The publication review [7] stated that the use of endplates on the Savonius wind rotor reached a maximum power coefficient of 0.462. In addition, the use of endplates can also strengthen the overall turbine structures. The author [8] researched the circular cylinder installation before the advancing blade and varying *ds/D*; 0.1, 0.3, and 0.5 with stagger angles of 0, 30, and 60 at *s/d* 0.95. The final study shows that the best Savonius rotor performance is obtained at *ds/D* = 0.5 diameter and 30° stagger angle at a TSR (λ) of 0.9. Turbine performance improved 41.18% higher than conventional turbine at staggered angle 30 and TSR (λ) 0.9. The researcher [9] conducted numerical and experimental tests on five rotors with two semicircular blades. Rotor blades were tested with overlap ratios of 0.15, 0.2, 0.3, and 0.5. The result is that the maximum performance increase is obtained at an overlap ratio of 0.15. Aspect ratio testing is also reported by [6], with the research variable aspect ratio $\beta = 0$–3.5. The results show that the turbine with an overlap ratio of $\beta = 0$ provides higher mechanical power than the one with overlap.

Based on these scientific reports related to the Savonius turbine, the rotor geometry was determined with two blades, single stage, without overlap, and using an endplate for further investigation. In addition, the placement of a circular cylinder with a staggered angle in front of the returning blade was further investigated to reach an area of study that previous researchers have not studied. So that aspects of research that are sustainable and renewable can be fulfilled in this study.

3 Numerical Setup

3.1 Geometry and Boundary Condition

The numerical simulation study is conducted by using the Finite Volume Method with ANSYS 19.1 software. The 2D Savonius model and boundary conditions were constructed using commercial software in a *.iges* file for later discretization (meshing).

Figure 1 presents the boundary condition of numerical simulation. The geometry of the Savonius rotor is split into two parts, a *fixed zone* and a *rotating zone,* to make it easier to define both the interfaces between the two zones.

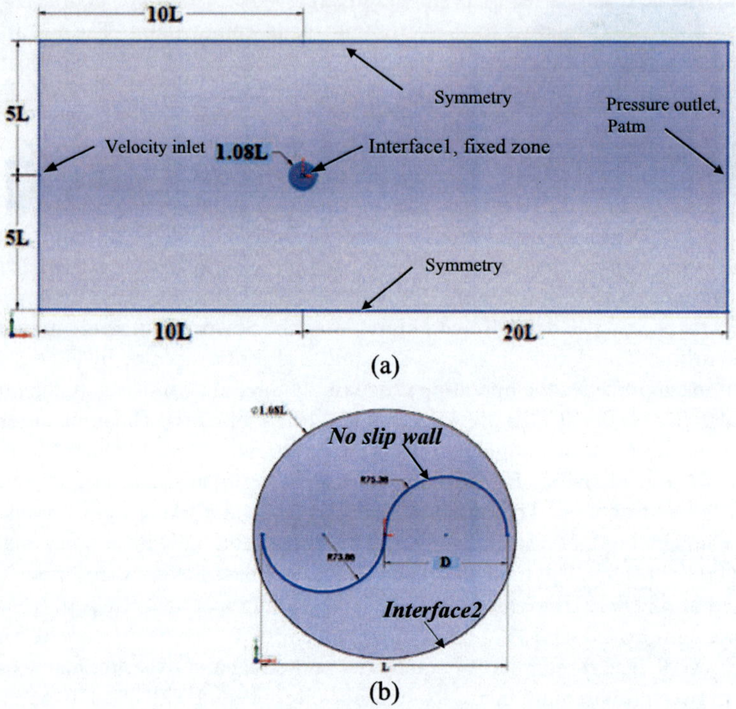

Fig. 1. The geometry construction and its boundary condition of current simulation: **a** fixed zone with *interface1*, and **b** rotating zone with *interface2*

The two solver setup interfaces are defined as *interface1* for the circular edge of the *fixed zone* and *interface2* for the circular edge of *the rotating zone* in the solver setup. Boundary condition inlet is determined as *velocity inlet*. The outlet is determined as a *pressure outlet,* while the upper and lower surfaces are *symmetrical*. A distance of 5 times the turbine diameter (5*L*) on each side is determined to avoid the *blockage effect* [10]. The outflow is extended by approximately 20 times the turbine diameter (20L) to reduce the *backflow effect* and allow fully developed airflow during the simulation. This procedure is in agreement with other researchers in their simulation work [10, 11].

3.2 Grid Independence

Figure 2 shows the Savonius wind turbine's mesh construction with cylinder circular staggered configuration angle $\alpha = 70°$. The Quadrilateral mesh is constructed with a vertical and horizontal density gradient. The gradient is distributed denser from fixed zone to turbine wall inside the rotating zone with varying its face sizing element size. The inflation, first layer thickness Δs, for the desired y^+ is also determined and set on the Savonius turbine and the cylinder circular wall. It was calculated using the following equation.

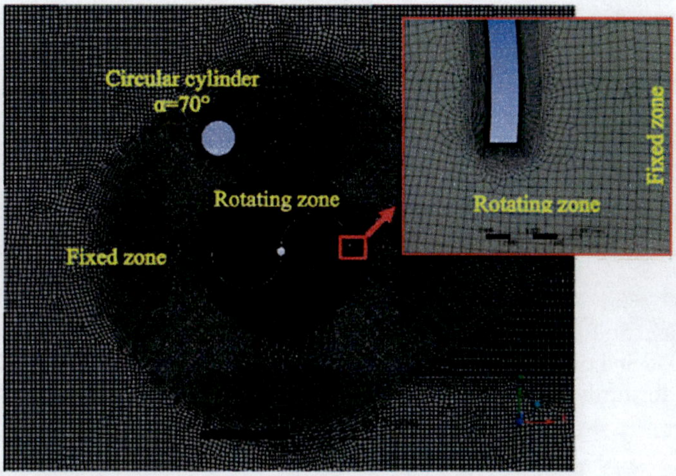

Fig. 2. Mesh construction at the fixed zone and rotating zone turbine Savonius with the installation of cylinder circular at staggered angle $\alpha = 70°$

$$\Delta s = \frac{y + v}{\sqrt{\frac{0.013 \times U_f}{Re^{\frac{1}{7}}}}} \tag{1}$$

Uf is expressed as frictional velocity, calculated based on shear stress rate at wall τ_w.

Table 1 shows grid independence, which is one of several essential steps during the numerical investigation. Seven meshes were assembled with an increment of mesh cells varying in their both fixed and rotary zone face sizing, which was determined to start from 450,000 up to 840,000 mesh cells. The best mesh quality is also determined by the value of $y^+ < 1$, and based on Table 1, mesh C and E are compliant. Furthermore, both the average aspect ratio and maximum skewness can reach below 1.5 and 0.9. respectively, and this is indicated the determined criterion could be compliance by the meshes.

Table 1 also shows that starting from mesh C, the change in the value of *Cm* is getting smaller, or in other words, there is no significant change as the number of mesh cells increases. Additionally, the grid independence was calculated within the tips-speed ratio 0.6 at Reynolds Number $Re = 105{,}000$, and the freestream velocity was determined at $U_\infty = 5$ m/s.

Further analysis, Fig. 3 presented the coefficient of moment data for each mesh configuration. This graph shows that mesh C to mesh G gives a constant coefficient

Table 1. *Grid independence* analysis for 2D conventional Savonius wind turbine, TSR = 0.6 *Re* = 105,000 at freestream velocity $U_\infty = 5$ m/s

Type of mesh	Number of cells	Average aspect. ratio	Max skewness	y^+	Average *Cm*
Mesh A	454,377	1.12	0.65	1.67	0.162
Mesh B	493,381	1.15	0,69	1,25	0.166
Mesh C	554,404	1.48	0.71	0.17	0.164
Mesh D	625,833	1.09	0.87	1.15	0.164
Mesh E	669,985	1.39	0.71	0.48	0.164
Mesh F	780,105	1.11	0.76	1.11	0.164
Mesh G	840,826	1.11	0.87	1.25	0.164

moment value. So for the efficiency and effectiveness of the simulation progress, reduce hardware load and time consumed, mesh C with a minimum number of 554,000 cells is maintained for further simulation work [12, 13].

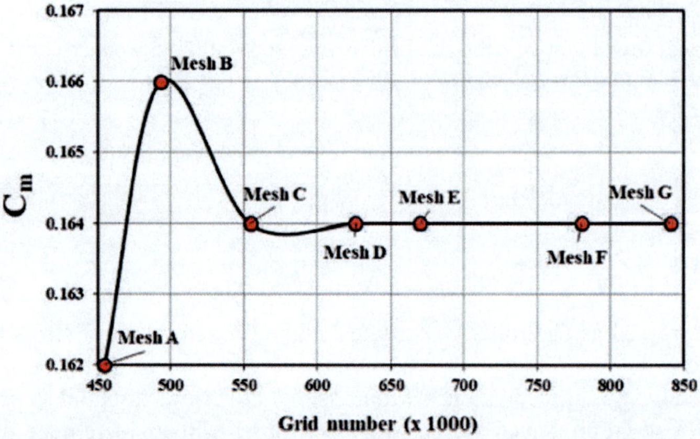

Fig. 3. Graphic *coefficient torque* as a function of *grid number* at TSR = 0.6 Re = 105,000

Figure 4 presents the graphic plot current numerical coefficient of power result, compared with 3D simulation data [14] and experimental data from [15] as a function of tip-speed ratio (λ). The current 2D simulation results show good agreement with the experimental data with a deviation rate of less than 5% for each tip-speed ratio within interval $0.6 \leq TSR \leq 0.7$. However, the result current 2D simulation also closed to the 3D simulation coefficient of power result with an erroneous less than 5% for all tip-speed ratio at $0.6 \leq TSR \leq 1.0$. Hence, the current numerical solver setup is valid and verified for further investigation in this study.

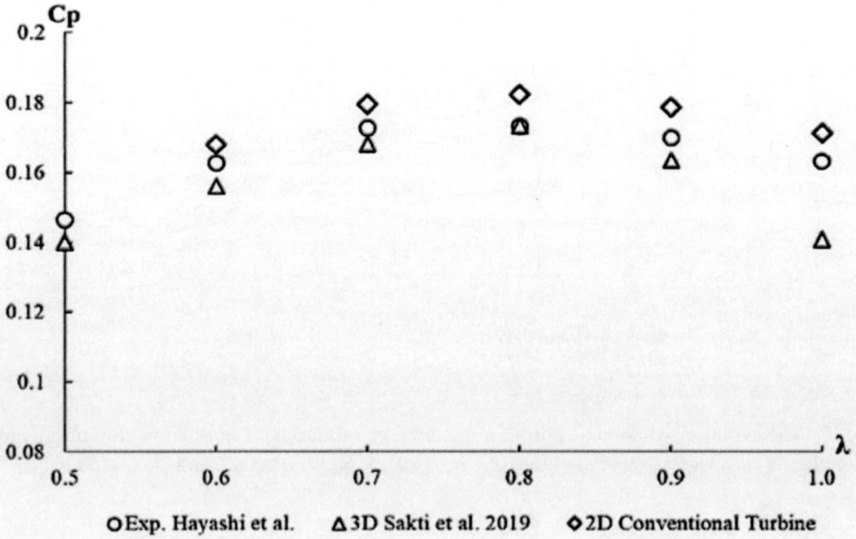

Fig. 4. The coefficient of power comparison between current numerical result and literature experimental data at tip-speed ratio $0.6 \leq \text{TSR} \leq 1.0$, $Re = 105{,}000$, and $U_\infty = 5$ m/s

4 Results and Discussion

4.1 The Different Stagger Angle Comparison

Figure 5 shows the coefficient of moment conventional Savonius turbine and turbine with the installation of staggered cylinder circular at Reynolds number Re = 105,000. The coefficient moment tends to increase by installation staggered cylinder circular within $10° < \alpha < 70°$, and falling when set on staggered angle $\alpha = -10°$, 0°, and 90°. This coefficient moment evolution proved that the staggered cylinder circular could improve the performance of turbine Savonius with the optimum angle at $\alpha = 70°$.

Figure 6 shows the coefficient of power conventional Savonius turbine and turbine with the installation of a staggered circular cylinder at Reynolds number $Re = 105{,}000$ freestream velocity at $U_\infty = 5$ m/s. The peak coefficient of power C_P tends to increase within the staggered angle $10° < \alpha < 70°$, and back to decrease at $\alpha = 90°$. The coefficient power C_P drop below conventional turbine Savonius when set on a staggered angle $\alpha =$ 0° and $\alpha = -10°$.

The conventional Savonius turbine has peak coefficient power at $C_p = 0.095$ and increased to $C_p = 0.103$ with staggered cylinder circular installed at $\alpha = 70°$. Based on this increased value, it can be concluded that the staggered circular cylinder configuration can strengthen the Savonius turbine performance by 9%. Furthermore, the graphic also noticed that maximum C_p was obtained at the tips-speed ratio $TSR = 0.8$.

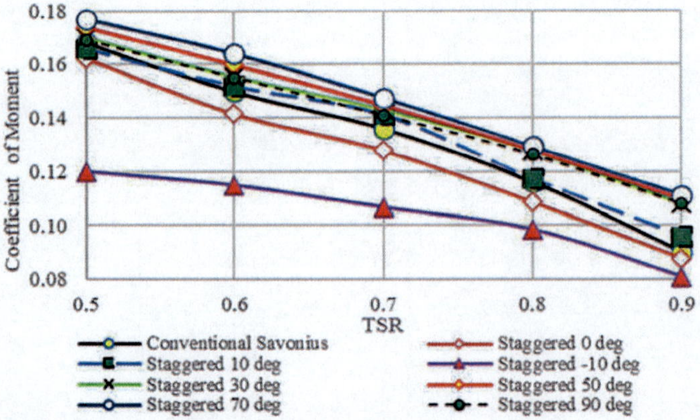

Fig. 5. The evolution of Savonius turbine's coefficient of moment *Cm* as a function of tip-speed ratio and its various staggered angle $-10 < \alpha < 90°$, at Re = 105,000, and $U_{\infty} = 5$ m/s

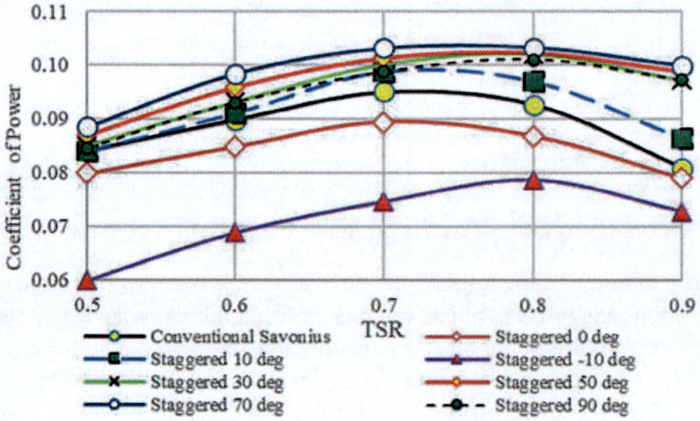

Fig. 6. The evolution of Savonius turbine's coefficient of power *Cp* as a function of tip-speed ratio and its various staggered angle $-10 < \alpha < 90°$, at Re = 105,000, and $U_{\infty} = 5$ m/s

5 Conclusion

Based on the validation simulation numerical method, it can be concluded that the numerical solver setup method can reach the experimental data and stated that verified and valid for this investigation. Furthermore, consideration parameters *Cm* and *Cp* proved that installing a staggered cylinder circular strengthens the turbine performance by 9% at TSR = 0.8 and $\alpha = 70°$. However, further analysis including pressure contour, velocity contour, and static torque of the turbine should be performed to reveal and clear the secret of this phenomenon.

References

1. Saad AS (2020) Performance enhancement of twisted-bladed Savonius vertical axis wind turbines. Energy Conv Manag 209. https://doi.org/10.1016/j.enconman.2020.112673
2. Singha S, Saini RP (2020) Performance analysis of a modified Savonius hydrokinetic turbine. Presented at the (2020)
3. Yuwono T, Sakti G, Nur Aulia F, Chandra Wijaya A (2020) Improving the Savonius wind turbine's performance by installing a circular cylinder upstream of returning turbine blade: Improving the performance of Savonius wind turbine. Alex Eng J 59:4923–4932. https://doi.org/10.1016/j.aej.2020.09.009
4. Lee S-J, Lee S-I, Park C-W (2004) Reducing the drag on a circular cylinder by the upstream installation of a small control rod. Fluid Dyn Res 34. https://doi.org/10.1016/j.fluiddyn.2004.01.001
5. Saha UK, Thotla S, Maity D (2008) Optimum design configuration of Savonius rotor through wind tunnel experiments. J Wind Eng Ind Aerodyn 96. https://doi.org/10.1016/j.jweia.2008.03.005
6. Mahmoud NH, El-Haroun AA, Wahba E, Nasef MH (2012) An experimental study on improvement of Savonius rotor performance. Alex Eng J 51. https://doi.org/10.1016/j.aej.2012.07.003
7. Kang C, Liu H, Yang X (2014) Review of fluid dynamics aspects of Savonius-rotor-based vertical-axis wind rotors. Renew Sustain Energy Rev 33. https://doi.org/10.1016/j.rser.2014.02.011
8. Setiawan P (2019) Effect of a circular cylinder in front of advancing blade on the Savonius water turbine using transient simulation. Int J Mech MechatronS Eng 19:151–159
9. Nasef MH, El-Askary WA, Abdel-Hamid AA, Gad HE (2013) Evaluation of Savonius rotor performance: static and dynamic studies. J Wind Eng Ind Aerodyn 123. https://doi.org/10.1016/j.jweia.2013.09.009
10. Mohamed MH, Janiga G, Pap E, Thévenin D (2011) Optimal blade shape of a modified Savonius turbine using an obstacle shielding the returning blade. Energy Convers Manage 52:236–242. https://doi.org/10.1016/j.enconman.2010.06.070
11. Hassanzadeh AR (2013) Comparison of conventional and helical Savonius marine current turbine using computational fluid dynamics. World Appl Sci J 28:1113–1119. https://doi.org/10.5829/idosi.wasj.2013.28.08.1385
12. Alom N, Borah B, Saha UK (2018) An insight into the drag and lift characteristics of modified Bach and Benesh profiles of Savonius rotor. Energy Proc 144:50–56. https://doi.org/10.1016/j.egypro.2018.06.007
13. Saeed HH (2019) Numerical study of improving Savonius turbine power coefficient by various blade shapes. Alex Eng J 58:429–441. https://doi.org/10.1016/j.aej.2019.03.005
14. Sakti G (2018) Investigasi Kinerja Turbin Angin Savonius Dengan Metode Pengganggu Tipe I-65^0 Sebagai Kontrol Pasif Aliran Pada Sudu Returning-Investigation Of Savonius …. repository.its.ac.id
15. Hayashi T, Li Y, Hara Y (2005) Wind tunnel tests on a different phase three-stage Savonius rotor. JSME Int J Ser B 48:9–16. https://doi.org/10.1299/jsmeb.48.9

Local Stress Analysis at Critical Tubular Joint of Offshore Wind Turbine Jacket Structure

Muhamad Ferdian Hendrawan, Rudi Walujo Prastianto(✉), and Daniel Muhammad Rosyid

Department of Ocean Engineering, Institut Teknologi Sepuluh Nopember, Surabaya 60111, Indonesia
rudiwp@oe.its.ac.id

Abstract. Energy is a basic need that cannot be separated from human's life. Almost all sectors in life require energy to meet human needs. One of the renewable energy sources is wind energy. Offshore wind turbine is a renewable energy source that recently has developed rapidly. Wind turbines built offshore have several advantages over onshore. To support the offshore wind turbine, there are several structural options that can be used. The jacket structure can be used for medium to deep seas. The jacket structure that supports the offshore wind turbine is composed of steel tubular members that are joined together by welding will then receive various loads acting on the jacket structure that finally will dictate the integrity of the structure. The purpose of this study is to investigate local stress distribution of critical tubular joint of an offshore wind turbine jacket by using finite element model. Normally, there are two conditions of operation on wind turbine, which are operating when the environmental conditions are normal, and parked when the environmental conditions are extreme. The result showed interesting local stress distribution due to all three types of loading which are axial, in-plane bending moment (IPB), and out-of-plane bending moment (OPB) loading conditions.

Keywords: Renewable energy · Offshore wind turbine · Jacket structure · Tubular joint · Local stress analysis

1 Introduction

There are several types of offshore wind turbine support structures, including gravity base, monopile, tripod, jacket, and tension leg. In general, the selection of the type of support for offshore wind turbines depends on the depth, type of soil and environmental conditions in the form of wind, currents, and waves at the installation location. Jacket is a three-dimensional frame structure composed of steel tubular members [1]. The jacket is also the most economical structure in terms of the use of steel, but there are other factors such as storage, logistics, and installation.

To build an offshore wind turbine, a support structure is needed to hold the turbine above the water surface. There are two types of the support structures, namely floating and fixed. In his study, Wu [2] discusses the selection of the right support structure

I. Tolj et al. (eds.), *Recent Advances in Mechanical Engineering*, Lecture Notes in Mechanical Engineering, https://doi.org/10.1007/978-981-19-0867-5_29

for offshore wind turbines. The support structure of the wind turbine costs 20–30% of the total cost. Economically, depths above 50 m are not efficient enough to use fixed structures so that floating structures are more considered.

The jacket structure that supports the offshore wind turbine is composed of steel tubular members that are joined together by means of welding that created what so called tubular joints. These tubular joints will be the final elements of the structure in receiving various loads acting on the jacket structure. Environmental loads will produce cyclic loads on the tubular joints that in turn produce various forces such as axial, shear, and moment on the tubular joints that finally dictate the fatigue life of the jacket structure. Therefore, the objective of this study primarily is to analyze local stress distribution at the critical tubular joint for ensuring the integrity of the structure during its operation.

According to IEC 61400-1 standard, two loading conditions considered on the analysis of offshore wind turbine are operating under normal environmental condition and parking under extreme environmental conditions [3]. These two conditions are the most common for a structural analysis of the offshore wind turbine.

2 Literature Review

Researches on the offshore wind turbine jacket structure are still relatively rare even though offshore wind turbines have grown rapidly in the last few decades. There are several studies on static, dynamic, and local analysis of the offshore wind turbine jacket structure.

Study on a floating offshore wind turbine structure was done by Prastianto et al. [4] regarding the mooring analysis in operating condition due to heave, roll, and pitch motions. Lai et al. [5] conducted research on the dynamic analysis of the offshore wind turbine jacket structure of 5 MW OWTG FAST under extreme conditions by considering wind, wave, and current loads in the Taiwan Sea.

In their study, analysis was carried out on the strength of a tubular K-joint of the jacket structure under the loads of axial force, in-plane and out-of-plane moments. The purpose of this study was to obtain the largest load until the maximum joint deformation occurred [6]. Moreover, a numerical work on the stress distribution along weld toes of tubular KT and KDT joints under axial and in-plane bending moments loads was done [7].

3 Method of Analysis

3.1 Geometrical Modeling

Global stress analysis (a global model of the jacket structure can be seen in Fig. 1) by means of in-place analysis is done for two loading conditions (operating and parking) which will then produce stresses on each member of the jacket structure. The critical joint is taken from the tubular connection that has the member with the largest stress indicated by largest Unity Check (UC). In this case, the largest UC in both operating and parking conditions occurred at the same tubular connection.

Fig. 1. Model of the jacket structure

The tubular joint in the jacket of offshore wind turbine with largest stress is to be analyzed in this case. The data used in this local analysis are geometry of the members (see Table 1) and loading data (see Tables 2 and 3) generated from in-place analysis. The chord and brace of the joint have same material of ASTM-992 with the yield strength, tensile ultimate strength, and Young modulus of the material are 50 ksi (≈345 MPa), 65 ksi (≈448 MPa), and 29,000 ksi (≈199.95 GPa), respectively.

Table 1. Geometry of the tubular joint members

Joint member	Thickness (in)	Outside diameter (in)	Inside diameter (in)	Length (in)
Chord 1	1.25	38.5	36	560
Chord 2	1.25	38.5	36	560
Brace 1	0.75	23	20.5	438
Brace 2	0.75	23	20.5	438
Brace 3	0.75	23	20.5	420
Brace 4	0.75	23	20.5	420

The tubular joint consists of one chord with 4 braces and to be modeled based on geometry data by using AutoCAD 3D software (see Fig. 2). Next, the tubular joint model will be exported to finite element software for local stress analysis. In addition, the weld model also is added to the joint model, because the intersection between the two tubulars

Table 2. Load data from the in-place analysis (operating condition)

Member	Group	UC	Axial (kips)	IPB (kips.in)	OPB (kips.in)
Chord 1	JL	0.388	−6.74	1269.20	223.22
Chord 2	JL	0.411	−7.12	1394.25	164.64
Brace 1	XB	0.077	−0.65	303.59	−60.99
Brace 2	XB	0.063	0.06	430.74	−17.02
Brace 3	XB	0.079	0.35	−470.61	−142.01
Brace 4	XB	0.109	−0.41	−663.09	98.39

Table 3. Load data from the in-place analysis (parked condition)

Member	Group	UC	Axial (kips)	IPB (kips.in)	OPB (kips.in)
Chord 1	JL	0.487	−8.47	1541.00	237.93
Chord 2	JL	0.523	−9.03	1781.03	220.75
Brace 1	XB	0.101	−0.93	373.97	−78.10
Brace 2	XB	0.079	−0.11	529.59	−20.57
Brace 3	XB	0.099	0.57	−532.61	−178.88
Brace 4	XB	0.117	−0.32	−769.50	115.63

(brace and chord) will form intersection line, where the stress concentration occurs in the weld toes area along the intersection line. The weld size along the intersection satisfies the AWS specifications [8].

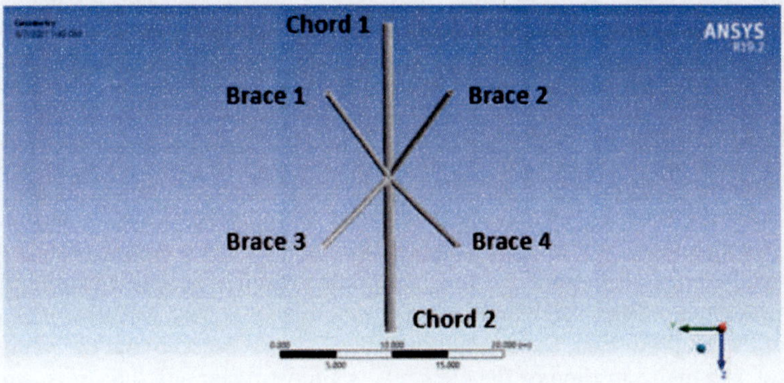

Fig. 2. The geometry of the tubular joint model

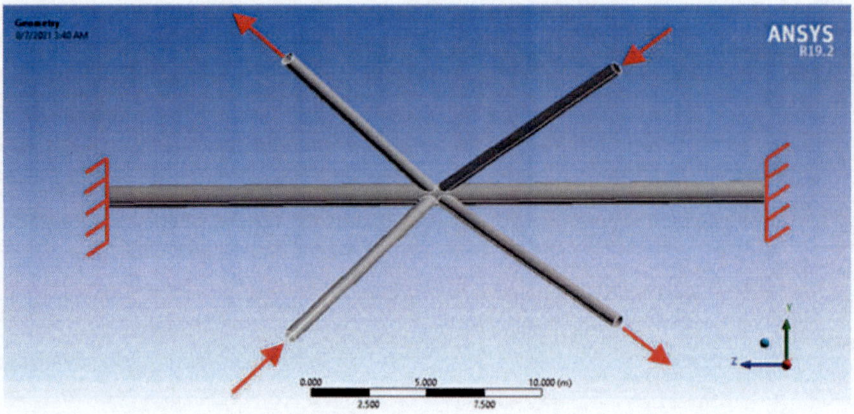

Fig. 3. Sample of the boundary and loading conditions for the model

3.2 Meshing and Analysis

Prior to do the local stress analysis, finite element meshing of the model should be generated as well as the support and loading conditions. The boundary condition used for the model is fixed support at both ends of the chord. Figure 3 shows the model with axial forces applied to all the braces. The element used for the model is a solid three-dimensional element with linear element type of tetrahedron. Meshing strategy for the model is performed with meshing size is made smaller (fine mesh) on regions around the brace-chord intersection line. While in regions far from the connection area the meshing size is made larger (coarse mesh) as can be seen in Fig. 4.

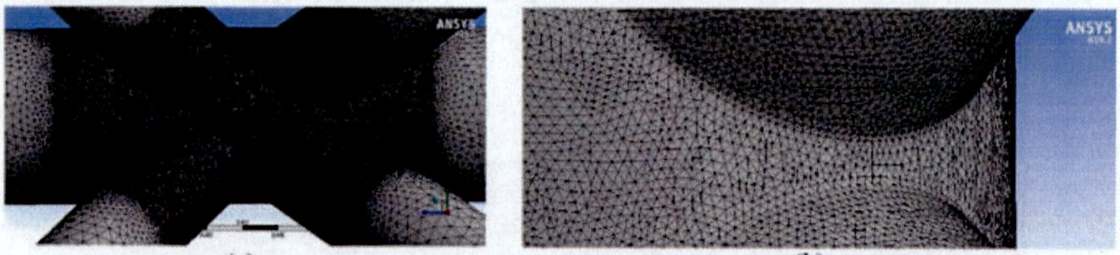

Fig. 4. **a** Meshing of the tubular joint model, **b** detail meshing around the weld area

For meshing sensitivity analysis, the size of meshing is iteratively changed until the stress generated at a particular location becomes almost constant. This sensitivity analysis should be done to obtain a consistent stress result especially near the weld toes area as presented in Fig. 5 and Table 4.

From the graph in Fig. 5 and result in Table 4, it can be seen that the stress within a range of 1 until 2.3 million of the elements is almost linear and the stress error value obtained is low (still within an acceptable range of error). Then, meshing of the model is selected with the number of elements of 2,301,662.

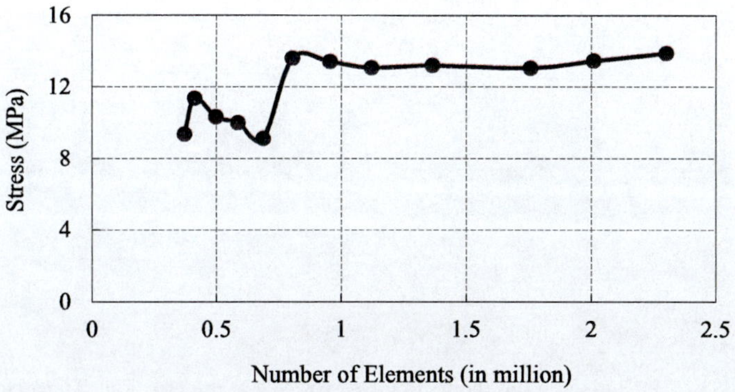

Fig. 5. A graph of the meshing sensitivity analysis of the model

Table 4. Result of the meshing sensitivity analysis

Element size (in)	Number of elements	Stress (Pa)	Stress (MPa)	Change (%)
0.8	371,402	9.34E + 06	9.34E + 00	
0.76	412,046	1.13E + 07	1.13E + 01	17.61
0.72	498,111	1.03E + 07	1.03E + 01	10.02
0.68	585,266	9.98E + 06	9.98E + 00	3.29
0.64	688,113	9.11E + 06	9.11E + 00	9.56
0.6	804,667	1.36E + 07	1.36E + 01	32.95
0.56	954,640	1.34E + 07	1.34E + 01	1.38
0.52	1,121,265	1.30E + 07	1.30E + 01	2.68
0.48	1,364,786	1.32E + 07	1.32E + 01	1.00
0.44	1,755,944	1.30E + 07	1.30E + 01	1.12
0.42	2,009,451	1.34E + 07	1.34E + 0 1	2.94
0.4	2,301,662	1.39E + 07	1.39E + 01	3.07

4 Results and Discussion

The local stress analysis is done to determine the stress distribution that occurred in the brace-chord intersection line. Among the loading conditions of axial load, in-plane bending moment (IPB), and out-of-plane bending moment (OPB) as outlined in Tables 2 and 3, the hot spot stress of 29.3 MPa occurred due to the IPB loading at the weld toe of Brace 4 as shown in Fig. 6.

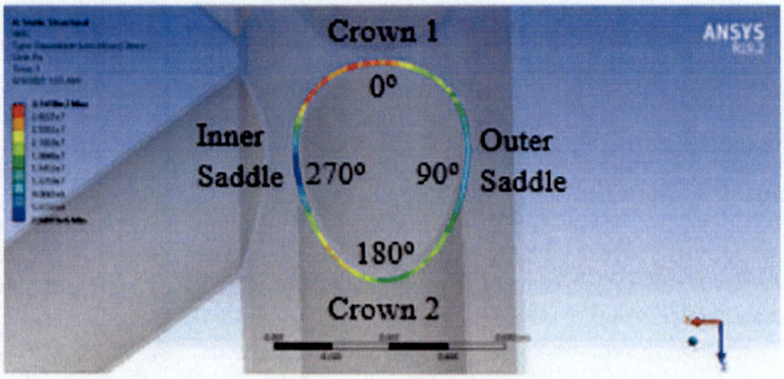

Fig. 6. Hot spot stress at the weld toe of Brace 4 at the position of Crown 1

Even though the stress distribution occurred in the same pattern but in parking condition the stresses at the joint were slightly larger than that in operation condition as shown in Fig. 7. In these cases, two lowest stresses of about 7.51 MPa and 4.42 MPa occurred at points of Outer Saddle (90°) and Inner Saddle (270°), respectively.

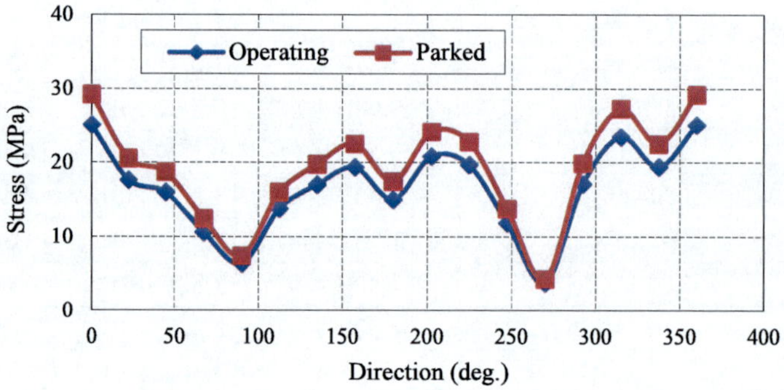

Fig. 7. Stress distribution at the Brace 4 for operating and parked conditions

Meanwhile, the lowest hot spot stresses occurred for the OPB load case. The hot spot stresses occurred in the weld toe of Brace 1 at locations of the Inner Saddle and Outer Saddle points as depicted in Fig. 8 for both cases of operation and parked conditions.

Figure 9 shows pattern of the stress distribution along the weld toe of Brace 1 for the cases of operation and parked conditions. The hot spot stresses of 5.81 MPa and 6.16 MPa occurred when the turbine was in operating condition. For the parked condition, the hot spot stresses occurred are 7.37 MPa and 7.81 MPa which are larger than the operating condition.

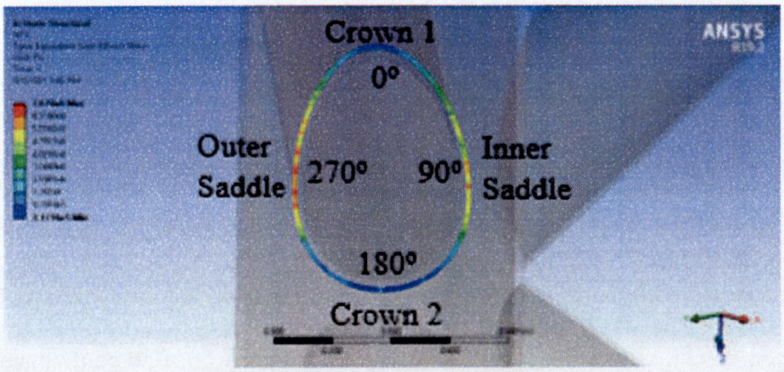

Fig. 8. Hot spot stresses at the weld toe of Brace 1 at the position of Inner Saddle and Outer Saddle points

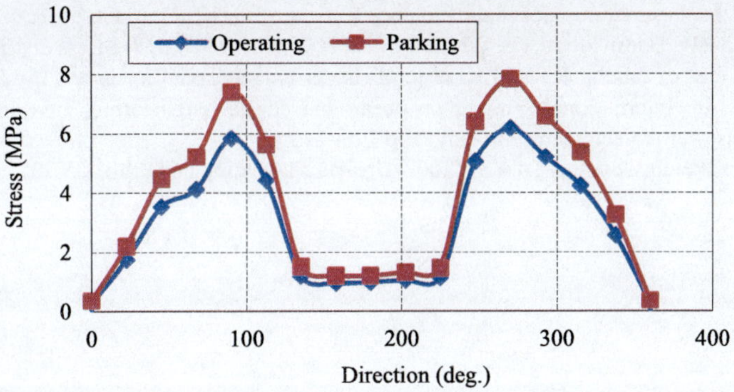

Fig. 9. Comparison of the stress distribution at the Brace 1 for operating and parked conditions

5 Conclusions

The wind has an impact on the strength of the jacket structure of the offshore wind turbine. In the present study, it was found that when the turbine is in parking conditions during a storm event the stresses occurred at the weld line of the tubular joint are always higher compared to the operating conditions at a normal sea environment.

In more detail, among the loading conditions that have been applied to the analysis which are axial, IPB, and OPB loadings, the largest hot spot stresses occurred for the case of IPB loading. Meanwhile, the lowest hot spot stresses occurred for the OPB load condition.

Moreover, different loading modes imposed on the braces caused different braces where the hot spot stresses occurred. In addition, each loading condition produced a different location where the hot spot stresses occurred at the weld toes along the length of the brace–chord intersection line. Overall, the hot spot stresses that occurred in the critical tubular joint are still under about 9% of the yield strength of the material which is 345 MPa.

References

1. Prastianto RW, Hadiwidodo YS, Fuadi IF, Naibaho MBYP, Rahardianto PN (2017) Stress concentration factor distribution of inclined brace in multiplanar offshore tubular double kt joints. In: International conference on maritime sciences and advanced technology. Denpasar, pp 480
2. Wu X (2019) Foundations of offshore wind turbines. In: Renewable and sustainable energy reviews, pp 379–393
3. International Electrotechnical Commission (IEC) (2005) Wind turbines standard: IEC 61400-1
4. Prastianto RW, Ramzi M (2020) Mooring analysis of SPAR type floating offshore wind turbine in operation condition due to heave, roll, and pitch motions. In: IOP conference series earth and environmental science. Makassar
5. Lai WJ et al (2016) Dynamic analysis of jacket substructure for offshore wind turbine generators under extreme environmental conditions. Appl Sci
6. Chandran AS, Arathi (2016) Static strength analysis of a tubular K-joint of an offshore jacket structure. In: International journal of science and research (IJSR). India, pp 708–713
7. Prastianto RW, Hadiwidodo YS, Widhestomo SW, Yazhahir R (2018) Stress distribution along the weld toes of tubular KT and KDT joints under balance axial loads and in-plane-bending moments. In: International seminar on ocean and coastal engineering, environmental, and natural disaster management. Surabaya, pp 170–175
8. American Welding Society (AWS) (2002) Structural welding code. In: AWS D1.1. USA

Numerical Study of Temperature Distribution and CO_2 Forming of Gaseous Tangential Firing Furnace with O_2 Level Variations in Air Combustion

Muhammad Deni Indrawan[1(✉)], Bambang Sudarmanta[2], and Atok Setiyawan[2]

[1] Graduate Program of Mechanical Engineering, Institut Teknologi Sepuluh Nopember, Surabaya 60111, Indonesia
m.deni@ptpjb.com

[2] Department of Mechanical Engineering, Institut Teknologi Sepuluh Nopember, Surabaya 60111, Indonesia

Abstract. Oxyfuel combustion is considered to be one of the most effective methods to improve thermal efficiency for high temperature processes such as boiler water of steam power plant. In the oxyfuel combustion, the combustion characteristics such as temperature and CO_2 emission are largely influenced by the process parameters such as the mixing between fuel gas and oxygen and the nitrogen content in the oxidant and fuel. A numerical simulation of 200 MW gaseous tangential firing boiler with varying O_2 content 25%, 30%, 35% is presented. Simulations demonstrate the characteristics of temperature distribution and CO_2 form rate in the furnace. The results show that by increasing O_2 percentage in air combustion there is an increase in temperature level and CO_2 forms rate by 6.8% and 13.92% in Oxy25, 16.9% and 15.02% in Oxy30, and 19.14% and 15.72% in Oxy35 compared to airfuel combustion.

Keywords: Oxyfuel · Gas fired furnace · CFD

1 Introduction

Climate change due to the increase in greenhouse gases caused by industry has made scientists work hard to find methods that can reduce the increasing amount of CO_2 gas in the atmosphere. One method used is oxyfuel combustion where the amount of O_2 content in the combustion air is increased so that more CO_2 gas as a result of combustion can be captured and stored so that it does not escape into the atmosphere as well-known as Carbon Captured and Storage (CCS) technology [1]. Fande and Joshi [2] conducted research in tangential firing boiler using CFD software to determine the temperature, velocity and thermal characteristics of 660 MW boiler firing medium coal. Results appeared relations among temperature, O_2, and CO_2 mass fraction clearly demonstrated based on calculated distributions and showed good agreement with measured and reported data. Azeez et al. [3] used CFD simulations on combustion process of 120 MW tangential furnace

I. Tolj et al. (eds.), *Recent Advances in Mechanical Engineering*, Lecture Notes in Mechanical Engineering, https://doi.org/10.1007/978-981-19-0867-5_30

which has three different levels burners. The temperature level has lower value at the bottom area and then increase at the combustion area and approximately constant at the upper zone of boiler. Himachandra and Prasad [4] investigated pulverizer coal boiler. The tangential firing boiler generates a swirling fire ball at the center of the combustion zone which is stronger at the lower level of boiler height. The maximum temperature of 1850 K is found to exist at the central part of the furnace. Ben Mansour et al. [5] used CFD simulation on oxyfuel combustion in a 600 MW gaseous fuel tangentially fired boiler. Compared to airfuel combustion, oxyfuel consisting of O_2 and CO_2 mixed will lower the temperature with increasing CO_2 content due to higher heat capacity.

This research will study the characteristics of combustion in the gas fired furnace, not using pure O_2 as oxyfuel, but varying O_2 content in air combustion (oxidant) rates of 25%, 30%, 35% to captures the temperature distribution also CO_2 forming.

2 Methodology

2.1 Mathematical Model

The mathematical model applied is based on the commercial CFD (Fluent) where the governing equations for the gas flow in tensor notation are:

Continuity:

$$\frac{\delta(\rho Y_i)}{\delta t} + \frac{\delta(\rho U_j Y_i)}{\delta x_j} = -\frac{\delta(J_i)}{\delta x_j} + R_i + S_i \tag{1}$$

Momentum:

$$\rho \frac{DU_i}{Dt} = -\frac{\delta \rho}{\delta x_i} + \frac{\delta}{\delta x_j}\left((\mu + \mu_t)\left(\frac{\delta U_i}{\delta x_j} + \frac{\delta U_j}{\delta x_j}\right)\right) + S_f \tag{2}$$

Energy:

$$\frac{\delta \rho H}{\delta t} - \frac{\delta P}{\delta t} + \frac{\delta \rho U_j H}{\delta x_j} = \frac{\delta}{\delta x_j}\left(\lambda \frac{\delta T}{\delta x_j} + \frac{\mu_t}{\sigma_t}\frac{\delta H}{\delta x_j}\right) + S_h \tag{3}$$

The continuity equation predicts the local mass fraction of each species, Y_i, in a mixture. R_i is the net rate of production of species by chemical reaction, and S_i is the rate of creation by addition from the dispersed phase. J_i is the diffusion flux of species i. The eddy-dissipation model is used to calculate R_i.

2.2 Furnace Geometry

The furnace has following dimensions: X = 10,810 mm, Y = 20,535 mm, and Z = 10,098 mm and a volume of 2,430 m^3. The tangential angles of combustion burners are 38°/52° at corner A & C and 43°/47° at corner B & D with tilting angle = 0° as shown in Fig. 1, and each corner has 4 elevation of burners as shown in Fig. 1a.

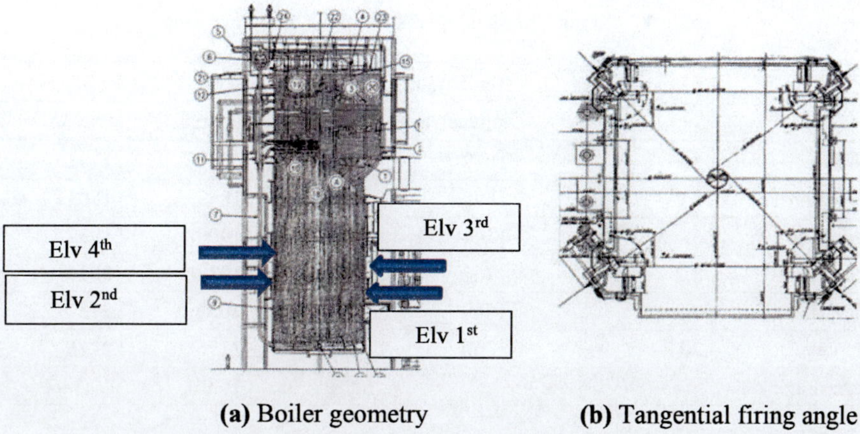

Fig. 1. **a** Boiler geometry. **b** Tangential firing angle

2.3 Boundary Conditions

Boundary conditions at the inlet are mass flow, temperature, viscosity, and density. The mathematical model based on the commercial CFD code where the gas flow is described by the mass, momentum, and energy equations. The standard k-ε turbulence model is used to simulate the turbulence flow and combustion process, non-premix model as species model and also probability density function (PDF) are used. The combustible of the fuel element and compounds in the fuel with all the oxygen requires high temperatures to ignite the constituent, mixing (turbulence) to provide intimate oxygen-fuel contact and sufficient time to complete process. In this study, the fuel and the oxygen entered furnace separately or so called non-premix combustion type. The equation of stoichiometric equation combustion for hydrocarbon can be written as:

$$C_mH_n + \left(m + \frac{4}{n}\right)O_2 \rightarrow mCO_2 + \left(\frac{n}{2}\right)H_2O \tag{4}$$

The amount of combustion air (oxidant) demand is carried out by calculating the conditions when operating normally (airfuel case at 200 MW) and then converting it into the oxyfuel cases shown in Table 1, given by following Eq. (4):

Table 1. Fuel gas and oxidant demand

Case	% O_2	Unit	Fuel gas rate	O_2 rate	Air flow (Oxidant) rate
Airfuel	21	kg/s	9,07	36,4	173,33
Oxy25	25	kg/s	9,07	36,4	145,59
Oxy30	30	kg/s	9,07	36,4	121,33
Oxy35	35	kg/s	9,07	36,4	103,99

Table 2. Average temperature distributions on burner levels

Z-Center	Elevation (m)	Temperatur (K)			
		Airfuel	Oxy25	Oxy30	Oxy35
Bottom	1.5	1652	1727	1875	1733
Elevation 1	5.2	1678	1699	1849	1801
Elevation 2	6.4	1744	1808	1900	1937
Elevation 3	8.2	1766	1861	1934	2014
Elevation 4	9.8	1849	1969	2082	2132
SH Level	20	1860	1988	2176	2216
Percentage			6.8%	16.9%	19.14%

3 Result and Discussion

3.1 Temperature Distributions

Figure 2 reveals temperature distribution contours of the all four cases at the vertical plane (Z-center). Generally, the highest temperature values have appeared above the burner area and slightly decrease at the middle of the furnace. The bottom area has lowest temperature value due to distance issue form the burners. In the airfuel combustion case, the combustion gases have average temperature of 1800 K, while in the Oxy35 case, the temperature increases rapidly to about 2100 K. This increasing in temperature values is mainly due to the increase of O_2 percentages in the mixture (Fig. 3).

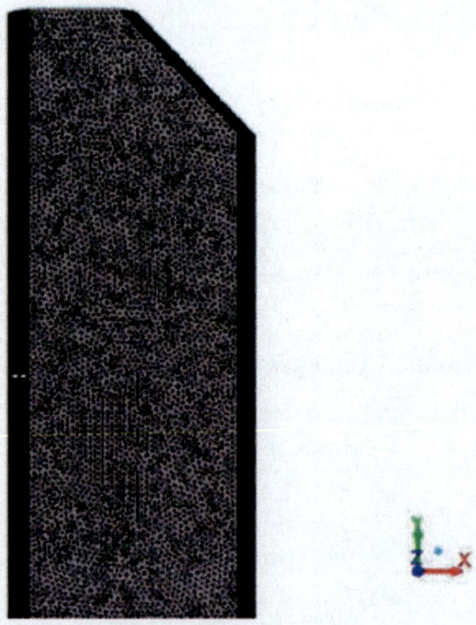

Fig. 2. Furnace model

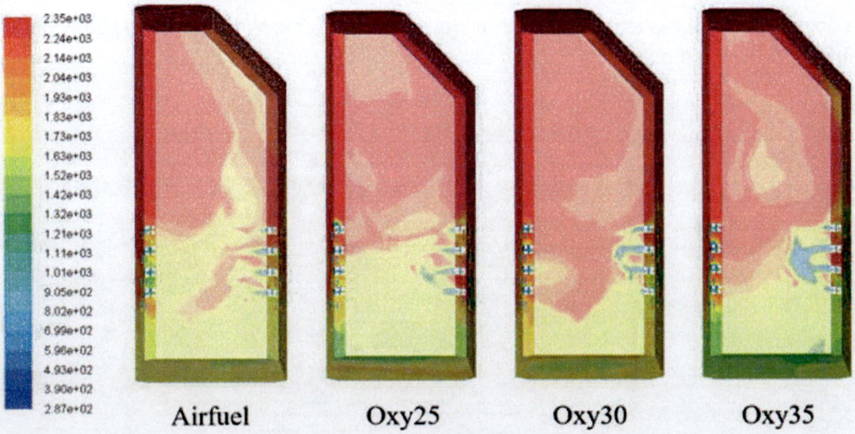

Fig. 3. Temperature distributions at Z-center view

The average temperature increase of Oxy25, Oxy30, and Oxy35 cases compare to Airfuel combustion 6.8%, 16.9%, and 19.14% is shown in Table 1 (Fig. 4)

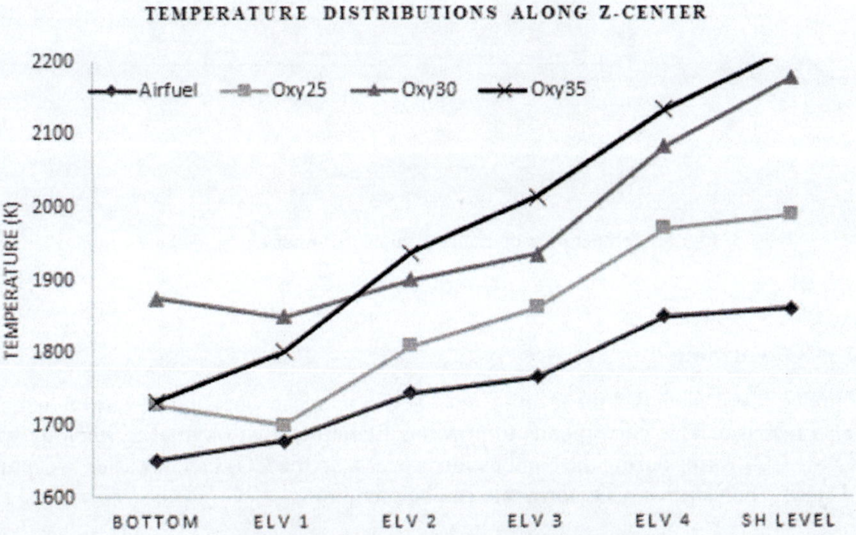

Fig. 4. Temperature values along Z-center

For all combustion cases, the gaseous fuel will dissipate gradually as it moves away from the burner nozzle therefore the highest concentration is located near the burner. The tangential angle of burners made fire ball in the middle of the burner but the highest temperature is beneath near the wall as shown in Fig. 5. Due to the tangential angles of the nozzle burners (as shown earlier in Fig. 1b), the flame will approach the walls and resulting in low isothermal circles in the center of the furnace (Fig. 6).

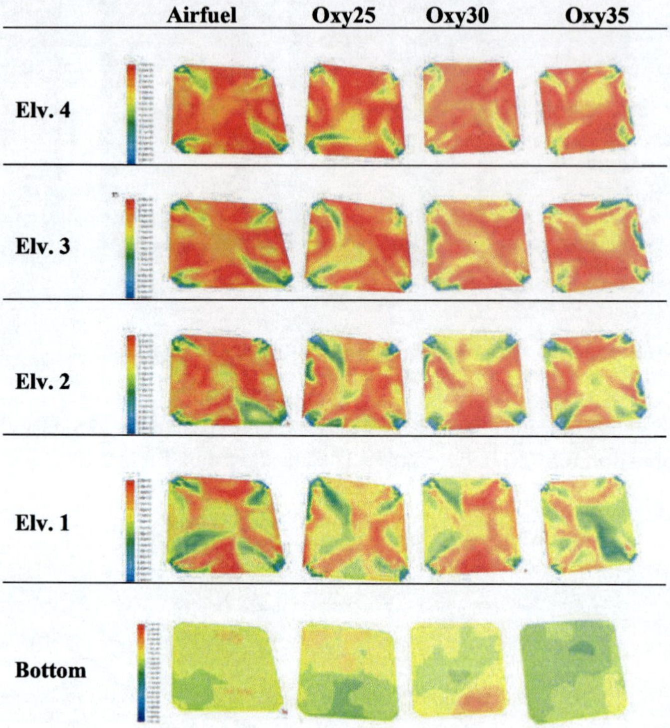

Fig. 5. Temperature contours from each burner along Y-plane

3.2 CO_2 Forming

From Fig. 5, it is clear that the forming of CO_2 will increase when the O_2 rate on oxidant is also increase. This corresponds to previous literatures that oxyfuel technology will increase CO_2 forms during the combustion process. So the CO_2 can be easier to capture and store. The value on CO_2 form rate can be seen in Fig. 7.

4 Conclusion

The results show that by increasing O_2 percentage in air combustion there is an increase in temperature level and CO_2 forms rates by 6.8% and 13.92% in Oxy25, 16.9% and 15.02% in Oxy30, and 19.14% and 15.72% in Oxy35 compared to airfuel combustion. High values of temperature in the Oxy25 case can help in increasing the efficiency of the burner by increasing the heat characteristics. However, this is applicable for the highly sustainable materials of which the combustors are made.

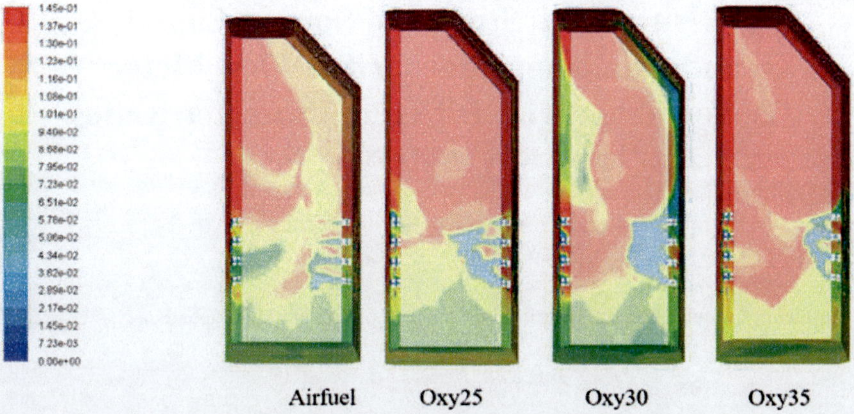

Fig. 6. CO_2 distributions along Z-center

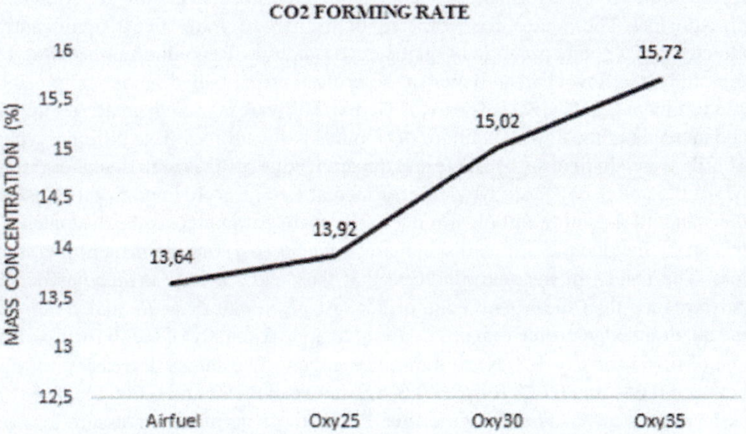

Fig. 7. CO_2 Forming rate of air fuel and oxyfuel combustion

References

1. https://www.climatecouncil.org.au/
2. Fande A, Joshi PV, Gabhane S, dan Sahu MK (2014) CFD Investigations of combustion in 660 MW tangential fired boiler. Int J Eng Res Technol (IJERT) 3(1)
3. Azeez W, Harsil H, Yusoff Z (2015) CFD investigation of effect of tilt angle to flow and combustion in 120 MW Gas-fired full scale industrial boiler. Int J Sci Technol Res 4(12)
4. Himachandra JJ, Prasad LL (2017) CFD analysis of 210 MW tangential fired boiler. Int J Sci Eng Technol 6(1)
5. Ben Mansour et al (2017) Oxy-fuel combustion in a 600 MW gaseous fuel tangentially fired boiler, pubs.acs.org/EF

Numerical Studies of Square Edge and Quadrant Edge Orifice Flow Meter Performance with Different Diameter Ratios and Reynolds Numbers

Sutardi(✉) and L. Z. Reinaldhy Aditya

Department of Mechanical Engineering, Institut Teknologi Sepuluh Nopember, Surabaya 60111, Indonesia
sutardi@me.its.ac.id

Abstract. Orifice plate flow meter still plays significant role in industrial process flow measurements and has many variations in shapes and dimensions. Each orifice type has different performances such as coefficient of discharge and permanent pressure loss. The value of the coefficient of discharge of the orifice is significantly affected by its geometry, ratio of diameter (β) and flow Reynolds number (Re). In this study, the flow characteristics in the orifices were studied numerically using Ansys Fluent 6.3.26 CFD software. A Gambit software is used to create geometry and meshing of the flow domain for 3D models of the orifices. Two different types of orifice are studied: square edge and quadrant edge orifices with diameter ratios (β) of 0.5 and 0.7. The study was performed at $Re = 5 \times 10^3$ and 5×10^4 using the standard *k*-epsilon turbulent model. The main parameters to be evaluated in this study include the orifice discharge coefficient (C_d) and permanent pressure loss. The results of the study indicate that the quadrant edge orifice has better performance than the square edge orifice. At given values of Re and β, C_d of the quadrant edge orifice remains constant at approximately of 0.746 for $\beta = 0.5$ and of 0.765 for $\beta = 0.7$. Next, the square edge orifice shows decreasing in C_d between 0.644 and 0.628 for $\beta = 0.5$, and between 0.693 and 0.646 for $\beta = 0.7$ as Re increases. Quadrant edge orifice has a lower permanent pressure loss of approximately 3.25% than that for square edge orifice.

Keywords: Orifice flow meter · Square edge · Quadrant edge · Diameter ratio · Coefficient of discharge · Permanent pressure loss

1 Introduction

In the process industries, the flow meter is a vital operational instrument. One type of the flow meter that is widely used in industry is the orifice plate. The orifice plate is generally made with a sharp angle and is installed concentrically in pipe. The orifice plate has many advantages, such as geometrically leak proof, economical, no moving parts, and suitable for handling large flow rates [1]. Orifice plates can also be used as energy dissipating tool in the flow [2]. However, the level of measurement accuracy of

I. Tolj et al. (eds.), *Recent Advances in Mechanical Engineering*, Lecture Notes in Mechanical Engineering, https://doi.org/10.1007/978-981-19-0867-5_31

the orifice plate flow meter is low [3]. The general geometry of the orifice plate can be obtained in ISO 5167-2 [4]. There are also many shapes of orifice plate application in the engineering, such as concentric shape with square edge, quadrant edge, and conical.

The main concern of the use of the orifice plate as the flow measuring device is the high pressure loss due to the flow separation from the sharp edges of the plate. The pressure downstream of the orifice drop to the minimum at the vena contracta, and it is expected to recover as the flow goes farther downstream. In practical, the pressure recovery is not as high as expected. The non-recovery of this pressure is called permanent pressure loss. The amount of permanent pressure loss is a major concern in the selection of the geometry and diameter ratio of the orifice plate.

Hollingshead [5] conducted numerical study on several types of differential pressure flow meters, one of which is the standard concentric orifice plate (i.e. square edge) to evaluate the discharge coefficient (C_d) of the flow meter as a function of flow Reynolds number (*Re*). The study showed that at *Re* between 100 and 300, the value of C_d increases, while at $Re > 300$, the value of C_d continues to decrease as *Re* increases. Britton and Stark [6] studied experimentally the quadrant edge orifices and square edge orifices at low *Re*, small diameter ratios between 0.25 and 0.54, and pipe diameters of 3 in and 4 in. In general, the quadrant edge orifice results in almost constant C_d, while C_d for the square edge orifice is significantly affected by the flow Reynolds number.

The main purpose of the present study is to evaluate the characteristics of the orifice flow meter using numerical simulation. Some parameters to be evaluated include discharge coefficient (C_d) as a function of flow Reynolds number and diameter ratio (β) of the orifice plates and the effect of the edge shape of the orifice plate to the C_d. Another flow parameter to be evaluated is the permanent pressure loss as a function of the orifice plate geometry and flow Reynolds number.

2 Methods

This study was conducted for orifice plate types: square edge orifice and quadrant edge orifice. The geometries of the orifice plates used in this study are shown in Figs. 1 and 2. Some definitions shown in Figs. 1 and 2 include $D = 2R =$ pipe diameter, $d =$ diameter of the orifice plate, and $\beta = d/D =$ orifice plate diameter ratio.

This study uses Ansys Fluent 6.3.26 commercial software along with the Gambit 2.4.6 software to create geometry and meshing of the flow simulation domain. The 3D CFD model is used to analyse the flow characteristics in the *x*, *y*, and *z* coordinates. This research was conducted using several orifice parameters including orifice plate geometry, orifice diameter ratio (β), and the flow Reynolds number (*Re*) as shown in Table 1. The study was performed at $Re = 5 \times 10^3$ and 5×10^4 using the standard *k*-epsilon turbulent model.

This simulation domain of present study is shown in Fig. 3 with a value of *L*/*D* ratio = 20 [7]. All boundary conditions for the simulation are also shown in Fig. 3. The lengths of upstream and downstream pipes are based on the study of Shah et al. [7]. The meshing of the square edge orifice plate is shown in Fig. 4, while the meshing of the side view of the quadrant edge orifice plate is shown in Fig. 5. Grid independency test performed by Reinaldhy [8] suggested that the number of grids used in this simulation

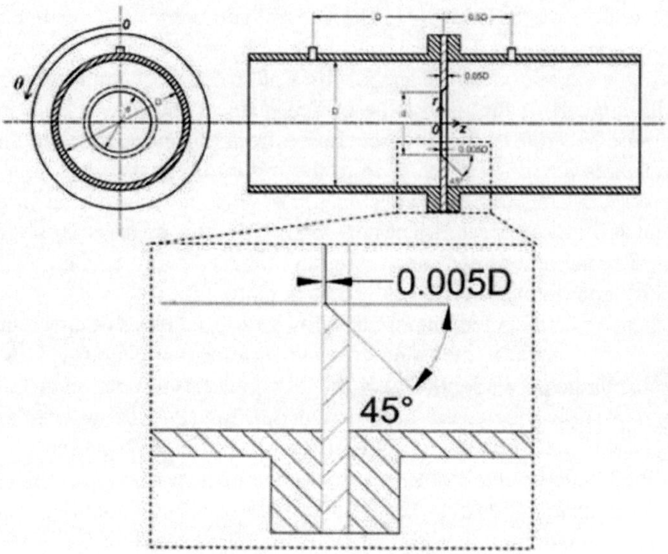

Fig. 1. Geometry of square edge orifice

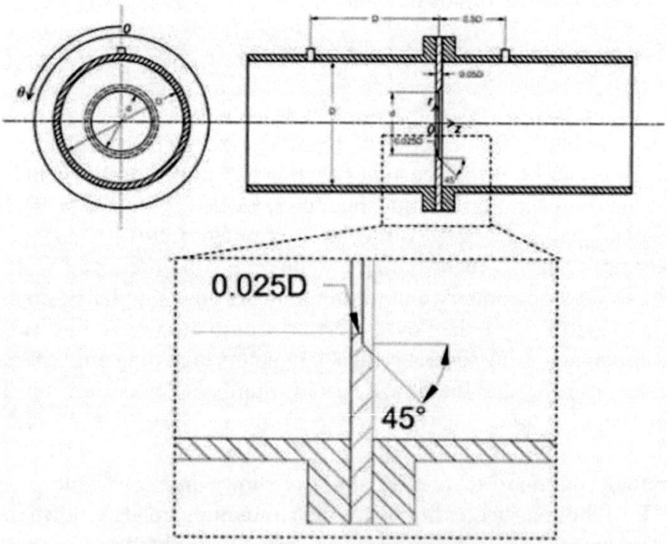

Fig. 2. Geometry of quadrant edge orifice

Table 1. Research parameters for orifice plate study

Orifice	Diameter ratio (β)	*Re*
Square edge	0.5	5×10^3
		5×10^4
	0.7	5×10^3
		5×10^4
Quadrant edge	0.5	5×10^3
		5×10^4
	0.7	5×10^3
		5×10^4

was more than 1.2×10^6. Next, based on his study, the y^+ values were in the range of 15 and the convergence criteria based on the residual level was less than 1×10^{-5}. The parameters used in this study are shown in Table 2.

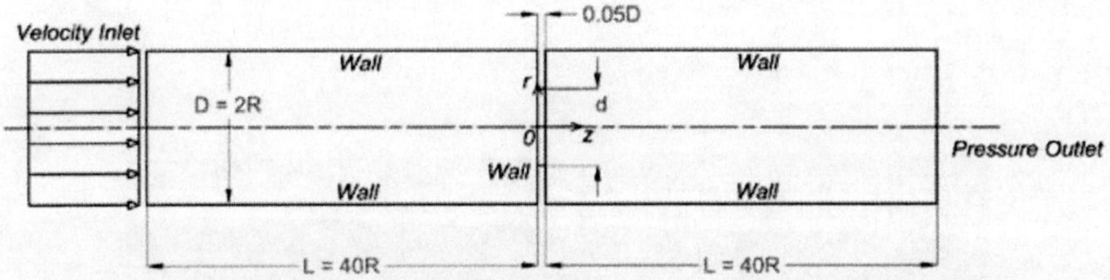

Fig. 3. Simulation domain for orifice plate in CFD analysis with all dimensions

3 Results and Discussion

Figure 6 shows the velocity at the pipe centreline at $Re = 5 \times 10^3$. The maximum centreline velocity for the square edge orifice is higher than that for the quadrant edge orifice for both β (0.5 and 0.7). The quadrant edge orifice results in approximately of 12.2% maximum centreline velocity lower than that for the square edge orifice. This is probably due to the flow separation at the neck of the quadrant edge orifice is delayed up to the end of the orifice neck because of its shape, the bevel-like geometry. Delayed flow separation for the case of quadrant edge orifice causes the flow has enough momentum so that the area of the *vena contracta* is larger than that for the square edge orifice.

Figure 7 shows velocity contour of flow passing through the square edge and quadrant edge orifices for $\beta = 0.5$ and 0.7 and at the flow Reynolds number (*Re*) $= 5 \times 10^3$. The flow structures indicate that there is high velocity flow region at the centre part of the pipe, where the square edge orifice with $\beta = 0.5$ generates the maximum absolute velocity compared to the other orifice geometries. The values of these maximum velocities in

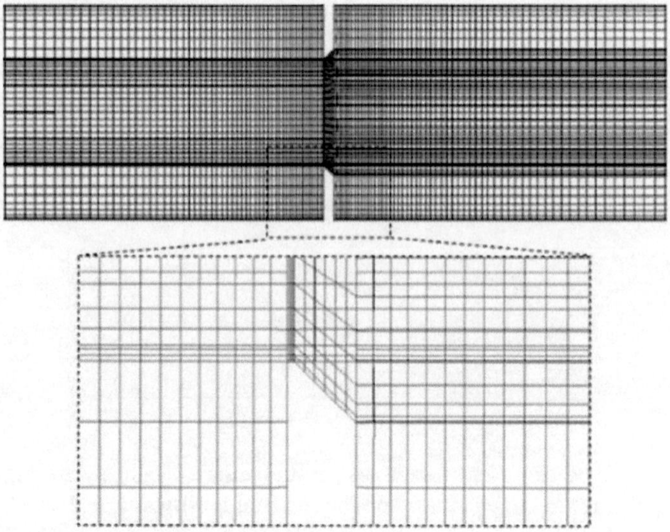

Fig. 4. Meshing of the square edge orifice plate ($\beta = 0.5$)

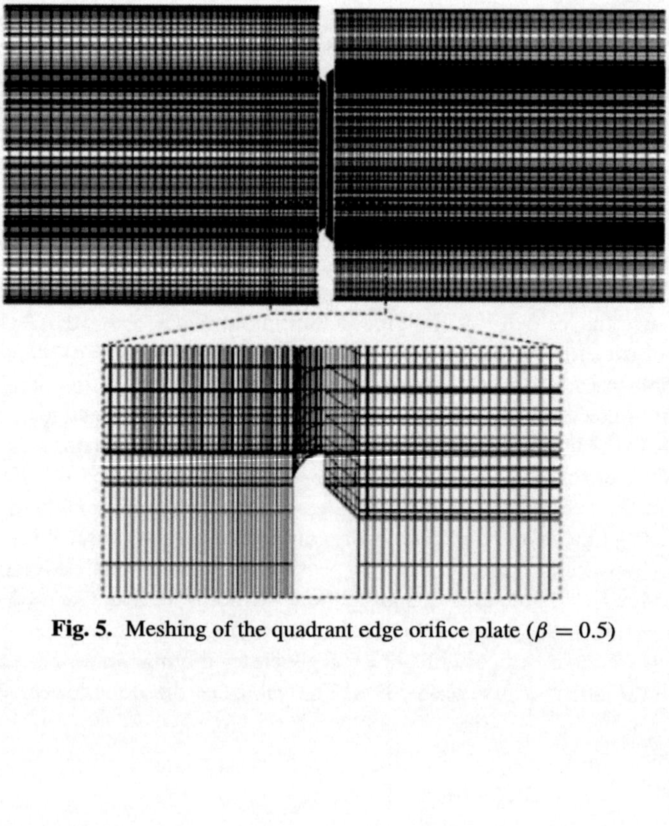

Fig. 5. Meshing of the quadrant edge orifice plate ($\beta = 0.5$)

Table 2. Simulation parameters

Re	*D* (mm)	*L* (mm)	Velocity inlet (m/s)	Pressure outlet (Pa (g))
5×10^3	12.7	254	0.394	0
5×10^4	12.7	254	3.941	0
	38.1	762	1.315	0

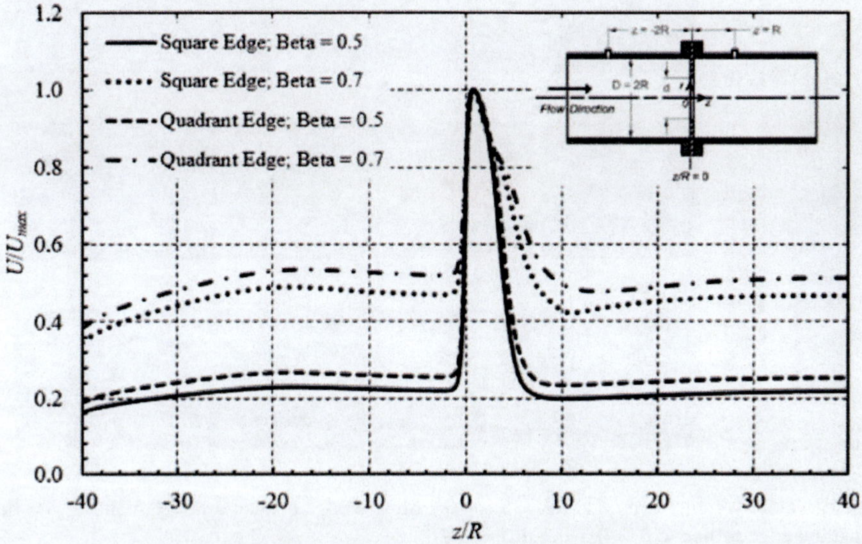

Fig. 6. Centreline velocity distribution for square edge and quadrant edge orifices at $Re = 5 \times 10^3$

terms of non-dimensional parameters are shown in Fig. 6, where the locations of these maximum velocities are approximately at $z/R = 0.85$ (or $z = 0.425D$) for all geometries.

The pressure distribution for the square edge orifice is different from that of the quadrant edge orifice, as shown in Fig. 8 for $Re = 5 \times 10^3$ and Fig. 9 for $Re = 5 \times 10^4$. The quadrant edge orifice produces a maximum pressure drop of approximately 25.7% lower compared with that of for the square edge orifice for both β (0.5 and 0.7) and both *Re* (5×10^3 and 5×10^4). This is probably that the quadrant edge orifice has a lower maximum centreline velocity than the square edge orifice. Figure 9 also shows that the difference in the pipe diameter or velocity at the inlet section does not affect the pressure distribution through the orifice as long as the Reynolds number (*Re*) is kept constant. The difference in the pipe diameter or velocity that performed in this study is just merely based on the consideration of practical conditions of the average velocity of approximately is 1.32 m/s and the pipe diameter is 38.1 mm (~1.5 in).

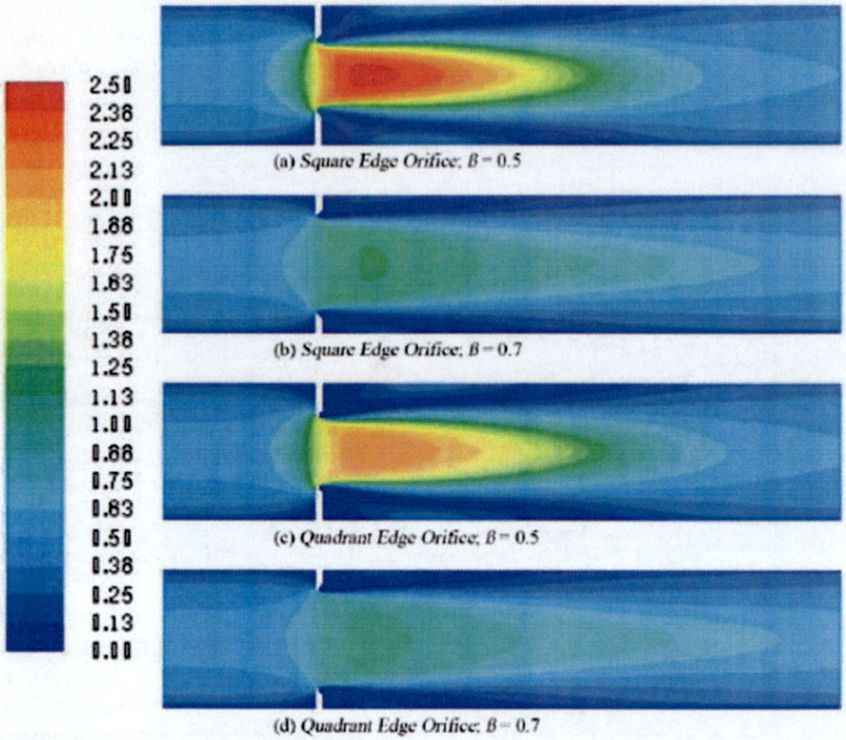

Fig. 7. Velocity contour (m/s) at $Re = 5 \times 10^3$. Square edge orifice: **a** $\beta = 0.5$ and **b** $\beta = 0.7$; Quadrant edge orifice: **c** $\beta = 0.5$ and **d** $\beta = 0.7$

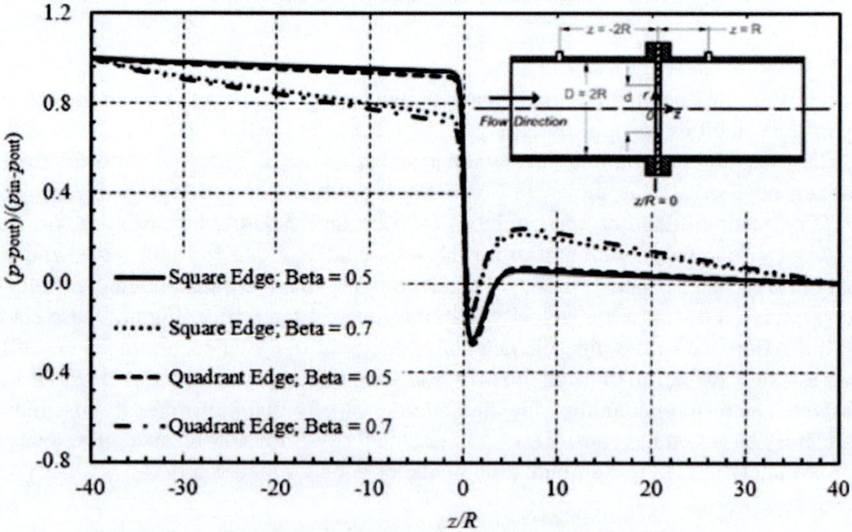

Fig. 8. Distribution of orifice centreline static pressure at $Re = 5 \times 10^3$

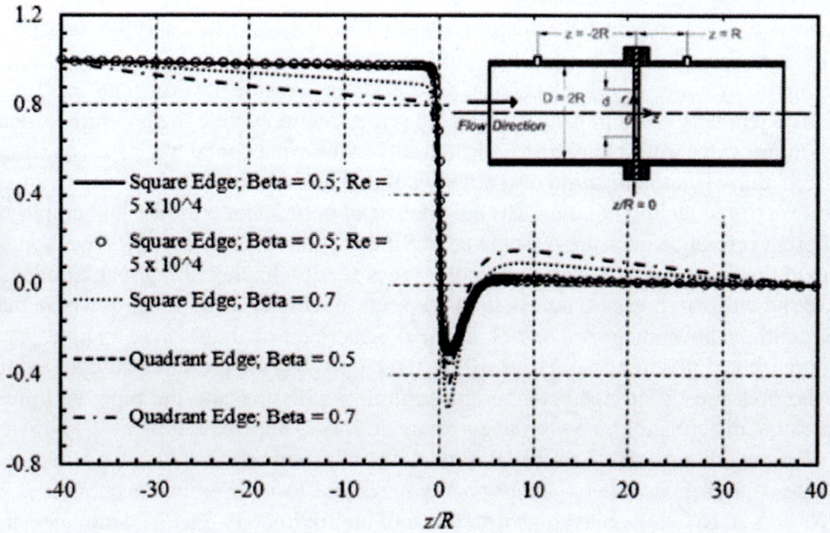

Fig. 9. Distribution of orifice centreline static pressure at $Re = 5 \times 10^4$

Table 3 shows the results of permanent pressure loss for each orifice geometry. The permanent pressure loss is expressed as a percentage of the maximum pressure drop. The numerical values shown in Table 3 were calculated using the following equation:

$$C_d = \frac{\text{Permanent pressure loss}}{\text{Maximum pressure drop}}(\%) = \frac{p_{\text{upstream }(-2z/R)} - p_{\text{recovery point}}}{\Delta p_{\max}} \times 100\%, \quad (1)$$

Table 3. Permanent pressure loss with the centreline and wall pressures as the reference value

Orifice	β	*Re*	Permanent pressure loss/Maximum pressure drop (%)		
			Miller [9] (f (β))	Centreline	Wall
Square edge	0.5	5×10^3	73.0	74.7	73.4
		5×10^4		74.6	73.6
	0.7	5×10^3	52.2	59.6	56.1
		5×10^4		55.2	52.9
Quadrant edge	0.5	5×10^3	–	71.5	70.3
		5×10^4		71.0	70.0
	0.7	5×10^3	–	56.3	52.9
		5×10^4		52.6	49.7

where $p_{upstream\ (-2z/R)}$ is the static pressure at $z/R = -2$, $p_{recovery\ point}$ is the static pressure at recovery point, downstream of the orifice, $\Delta p_{max} = p_{upstream\ (-2z/R)} — p_{vena\ contracta}$ = maximum pressure drop, and $p_{vena\ contracta}$ is static pressure at *vena contracta.*

The recovery point indicated in Eq. (1) is the location where the pressure raise to the maximum value after being drop considerably at the vena contarcta. Based on the study of Reinaldhy [8], the location of recovery point is just a few distance downstream of the location of reattachment point. The calculation of permanent pressure loss employs two different reference pressures: (i) the centreline pressure and (ii) the pipe wall pressure. Theoretically, the flow for at a particular cross section having straight streamlines will have the uniform pressure across the flow section. However, in the case where there is a secondary flow, such as a vortex at the downstream of the orifice, a slight pressure difference at a given cross section will be developed. At the present study, the existence of the pressure difference between the centreline pressure and the pipe wall pressure yields the difference in the permanent pressure loss is approximately of 2.1%.

Permanent pressure loss for the quadrant edge orifice is lower than that for the square edge orifice. The largest permanent pressure loss difference occurs for $\beta = 0.5$ at $Re = 5 \times 10^4$ with an average difference of approximately 3.61%, while the smallest permanent pressure loss difference occurs for $\beta = 0.7$ at $Re = 5 \times 10^4$ with a difference of approximately 2.96%. This happens because the flow separation at the quadrant edge orifice can be able to be delayed as it has been discussed regarding to the pressure distribution (Figs. 8 and 9).

The values of the permanent pressure loss are also compared with the results Miller [9] based on the empirical correlation that correlates the permanent pressure loss and the diameter ratio (β). Miller [9] suggested that the permanent pressure loss is a function of β, while *Re* does not play a significant role to the permanent pressure loss across the orifice. The present simulation results also show that the difference in the permanent pressure loss based on analytical calculations is approximately of 0.29% and at $\beta = 0.7$ at present study is approximately of 3.61%.

Table 4 shows discharge coefficient (C_d) of each orifice geometry. The table also presents comparison with data based on analytical calculations for standard orifices or square edge orifices [4]. Present C_d calculation uses following equation:

$$C_d = \frac{Q\sqrt{1-\beta^4}}{A_0\sqrt{\frac{2\Delta p_{\max}}{\rho}}}, \tag{2}$$

where Q is the fluid volumetric flowrate, A_0 is the cross sectional area of main pipe, and ρ is the fluid density. The average present C_d results is approximately within 2% difference compared to that of obtained from the calculation of ISO [4].

Table 4 also shows that the increase in *Re* causes C_d of the square edge orifice decreases. On the contrary, the increase in *Re* does not affect significantly C_d of the quadrant edge orifice where C_d is relatively constant at approximately of 0.8% difference. For both the square edge and quadrant edge orifices, C_d increases as the diameter ratio (β) increases for the two *Re* investigated in this study. Present simulation results also show that the quadrant edge orifice exhibits larger discharge coefficient (C_d) compared

Table 4. Discharge coefficient (C_d) with centreline and wall pressures as reference values

Geometry	β	*Re*	Discharge coefficient (C_d)		
			ISO (−2z/R; 1z/R) [4]	Centreline	Wall
Square edge orifice	0.5	5×10^3	0.630	0.644	0.639
		5×10^4	0.608	0.633	0.628
	0.7	5×10^3	0.670	0.693	0.673
		5×10^4	0.618	0.660	0.646
Quadrant edge orifice	0.5	5×10^3	–	0.747	0.741
		5×10^4	–	0.751	0.746
	0.7	5×10^3	–	0.781	0.757
		5×10^4	–	0.772	0.751

with C_d of the square edge orifice. The discharge coefficient (C_d) of the quadrant edge orifice is in average 16% larger that of C_d of the square edge orifice.

4 Conclusion

The simulation study of the flow passes through two different geometry of orifice plates has been done. The two different geometry of orifice plates are the square edge and quadrant edge orifices with a diameter ratio (β) of 0.5 and 0.7. Two Reynolds numbers were used in the present study: $Re = 5 \times 10^3$ and 5×10^4. Based on the present study, some conclusions can be drawn as follows:

(1) Permanent pressure loss of the quadrant edge orifice is lower than that of the square edge orifice. The largest permanent pressure loss difference occurs for $\beta = 0.5$ and $Re = 5 \times 10^4$ with an average difference of approximately 3.6%. Aside, the smallest permanent pressure loss difference occurs for $\beta = 0.7$ and $Re = 5 \times 10^4$ with an average difference of approximately 2.96%. In average, the permanent pressure loss of the quadrant edge orifice is about 3.25% lower than that of the square edge orifice.

(2) The increase in the flow Reynolds number causes the discharge coefficient (C_d) of the square edge orifice become smaller. On the contrary, the greater the Reynolds number does not change significantly affect the discharge coefficient (C_d) of the quadrant edge orifice, where C_d is approximately constant with a slight difference of approximately 0.8%.

(3) The quadrant edge orifice produces larger discharge coefficient (C_d) than that of the square edge orifice of approximately 16%.

References

1. Singh V, Tharakan TJ (2016) Numerical simulations for multi-hole orifice flow meter. Flow measurement and instrumentation, https://doi.org/10.1016/j.flowmeasinst.2015.08.004. Last Accessed 29 Dec 2016
2. He N, Zhao Zx (2010) Theoretical and numerical study of hydraulic characteristics of orifice energy dissipator. Water Sci Eng 3(2):190–199
3. Boom R J (1991) Flow meter evaluation. ISA Annual Convention): ISA conference paper, pp #91–0509, USA
4. ISO: Measurement of fluid flow by means of pressure differential devices inserted in circular cross-section conduits running full - Part 2: Orifice plates (ISO 5167-2:2003): CEN, Brussels (2003)
5. Hollingshead CL (2011) Discharge coefficient performance of venturi, standard concentric orifice plate, V-Cone, and wedge flow meters at small reynolds numbers. Utah State University, Utah
6. Britton CL, Stark ST (1986) New data for the quadrant-edge orifice. International symposium on fluid flow measurement
7. Shah MS, Joshi JB, Kalsi AS, Prasad C, Shukla DS (2014) Analysis of flow through an orifice meter: CFD simulation. Chem Eng Sci 71(2012):300–309
8. Reinaldhy ALZ (2017) Numerical studies of square edge and quadrant edge orifice flow meter performance with diameter ratio variation, Final Project, ITS, Surabaya
9. Miller RW (1996) Flow measurement engineering handbook, 3rd edn. McGraw-Hill, New York

Numerical Study of the Effect of Excess Air Variation and Biomass Amount at Co-Firing Process for Tangential Coal Pulverized Boiler

Aryo Wicaksono(✉) and Bambang Arip Dwiyantoro

Department of Mechanical Engineering, Institut Teknologi Sepuluh Nopember, Surabaya 60111, Indonesia
bambangads@me.its.ac.id

Abstract. There is a power plant with a capacity of 3 × 315 MW which has a tangential pulverized boiler type. This power plant uses an LRC and MRC Coal type for the fuel to produce steam which is used to rotate steam turbine that coupled with a generator to produce electric power. The Indonesian government is striving used biomass to decrease using coal fuel until 23%. Previous sub-critical power plant not designed to be a co-firing fuel, and the best way to implement that system is using the direct co-firing method. This power plant has been an experiment using biomass until 1%, and there is no significant effect for the equipment. For the effectiveness combustion still need to do research. In this research are using 3 types fuels, 97.5% coal and 2.5% biomass, 95% coal and 5% biomass, 92.5% coal and 7.5% biomass. And it is also using excess air variation of 15%, 17%, and 20%. Efficiency's formula was carried out to find optimum variables with biomass and excess air at this power plant that the CFD numerical analysis was carried out to find an optimum model to analyze in more detail the effect to temperature contour, velocity particle combustion. In the end, we obtained that the efficiency biomass is still underusing coal but 97.5% biomass is the best for efficiency between the other variable.

Keywords: Pulverizer · Coal pulverized · Excess air · Biomass · Co-firing · CFD

1 Introduction

Sub-critical tangential pulverized boiler type has 20 coal burners, 12 HSD burners, 4 corners, negative pressure in the furnace, one furnace, one reheat, one steam drum, using bottom ash. There are wall pipes around the combustion chamber, water goes to the steam drum trough inside the wall pipes which are heated by fire from a coal burner. Boiler model DG1025/17.4-II 13 from dong fang boiler group co., Ltd. It designs at BMCR operating superheated steam-rated evaporation until 1025 t/h, outlet pressure 17.4 MPa, and outlet pressure until 541 °C. For reheated steam, flow 839.4 t/h, inlet/outlet pressure 3.74/3.58 MPa, and inlet/outlet temperature 329/541 °C. The heat transfer process also occurs in furnaces even though the fuel is biomass, and it is a reference to Sudarmanta et al. [1] conducting an experimental study of the effect of two stages gasifying agents

I. Tolj et al. (eds.), *Recent Advances in Mechanical Engineering*, Lecture Notes in Mechanical Engineering, https://doi.org/10.1007/978-981-19-0867-5_32

on biomass downdraft gasification on the gasifier performance. Maria et al. [2] classify 3 technologies for the configuration of biomass with coal in power plants, namely: direct co-firing, indirect cofiring, and parallel cofiring (see Fig. 1). Direct co-firing is a cheaper and easier method and is also commonly used in subcritical coal pulverized boiler. Direct Co-firing of coal and biomass has the advantage of efficiencies gained especially in large capacity plants and improved combustion based on the volatile nature of biomass, although there is potential to produce more ash deposits. Several studies have been researched the correlation between excess air and boiler efficiency. For example, Yaojie et al. [3] researched the Co-firing low-rank coal and biomass in a bubbling fluidized bed with waiting for excess air ratio and fluidization velocity found that increasing excess air would increase furnace temperature too and decreasing boiler efficiency. Paulo et al. [4] studied the effect of excess air variation in the co-firing between low-rank coal and biomass at bulling fluidized bed found that additional biomass will increase boiler temperature but when we add excess air ratio it will decrease the temperature inside the boiler. Jiusheng Shi et al. [5] have a conclusion about their studies, the higher the coefficient of excess air in the combustion chamber, the lower the efficiency of the boiler. And the excess air coefficient increases by 0.1, the boiler efficiency will decrease by about 0.5–0.7% therefore controlling excess air to increase boiler efficiency is highly recommended. The research was conducted to see the effect of biomass amount variation on boiler efficiency. Due to that condition, lab-scale data collection has been carried out for each variation in the amount of biomass SRF type [6].

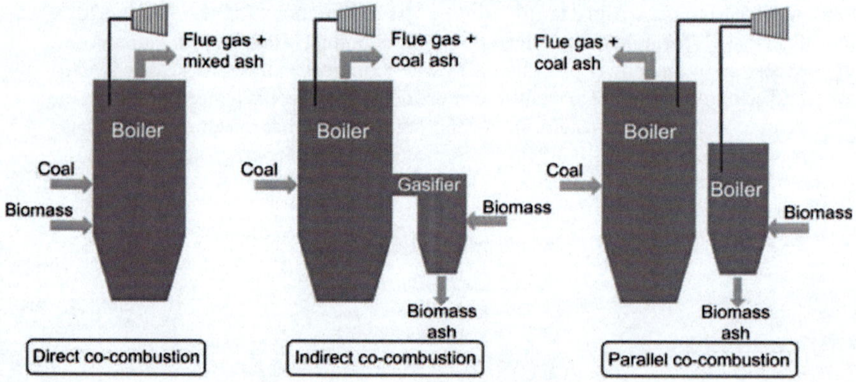

Fig. 1. Co-firing technology

2 Methodology

2.1 Heat Loss Boiler Efficiency

Due to in indirect method, each variable is calculated and produces data boiler losses and boiler efficiency compared to performance test data [7] and indirect method formula [8].

$$\text{Theoretical air required} = \frac{[(11,6 \times C) + \left\{34,8x\left(H_2 - \frac{O_2}{8}\right)\right\} + (4,35 \times S)}{100} \tag{1}$$

$$\text{Excess Air} = \frac{\mathrm{O_2}\%}{21 - \mathrm{O_2}\%} \times 100 \quad (2)$$

$$\text{Actual mass of air supplied(AAS)} = \left[1 + \frac{EA}{100}\right] \times \textit{theoretical air} \quad (3)$$

$$\dot{m}_{air}\text{total} = \text{AAS} \times \textit{fuel firing rate} \quad (4)$$

$$\begin{aligned}\text{Mass of dry flue gas} = {} & \textit{Mass of } CO_2 + \textit{Mass of } N_2 \textit{ content in the fuel} + \\ & \textit{Mass of } N_2 \textit{ in the combustion air supplied} + \\ & \textit{Mass of oxygen in flue gas}\end{aligned} \quad (5)$$

$$\text{Mass of dry flue gas} = \textit{Mass of dry flue gas}\left(\frac{kg}{kg}\textit{coal}\right) \times \textit{fuel firing rate} \quad (6)$$

Heat loss at boiler with indirect method:

$$\mathrm{L1} = \frac{\mathrm{m} \times C_p \times (\mathrm{T_f} - \mathrm{T}_a)}{\textit{GCV of Fuel}} \times 100 \quad (7)$$

$$\mathrm{L2} = \frac{9 \times H_2 \times [584 + C_p(\mathrm{T_f} - \mathrm{T}_a)]}{\textit{GCV of Fuel}} \times 100 \quad (8)$$

$$\mathrm{L3} = \frac{\mathrm{M} \times C_p \times (\mathrm{T_f} - \mathrm{T}_a)}{\textit{GCV of Fuel}} \times 100 \quad (9)$$

$$\mathrm{L4} = \frac{\mathrm{AAS} \times \textit{Humidity ratio} \times C_p(\mathrm{T_f} - \mathrm{T}_a)}{\textit{GCV of Fuel}} \times 100 \quad (10)$$

$$\mathrm{L5} = \frac{\%CO \times C}{\%CO + (\%CO_2)a} \times \frac{5744}{\textit{GCV Fuel}} \times 100 \quad (11)$$

$$\begin{aligned}\mathrm{L6} = \Bigg\{ & 0,548 \times \left[\left(\frac{T_s}{55,55}\right)^4 - \left(\frac{T_a}{55,55}\right)^4\right] + 1,957 \times \\ & (T_s - T_a)^{125} \times \sqrt{\frac{196,85V_m + 68,9}{68,9}}\end{aligned} \quad (12)$$

$$\mathrm{L7} = \frac{\textit{Total ash}\frac{\textit{collected}}{kg}\textit{ of fuel} \times \textit{GCV of fly ash} \times 100}{\textit{GCVoffuel}} \quad (13)$$

$$\mathrm{L8} = \frac{\textit{Total ash}\frac{\textit{collected}}{kg}\textit{ of fuel} \times \textit{GCV of bottom ash} \times 100}{\textit{GCV of fuel}} \quad (14)$$

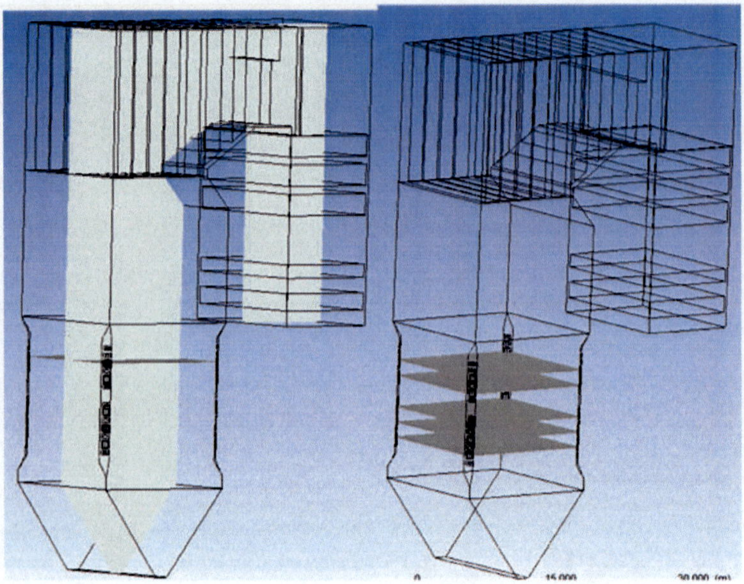

Fig. 2. Tangential pulverized boiler's geometry and data retrieval surface

2.2 Numerical Study

This study focuses on finding the effect of fuel-excess air alteration. The boiler model is consisting of 7 heat exchanger, panel superheater, platen superheater, final superheater platen reheater, final reheater, low-temperature superheater, and economizer. 13 mass input flow consist of 2 layers separated overfire air (SOFA), 3 layers secondary air, 3 layers oil burner, and 5 layers coal burner and is divided into 4 corners. By numerical analysis using CFD, then the model and equation developed. As shown in Fig. 2, this is the geometry model to find the best fuel variation in optimum excess air setting.

From the simulation produces temperature contour data rate CO and CO_2 at flue gas. The data is used as input for the L6 formula.

3 Result and Discussion

The result of temperature contour for each variation condition is shown in Fig. 3. Temperature contour shown that using 2.5% of biomass mixed by coal compared with 5% of biomass and 7.5% of biomass will produce more equal heat than 7.5% of biomass. The gross calorific value for each fuel makes the differences for each condition. Increasing excess air gradually has an impact on heat distribution higher value of the excess air, fire formed is getting closer to the middle furnace area.

From Fig. 4 it can be seen that the higher the excess air value makes lower the FEGT value. When combined with Fig. 3, it is concluded that the higher the excess air value, the more the fire will shift to the center of the furnace. It is related to Abdallah elorf [9]

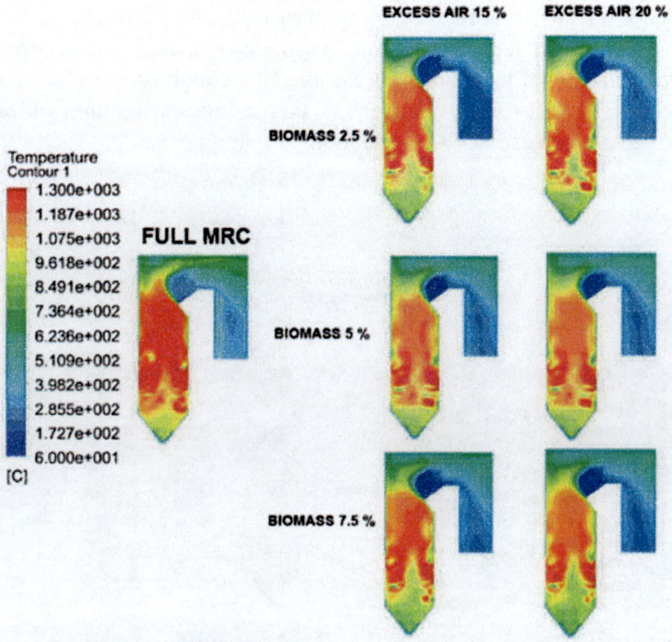

Fig. 3. Temperature contour at mid furnace surface

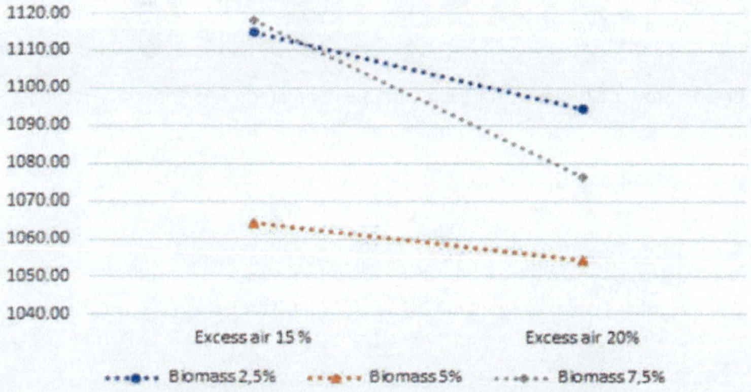

Fig. 4. Temperature furnace before entering superheater

studied, another effect of excess air is the visible length reduction of flame when excess air increases.

Figure 5 shows that increasing biomass causes velocity at the center of the furnace to decrease and move slowly to near the wall tube. The effect of increasing the excess air from 15 to 17% and 20% at stable biomass is causing velocity at near-wall tube increase. Figure 6 shows that increasing biomass causing % CO increasing too. When the excess air is increasing, every variation of biomass has the same reaction. %CO will decrease according to the increasing excess air. This condition is in line with research about excess air effect at CO by Wander [4]. Increasing the amount of biomass in the mixture leads to an increase in CO production.

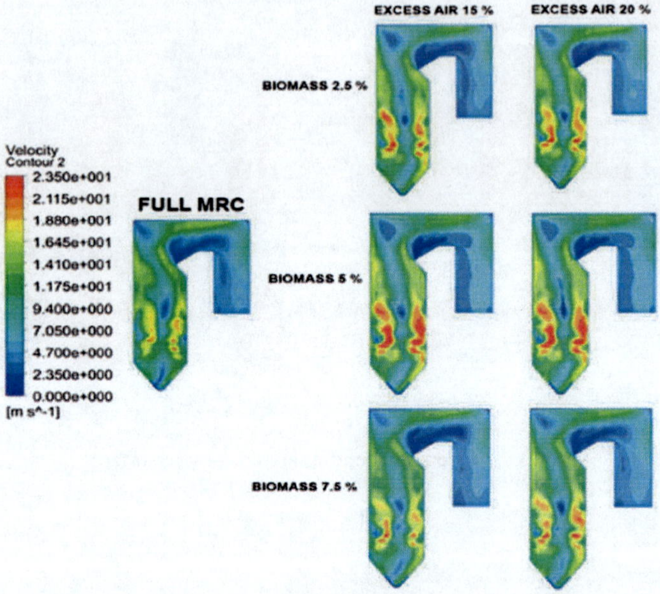

Fig. 5. Boiler Efficiency for each variable compared to performance test data

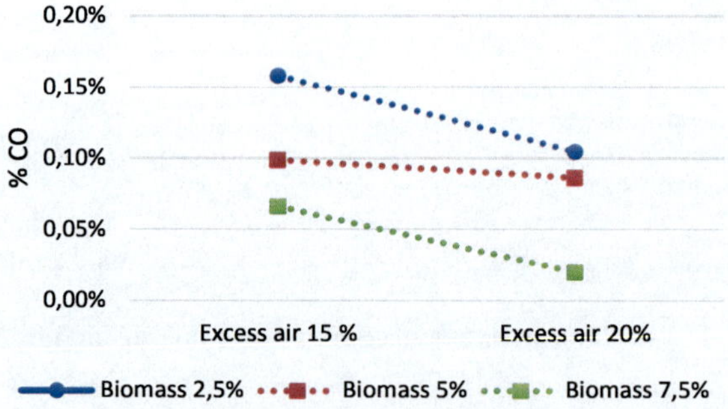

Fig. 6. % CO at flue gas

Based on the results of the CFD-post, the values of CO and CO_2 are obtained which are used as input parameters in the L5 formula and by ignoring the value of the change in unburnt carbon in bottom ash and fly ash, the boiler efficiency can be calculated by the indirect method of boiler losses. Figure 7 shows that, at 2.5% biomass mixed with MRC, 20% excess air gave the best boiler efficiency. At 5% biomass mixed with MRC, 15% excess air gave the best results, and at 7.5% biomass mixed with MRC 15% excess air was more efficient (Table 1).

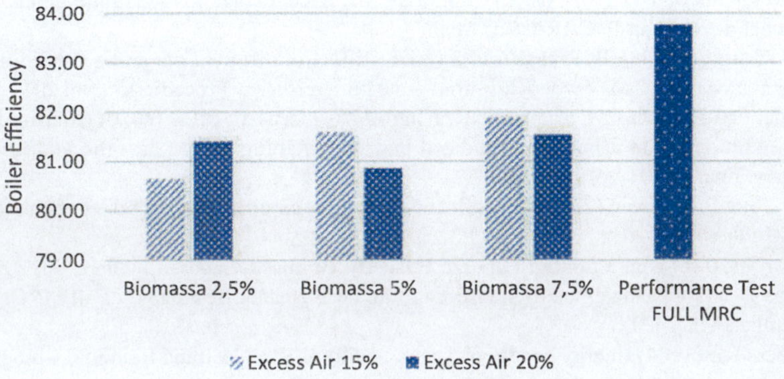

Fig. 7. Boiler efficiency

Table 1. Boiler efficiency by an indirect method

Excess air (%)	Biomass 2.5% (%)	Biomass 5% (%)	Biomass 7.5% (%)	Full MRC (%)
15	80.64	81.57	81.88	83.76
20	81.40	80.86	81.52	

4 Conclusion

1. With the increasing amount of biomass, the heat generated will decrease because of gross calorific value in each composition decrease too. When increasing excess air number will cause the fire that is formed to be more focused in the lower area.
2. The addition of excess air value from 15 to 20% generally causes an increase in the temperature of the burner layer. This causes the fire to be concentrated in the middle burner area.
3. The best efficiency is obtained when using 2.5% biomass and 20% excess air setting. While using 5% biomass, setting 15% for excess air is the best composition too, and when using 7.5% of the biomass, with 15% excess air setting will obtain the best efficiency.
4. The best number of amount biomass and excess air is 7.5% biomass using 15% excess air will obtain boiler efficiency until 81.88%. But it still needs more research for the pulverized effect when the amount of biomass was added.

References

1. Sudarmanta B, Gafur A, Saleh AR, Dwiyantoro BA, Sampurno (2018) The effect of two stage gasifying agent on biomass downdraft gasification to the gasifier performance. In: AIP conference proceedings, Vol 1983
2. Gill MV, Ruberia F (2010) Coal and biomass cofiring-fundamentals and future trends. Instituto Nacional del Carbon INCAR-CSIC Spain
3. Tu Y, Yang W, Siah KB, Prabakaran S (2018) Effect of different operating conditions on the performance of a 32 MW woodchip-fired grate boiler. Energy Procedia 158 Vol 2019:898–903
4. Wander PR, Bianchi FM, Caetano NR, Klunk MA, Indrusiak MLS (2020) Cofiring low-rank coal and biomass in a bubbling fluidized bed with waiting excess air ratio and fluidization velocity. Energy 203 Vol 2020:117882
5. Shi J, Peng F, Bingwen (2013) The influence of excess air coefficient of boiler efficiency. Trans Tech Public 409–410
6. Lontar T (2021) Kimia Lontar: Laporan Hasil Uji Biomasa. Laporan Januari, Vol 2021
7. Lontar T (2020) Manajemen Efisiensi: Laporan Performance Test dan uji Full MRC. Laporan Desember, Vol 2020
8. Loftness RL (1984) Energy handbook second edition. Von Nostrand Reinhold Company
9. Elorf A, Sarh B (2019) Excess air ratio effects on flow and combustion characteristics of pulverized biomass (olive cake). Case Stud Thermal Eng 13100367

The Coats-Redfern Models Kinetics Analysis of *Tetraselmis Chuii*—Titanium Dioxide Nanoparticle Blend During Combustion Process

Ardianto Prasetiyo[1], Sukarni[1,2](✉), and Poppy Puspitasari[1,2]

[1] Center for Renewable and Sustainable Energy Engineering (CRSEE), Department of Mechanical Engineering, Universitas Negeri Malang, Malang 65145, Indonesia
sukarni.ft@um.ac.id

[2] Center of Advanced Materials for Renewable Energy (CAMRY), Universitas Negeri Malang, Malang 65145, Indonesia

Abstract. The increasing demand for energy from fossil fuels and depleting reserves has made researchers look for efficient energy and guaranteed its availability. This paper intends to understand the effect of TiO_2 nanoparticles on *Tetraselmis chuii* (*T. chuii*) microalgae combustion using a thermogravimetric (TG) analyzer. The kinetic parameters were analyzed using the Coats-Redfern method on the ten models of reaction. The results showed that the combustion of *T. chuii* occurred in three stages, with the main combustion present at a range of 186.43–589.75 °C. The presence of TiO_2 plays as an accelerator which promoted the decomposition of lipids to the lower temperature and shortened the combustion time. However, the presence of TiO_2 increases the final residue. The adding of TiO_2 reduced the activation energy up to 11.82% (from 72.99 to 64.36 kJ.mol^{-1}) in the model of Chemical Reaction Order F1.5. This phenomenon confirmed that the presence of TiO_2 affects thermal decomposition and activation energy.

Keywords: *T. chuii* · Titanium dioxide · Nanoparticle · Combustion · Coats-Redfern

1 Introduction

Global dependence on fossil-based resources has received special attention due to the issue of resource limitation and its effect on the environment. The role of biom as an energy source is expected to increase in the coming years due to its renewable and zero CO_2 emissions [1]. Biom consists of three different levels involved the first, second, and third-generation [2, 3]. The first generation comes from edible raw materials and has been concerned about affecting food security [2, 4]. The second-generation source comes from inedible raw materials. Finally, the third generation comes from photosynthetic biom, where microalgae have been considered a third-generation feedstock.

The thermochemical conversion process of microalgae raw materials into fuel can be through gasification, pyrolysis, combustion, liquefaction, and torrefaction [5]. The

I. Tolj et al. (eds.), *Recent Advances in Mechanical Engineering*, Lecture Notes in Mechanical Engineering, https://doi.org/10.1007/978-981-19-0867-5_33

direct combustion process has the advantage of being able to convert solid raw materials into heat without being converted into other forms. The direct combustion method supports more than 97% of the world's energy production [6]. However, the combustion of *T.chuii* microalgae with additives of TiO_2 nanoparticles under atmospheric air at different heating rates has not been thoroughly explored.

The novelty of this work lies in the utilization of TiO_2 nanoparticles to the combustion of *T. chuii* microalgae, intending to pave the way for efficient energy production. Furthermore, this is the first time that TiO_2 nanoparticles used as a catalyst in the combustion of microalgae biom.

2 Experiment Procedures

2.1 Materials

The *T. chuii* microalgae have been provided by the Center for Development of Brackish Water Aquaculture of Jepara, Indonesia. The culturing was performed on a laboratory scale and harvested within eight days. First, the samples were precipitated and filtered. Subsequently, the sample was washed using purified water and soaked for six hours, and then sieved; this process was carried out two times. Finally, in order to remove moisture content, the sampled sediment was employed the drying process in an oven at 80 °C for 24 h. The dry sample was then crushed on a mortar and equalized with a 60 mesh sieve. About 10 mg of dry microalgae powder was used for each experiment.

2.2 Nanoparticle TiO_2

The titanium dioxide (TiO_2) nanoparticles (PA grade) get from Sigma Aldrich, with particle size <25 nm. About 5% of TiO_2 nanoparticles are blended with 10 mg of *T. chuii* in a mortar container then stirred by mechanochemical mixing processes to homogenize blends [7].

2.3 Experimental Procedure

The combustion experiment was performed on a thermal analyzer METTLER TOLEDO TGA/DSC1 to determine the sample combustion behavior. A samples weight of around 10.05 mg was laid in an Al_2O_3 ceramic crucible for the purpose of combustion in a thermogravimetric apparatus. The sample was heated up from ambient temperature to 1000 °C at two selected heating ramps of 20 and 25 $°C.min^{-1}$. To ensure sufficient oxidative air in the reaction chamber, atmospheric air has flowed at 100 ml min^{-1}. M loss was recorded and plotted with temperature during the testing process and then represented as the thermogravimetric (TG) and derivative thermogravimetric (DTG) curves.

2.4 Combustion Kinetic Analysis

The material degradation rate throughout combustion is stated by the basic equation:

$$\frac{d\alpha}{dt} = k(T)f(\alpha) \tag{1}$$

where

$$\alpha = \frac{m_o - m_i}{m_0 - m_f} \tag{2}$$

where m_o is the initial m, m_i is the m at time t(s), and m_f is the final m in mg. The dependence of reaction rate to the temperature is determined by:

$$k(T) = A\exp\left(-\frac{E_a}{RT}\right) \tag{3}$$

where A is a pre-exponential factor (min^{-1}), Ea is the activation energy ($kJ.mol^{-1}$), R is the universal gas constant (0008314 $kJ.mol^{-1}.K$), and T is the reaction temperature (K). For $\beta = dT/dt$, then Eqs. (1) and (3) re-written to be:

$$\frac{d\alpha}{dT} = \frac{A}{\beta}\exp\left(-\frac{E_a}{RT}\right)f(\alpha) \tag{4}$$

Integrating Eq. (4) resulted in a new equation as the following:

$$g(\alpha) = \int_0^\alpha \frac{d\alpha}{f(\alpha)} = \frac{A}{\beta}\int_{T_0}^{T} exp\left(-\frac{E_a}{RT}\right)dT \tag{5}$$

where g(α) represented the reaction model integration form. Due to the right side of the equation are not possible to solve by the analytical method; therefore, several approximate models are applied to address complex parts of this equation.

The fitting method of the Coats-Redfern is widely applied to evaluate the kinetic parameters by being subjected to the various selected $g(\alpha)$. The basic equation of the Coats-Redfern is given below:

$$ln\left[\frac{g(\alpha)}{T^2}\right] = ln\frac{AR}{\beta E_a}\left(1 - \frac{2RT}{E_a}\right) - \frac{E_a}{RT} \tag{6}$$

Activation energy is obtained by making plot between $ln\left[\frac{g(\alpha)}{T^2}\right]$ versus $1/T$ to generate the slope. The pre-exponential factor is attained from this intercept. It would be varied according to various reaction mechanisms and models adopted. The various model for solid-state degradation is listed in Table 1.

Table 1. Reaction mechanism with their differential form $f(\alpha)$ and integral form $g(\alpha)$

Reaction mechanism		Differential form $f(\alpha)$	Integral Form $g(\alpha)$
Symbol	Model name		
Reaction order			
F1	Reaction Order 1	$1-\alpha$	$-\ln(1-\alpha)$
F1.5	Reaction Order 1.5	$(1-\alpha)]^{3/2}$	$2(1-\alpha)]^{3/2}-1]$
Diffusion			
D1	Parabolic law	$1/2\alpha$	α^2
D2	Valensi equation	$-[\ln(1-\alpha)]^{-1}$	$\alpha+(1-\alpha)\ln(1-\alpha)$
D3	Ginstling-Broushtein equation	$3/2[(1-\alpha)^{1/3}-1]^{-1}$	$\left(1-\frac{2}{3}\alpha\right)-(1-\alpha)^{2/3}$
Nucleation and growth			
N1.5	Avrami-Erofeev equation of Nucleation and growth	$3(1-\alpha)[-\ln(1-\alpha)]^{2/3}$	$[-\ln(1-\alpha)]^{2/3}$
N2	Avrami-Erofeev equation of Nucleation and growth	$2(1-\alpha)[-\ln(1-\alpha)]^{1/2}$	$[-\ln(1-\alpha)]^{1/2}$
Phase of interfacial reaction			
S1	Shrinkage geometrical column	$2(1-\alpha)^{1/2}$	$1-(1-\alpha)^{1/2}$
S2	Shrinkage geometrical Spherical	$2(1-\alpha)^{2/3}$	$1-(1-\alpha)^{1/3}$
Power law			
P	Power law	1	α

3 Results and Discussions

3.1 The Combustion Behavior

Figure 1 represented the m loss (TG), and the rate in m loss (DTG) curves at heating rates of 20 and 25 °C.min^{-1}, respectively. The curve indicates that there are three stages of decomposition. In stage one, as indicated by a TG curve, the m was lost of up to 10.53% in correlation with the release of water as reported in some literature [8, 9], which occurs at room temperature until 191.15 °C.

In stage two, there are two broad basins on the DTG curve. In this stage, the material decomposition is divided into two zones. The first zone is in the temperature range of 191.15–305.00 °C with a weak peak at a temperature of 287.68 °C indicated the decomposition of protein and carbohydrates, followed by the second zone at a temperature range of 305–589.75 °C with a pronounced peak at temperature 319.88 °C indicated the decomposition of lipids [9, 10].

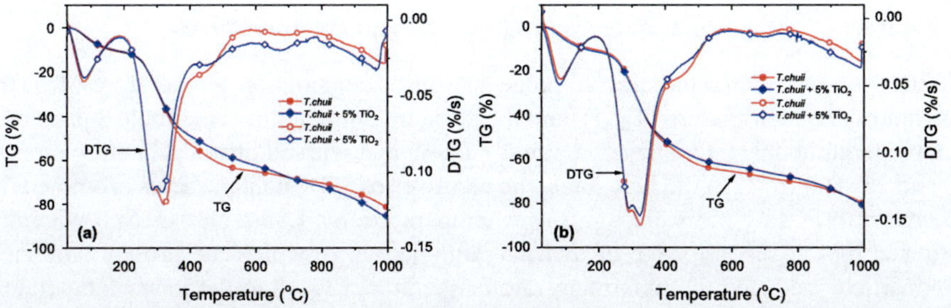

Fig. 1. TG-DTG *T.chuii* and *T.chuii* + TiO_2 at heating rate **a** 20 °C.min^{-1}, **b** 25 °C.min^{-1}

The third stage exceeds the temperature of 589.75–1000 °C, indicating the decomposition of the remaining compounds of decomposed biom to ash [11]. This stage is characterized by a flat slope of m loss curve (TG curve), indicating the m is lost slowly.

3.2 Effect of Different Heating Rate

Figure 1 and Table 2 show that different heating rates affect the temperature of thermal decomposition. Figure 1 reveals that a higher heating rate tends to push thermal decomposition to a higher temperature, making the extension of basin space of the stage of decomposition. This phenomenon revealed that at a certain temperature, a high heating rate affected the material to reach the decomposition temperature in a relatively short time. In shorter induction time, they were inducing a shift in the volatile decomposition to a higher temperature [8, 12].

Table 2. Combustion characteristic parameters of *T. chuii* without and with additives TiO_2

β (°C.min^{-1})	Samples	T_o (°C)	T_{max} (°C)	T_b (°C)	*M loss* (%)	DTG_{max}-1 (%/s)
20	*T. chuii*	188.41	319.88	586.97	−65.34	−0.1217
	T. chuii + TiO_2	186.43	315.89	580.19	−61.87	−0.1131
25	*T. chuii*	191.15	324.33	589.75	−65.58	−0.1538
	T. chuii TiO_2	189.27	322.19	587.24	−62.69	−0.1468

T_o: onset temperature and T_b: burnout temperature
T_{max}: the maximum rate temperature
DTG_{max}: the maximum m loss rate

3.3 Effect of TiO_2 Nanoparticles at Characteristic Temperatures

The temperature characteristics included the onset degradation temperature (T_o), the maximum rate temperature (T_{max}), and the burnout temperature (T_b). Table 2 presented the temperature characteristics of *T. chuii* without and with additives at heating rates of 20 and 25 °C.min^{-1}. In the first stage, the additives of TiO_2 nanoparticles promoted the process of releasing moisture to a lower temperature by 1.88–1.98 °C. Subsequently, followed by the second stage of thermal degradation, in which the presence of TiO_2 nanoparticles reduced the maximum rate temperature (T_{max}) to the lower temperature by 2.14–3.99 °C; this is followed by the completion stage of decomposition to a lower temperature by 2.51–6.78 °C. These phenomena justified that the additives of TiO_2 nanoparticles significantly affected the thermal degradation process to a lower temperature and a shorter decomposition process. Therefore, the thermal characteristics of the *T. chuii* microalgae sample could change due to the presence of TiO_2.

3.4 Kinetic Analysis

The kinetic parameters analyses were estimated using the Coats Redfern method that applied to the ten selected models. The most proper model is selected based on the highest linear fitting. Linear regression with a correlation coefficient (R^2) approach to 0.999 was considered a suitable model. Table 3 shows the kinetic parameters of *T. chuii* without and with additives that were resulted from an evaluation using the Coats-Redfern method.

The kinetic parameters were analyzed at heating rates of 20 and 25 °C min^{-1} in the main decomposition stage (Stage) II (186.43–589.75 °C), using various reaction models. It observed that all models gave the correlation coefficient (R^2) from 0.725 to 0.992, except the power-law model (P), which showed the worst linear regression with R^2 in the range of 0.596–0.683 for the sample without and with TiO_2. The Chemical Reaction Order 1.5 (F1.5) model can be considered a suitable model for *T. chuii* without and with additives TiO_2.

Table 3. The kinetics parameter of *T.chuii* without and with additives

Models	β (°C min^{-1})	*T. chuii*			*T. chuii* + TiO_2		
		Ea (kJ mol^{-1})	R^2	log A (min^{-1})	Ea (kJ mol^{-1})	R^2	log A (min^{-1})
F1	20	38.06	0.883	2.31	32.58	0.837	1.75
	25	38.52	0.887	2.32	37.91	0.873	2.29
F1.5	20	72.99	0.991	6.42	64.36	0.980	5.60
	25	73.76	0.992	6.45	73.10	0.992	6.42
D1	20	57.48	0.764	3.57	49.38	0.713	2.79
	25	58.18	0.768	3.61	57.08	0.752	3.53
D2	20	65.08	0.812	4.08	56.26	0.767	3.24
	25	65.86	0.816	4.12	64.70	0.801	4.04
D3	20	68.41	0.832	3.77	59.27	0.789	2.90
	25	69.21	0.836	3.81	68.04	0.821	3.73
N1.5	20	21.91	0.845	0.76	18.21	0.775	0.35
	25	22.21	0.849	0.77	21.80	0.831	0.74
N2	20	13.84	0.787	−0.10	11.02	0.683	−0.45
	25	14.05	0.793	−0.09	13.74	0.769	−0.11
S1	20	29.94	0.789	1.11	25.51	0.725	0.61
	25	30.33	0.795	1.13	29.74	0.776	1.08
S2	20	32.44	0.824	1.21	27.47	0.765	0.70
	25	32.85	0.828	1.23	32.25	0.811	1.19
P	20	23.55	0.677	0.67	19.43	0.596	0.26
	25	23.88	0.683	0.69	23.33	0.661	0.65

In the model chemical reaction order 1.5 (F1.5) at a heating rate of 20 °C.min^{-1}, the *T. chuii* + TiO_2 sample shows an activation energy value of 64.36 kJ.mol^{-1}; this value is 11.82% smaller than that of *T. chuii* of 72.99 kJ.mol^{-}. This is similar to the heating rate of 25 °C.min^{-1}; the *T. chuii* + TiO_2 sample shows an activation energy value of 73.10 kJ.mol^{-1}; this value is 0.89% smaller than the *T. chuii* that of 73.76 kJ.mol^{-1}. These results are in disagreement with the previous study of Sukarni et al. [13] that confirmed the additives of Al_2O_3 and $MgCO_3$ affected the increasing activation energy value up to 4.83 kJ.mol^{-1}.

The overall pre-exponential factor (log A) values of the samples with the additives of TiO_2 show a smaller value than without additives TiO_2, as illustrated in Table 3. The chemical reaction order 1.5 (F1.5) at a heating rate of 20 °C.min^{-1} indicated that log A value of *T. chuii* + TiO_2 was 5.60 min^{-1}; 12.77% smaller than the log A value of *T. chuii* that was 6.42 min^{-1}. A similar result was observed at the heating rate of 25 °C min^{-1}, in which the log A of *T. chuii* + TiO_2 was 6.42 min^{-1}; 0.46% smaller than the log A of *T. chuii* that was 6.45 min^{-1}. Usually, the pre-exponential factor value relates to the number of times the molecule will hit in the orientation needed to cause the reaction [14].

4 Conclusion

This study examined thermal decomposition, catalytic effects, and kinetic parameters of *T.chuii* without and with the additives of TiO_2 during the combustion process. It had observed that the thermal decomposition of microalgae occurs in three stages, with the main degradation taking place in the temperature range 186.43–589.75 °C (m loss range of 50–65%), known as the active combustion zone. The different heating rates affect the catalytic effectiveness during the combustion. The presence of TiO_2 promoted the m-loss rate in the main decomposition process toward lower temperatures. The thermo-kinetic evaluation showed that TiO_2 caused to reduce the activation energy by up to 11.82%.

Acknowledgements. The support from the scheme of Penelitian Tesis Magister with grand numbers 19.3.19/UN32.14.1/LT/2019 was acknowledged.

References

1. Babich IV, Van Der Hulst M, Lefferts L, Moulijn JA, Connor PO, Seshan K (2011) Catalytic pyrolysis of microalgae to high-quality liquid. Biomed Bioenergy 35:3199–3207. https://doi.org/10.1016/j.biombioe.2011.04.043
2. Kowthaman CN, Varadappan AMS (2019) Synthesis, characterization, and optimization of Schizochytrium biodiesel production using Na+-doped nanohydroxyapatite. Int J Energy Res 43:3182–3200. https://doi.org/10.1002/er.4387
3. Yew GY, Lee SY, Show PL, Tao Y, Law CL, Nguyen TTC, Chang JS (2019) Recent advances in algae biodiesel production: from upstream cultivation to downstream processing. Bioresour Technol Rep 7:100227. https://doi.org/10.1016/j.biteb.2019.100227
4. Singh D, Sharma D, Soni SL, Sharma S, Kumar Sharma P, Jhalani A (2020) A review on feedstocks, production processes, and yield for different generations of biodiesel. Fuel 262:116553. https://doi.org/10.1016/j.fuel.2019.116553
5. Sukarni S, Ardianto P, Retno W, Aloon EW, Poppy P (2020) Thermal characteristic of tetraselmis chuii combustion influenced by titanium dioxide (Tio2) nanoparticle. Key Eng Mater 851 KEM, 149–155. https://doi.org/10.4028/www.scientific.net/KEM.851.149
6. Demirbas A (2004) Combustion characteristics of different biom fuels. Prog Energy Combust Sci 30:219–230. https://doi.org/10.1016/j.pecs.2003.10.004
7. Cova CM, Luque R (2019) Advances in mechanochemical processes for biom valorization. BMC Chem Eng 1:1–12. https://doi.org/10.1186/s42480-019-0015-7
8. Sukarni S, Hamidi N, Yanuhar U, Wardana ING (2015) Thermogravimetric kinetic analysis of Nannochloropsis oculata combustion in air atmosphere. Front Energy. 9:125–133. https://doi.org/10.1007/s11708-015-0346-x
9. Agrawal A, Chakraborty S (2013) A kinetic study of pyrolysis and combustion of microalgae Chlorella vulgaris using thermo-gravimetric analysis. Bioresour Technol 128:72–80. https://doi.org/10.1016/j.biortech.2012.10.043

10. Peng W, Wu Q, Tu P (2001) Pyrolytic characteristics of heterotrophic Chlorella protothecoides for renewable bio-fuel production. J Appl Phycol 13:5–12. https://doi.org/10.1023/A:1008153831875
11. Azizi K, Keshavarz M, Abedini H (2017) Bioresource technology characteristics and kinetics study of simultaneous pyrolysis of microalgae Chlorella vulgaris, wood and polypropylene through TGA. Bioresour Technol 243:481–491. https://doi.org/10.1016/j.biortech.2017.06.155
12. Vamvuka D, Sfakiotakis S (2011) Combustion behaviour of biom fuels and their blends with lignite. Thermochim Acta 526:192–199. https://doi.org/10.1016/j.tca.2011.09.021
13. Sukarni S, Sumarli S, Nauri IM, Prasetiyo A, Puspitasari P (2019) Thermogravimetric analysis on combustion behavior of marine microalgae Spirulina platensis induced by MgCO3 and Al2O3 additives. Int J Technol 10:1174–1183. https://doi.org/10.14716/ijtech.v10i6.3611
14. Balasundram V, Ibrahim N, Kasmani RM, Hamid MKA, Isha R, Hasbullah H, Ali RR (2017) Thermogravimetric catalytic pyrolysis and kinetic studies of coconut copra and rice husk for possible maximum production of pyrolysis oil. J Clean Prod 167:218–228. https://doi.org/10.1016/j.jclepro.2017.08.173

Modeling and Sensitivity Analysis of Heat Transfer Enhancement in Double-Pipe Heat Exchanger Using Nanofluid

Avita Ayu Permanasari[1,2](✉), Muhammad Taufiq Affandi[2], Poppy Puspitasari[1,2], and Sukarni[1,2]

[1] Centre of Advanced Materials for Renewable Energy (CAMRY), Malang 65145, Indonesia
avita.ayu.ft@um.ac.id

[2] Department of Mechanical Engineering, State University of Malang, Malang, Indonesia

Abstract. Nanofluids have the property of being able to flow heat rapidly, so nanofluids are useful in many applications such as heat distribution, cooling, and various machine tools. The purpose of this research is to determine the heat transfer in the heat exchanger with the cold fluid $MnFe_2O_4$–water nanofluid at various input parameters, namely, number of tubes, cold fluid flow rate, and volume fraction of nanoparticles to determine the optimal input response. A sensitive analysis used in this study is the Response Surface Methodology (RSM) to determine the optimum value of the response that affects a factor. Computational Fluid Dynamics (CFD) method is also used in this research to determine heat transfer in fluid flow. The parameters used in this study include the volume fraction percentage of $MnFe_2O_4$ nanoparticles (0.025%; 0.05%; 0.075%), nanofluid flow as cold fluid (0.2 l/min; 0.4 l/min; 0.6 l/min), and the number of tubes in the heat exchanger for cold fluids (1 tube; 2 tubes; 3 tubes). The results showed that the input parameters have an influence on LMTD, overall heat transfer coefficient and heat transfer rate with a coefficient of determination (R^2) of 87.25%, 99.30%, and 94.18%. The heat characteristics also increased along with the addition of the volume fraction of $MnFe_2O_4$ nanoparticles and the nanofluid flow rate in the heat exchanger.

Keywords: Nanofluid · Heat exchanger · RSM · CFD · Optimization

I. Tolj et al. (eds.), *Recent Advances in Mechanical Engineering*, Lecture Notes in Mechanical Engineering, https://doi.org/10.1007/978-981-19-0867-5_34

1 Introduction

Heat exchangers in technological developments on the nanoscale (nanotechnology) began to grow more rapidly, both in the fields of industry and science [1–3]. Nanofluid is one of the fluids that influences heat exchangers because it has unique properties as a good heat conductor so that it becomes a topic of interest by researchers [4, 5]. In this study, previous research becomes a reference in determining steps in terms of concepts and theories. Shahrul et al. conducted experiments on Al_2O_3–water, SiO_2–water, and ZnO–water nanofluids and their application in a shell and tube heat exchanger. The heat transfer performance of nanofluids, Al_2O_3–water, SiO_2–water, and ZnO–water, was stabilized using polyvinylpyrrolidone (PVP) surfactant in a shell and tube heat exchanger [2]. From this study, the performance of nanofluids ZnO–water, Al_2O_3–water, and SiO_2–water with PVP surfactant increased, respectively, by 35%, 26%, and 12%. So, it can be said that nanofluids show the best performance in heat exchangers compared to conventional cooling fluids (water, oil, and ethylene glycol). Amani et al. [6] perform a thermal conductivity analysis of nanofluids. $MnFe_2O_4$–water is assisted by using a magnetic field. $MnFe_2O_4$–water was synthesized and prepared in nanoparticle concentrations of 0.25–3%. It was found that the thermal conductivity of the nanofluids greatly increased with the augmentation of the solid volume fraction. Dondapati et al. analyzed the thermal conductivity of Al_2O_3, SiO_2, SiC, CuO, and TiO_2 nanoparticles in liquid nitrogen (LN2) with a volume concentration of 3% in a microheat exchanger using computational fluid dynamics (CFD) simulations. CFD geometry and analysis of pressure drop and heat transfer were performed using ANSYS software. In the study, it is stated that the dispersion of nanoparticles in liquid nitrogen (LN2) is useful in microheat exchangers because it showed an increase in thermal conductivity caused by Brownian motion [7].

Fadakar Kourkah et al. [8] research on the performance of the double-pipe heat exchanger was analyzed using Response Surface Methodology (RSM) with various input parameters, namely, Reynolds number, torsion ratio (y/w), and SiO nanofluid concentration. The overall heat transfer coefficient outputs the response of the double-pipe heat exchanger. Experimental design is developed based on the Box–Behnken design method. The effect of vital input parameters and the interactions between them were investigated using Analysis of Variance (ANOVA) [9–11]. The optimum value of the overall heat transfer coefficient is 2732.59 ($W/m^2.K$) at the Reynolds number 19999.42, the nanoparticle concentration 0.5% w, and the bending ratio (y/w) 5.87. In the desired function approach, the value of R^2 is 0.937 which is proven to be able to predict the overall heat transfer coefficient of the double-pipe heat exchanger. Therefore, the above studies serve as the basis and comparison of this research. In addition, from previous studies, it can be concluded that nanofluids can be used as a replacement fluid in a cooling machine because it has good heat conductivity; the design of the heat exchanger also affects the effectiveness and optimization of heat transfer.

2 Methodology

2.1 Synthesis of $MnFe_2O_4$–water Nanofluid

The nanoparticle used in this study was $MnFe_2O_4$ (Sigma Aldrich). Nanoparticles have a clarity of 99% and an average particle size of less than 50 nm. These nanoparticles are the basic ingredients for the manufacture of nanofluids which will be synthesized with distilled water using a two-step method. The first step is that the nanofluids are stirred using a magnetic stirrer for 30 min. In the second step, the nanofluids were sonicated using a sonicator for 2 h. Variations in the volume fraction of nanoparticles used in this study were 0.025%, 0.05%, and 0.075%. The nanofluid was tested for thermophysical properties to see the value of thermal conductivity, specific heat, density, and viscosity of the $MnFe_2O_4$–water nanofluid. Thermal conductivity and specific heat of nanofluid were tested using KD2Pro. The nanofluid viscosity was tested using an NDJ-8S viscometer, and the nanofluid density was tested using a pycnometer. Table 1 shows the thermophysical properties of the nanofluids.

Table 1. Thermophysical properties of nanofluid

Volume of water (ml)	Mass of $MnFe_2O_4$ (gr)	Volume fraction of $MnFe_2O_4$ (%)	Thermal conductivity of nanofluid (W/m°C)	Viscosity of nanofluid (kg/m.s)	Density of nanofluid (kg/m^3)	Specific heat of nanofluid (J/kg°C)
300	0.075	0.025	0.625998	0.00097	1093.92	4095.95
300	0.015	0.005	0.638456	0.00102	1190.75	4012.9
300	0.225	0.075	0.650407	0.00108	1287.57	3929.85

2.2 Numerical Simulation Using CFD

Designing the heat exchanger geometry using Ansys Fluent CFD software. The initial steps taken in conducting the simulation are making a geometric design, meshing, and determining the boundary plane on the geometry. Figure 1 shows the geometry of the heat exchanger made with one tube, two tubes, and three tubes. Geometry design in the form of outlet pipe diameter 21.45 mm, pipe inlet diameter 7.45 mm, and pipe length 0.38 m as shown in Fig. 2. The second step is determining boundary conditions, where the stages are general, models, materials, cell zone conditions, boundary conditions, and iterations. In this case, the fluid flow in the pipe is steady. Geometry model using viscous laminar with the model in realizable state and near-wall treatment in standard wall functions. The material used in this simulation is $MnFe_2O_4$–water nanofluid as the cold fluid, while water is used as the hot fluid. Boundary conditions at the inlet which are applied in this simulation are mass flow rate of fluid, fluid temperature, and laminar flow. The results of the CFD simulation will produce the output temperature of the cooling fluid (nanofluid) and hot fluid (water).

2.3 Statistical Analysis Using RSM

RSM is a statistical method based on empirical models using experimental data. This method produces a polynomial function to model the curvature of the response in the

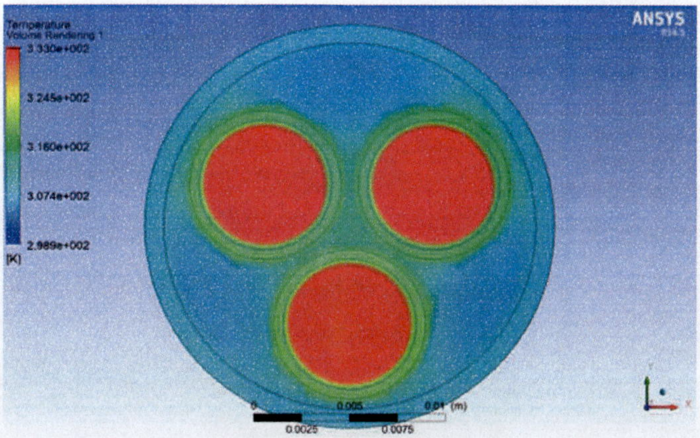

Fig. 1. Geometric design of the tube in the heat exchanger

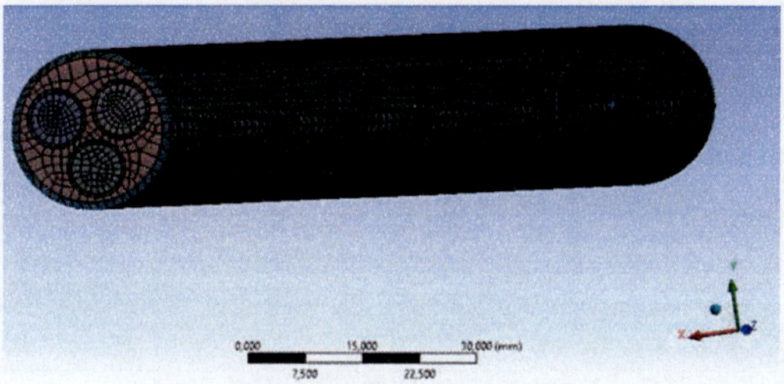

Fig. 2. Meshing of heat exchanger geometric design

optimum region and determine the factors (independent variable) to produce the optimum response (depending variable). RSM only concentrates on a certain level according to the selected experimental design. The experimental design consists of different combinations of levels (codes 1, 2, 3) of the independent variables as shown in Tables 2 and 3. The empirical model commonly used is a first-order polynomial. Equation 1 shows the form of a first-order polynomial in RSM based on a linear function, where β_0 is the constant term, β_i is the coefficient of the linear equation, k is the number of variables, x_i is the factor, and ε is the experimental residue. Equation 2 shows the form of a second-order polynomial. Second-order polynomial equations can significantly improve the optimization process. The Box–Behnken design uses second-order equations and provides a more efficient matrix design. The quality of the empirical model can be evaluated using the ANOVA. Determination of the contribution of a factor can be analyzed using the percentage contribution shown in Eq. 3. The percentage contribution is a function of the SS (Sum of Square) of the factor, resulting in a significant factor interaction which is an indication of the strength of the factor [12–14].

Table 2. Factors and levels used in RSM

Factor	Level		
	1	2	3
Pipe tube	1 tubes	2 tubes	3 tubes
Nanofluid flow	0.2 l/min	0.4 l/min	0.6 l/min
Nanoparticle volume fraction	0.025%	0.05%	0.075%

Table 3. Design of experiment using Box–Behnken method

Number of experiment	Pipe tube (A)	Nanofluid flow (B)	Nanoparticle volume fraction (C)
1	1	1	2
2	3	1	2
3	1	3	2
4	3	3	2
5	1	2	1
6	3	2	1
7	1	2	3
8	3	2	3
9	2	1	1
10	2	3	1
11	2	1	3
12	2	3	3
13	2	2	2
14	2	2	2
15	2	2	2

$$Y = \beta_0 + \Sigma_{i=1}^{k} \beta_i x_i + \varepsilon \quad (1)$$

$$Y = \beta_0 + \Sigma_{i=1}^{k} \beta_i x_i + \Sigma_{i=1}^{k} \beta_{ii} x_i^2 + \varepsilon \quad (2)$$

$$SS' = SS - (MSe \times D_f)$$

$$P = \frac{SS'}{SS_T} \times 100\% \quad (3)$$

3 Result and Discussion

This analysis was conducted to determine that the parameter factors have a significant effect on the response, namely, the characteristics of heat transfer. The value of ΔT_{LMTD} is the temperature difference between the inlet and outlet fluids of the heat exchanger. The ΔT_{LMTD} value is obtained from the temperature value generated from the CFD simulation results which are used to calculate heat transfer in the heat exchanger [15–17]. Based on the ANOVA for the ΔT_{LMTD} response shown in Table 4, the regression model of the entire model has a significant effect on the ΔT_{LMTD} response ($P < 0.05$) because the calculated F_{value} is greater than F_{table}. However, the volume fraction of nanoparticles ($P > 0.05$) did not have a significant effect on the response and the value of the coefficient of determination (R-squared) was 87.25% which indicated that the model could describe the ΔT_{LMTD} response, so it can be concluded that the model obtained is suitable for predicting the results of ΔT_{LMTD} on the optimum performance of the heat exchanger. In Fig. 3, the percentage of factor contribution to ΔT_{LMTD} can be seen that the number of tubes gives the largest contribution to ΔT_{LMTD} which is then followed by nanofluid flow rate and volume fraction of nanoparticles, which are 72.83%, 14.33%, and 0.09%, while the percentage of error contribution is 12.75% which is an external factor that influences and is explained by the number of the equation function of ANOVA. It can be said that the number of tubes in the heat exchanger has a great influence and contribution to the ΔT_{LMTD} [18, 19].

Figure 4 shows that the value of ΔT_{LMTD} calculated through experiments using CFD and the regression equation function is not much different and is almost close to the same value. Equation 4 shows the empirical model generated from the ANOVA, where A is the number of pipes, B is the flow rate, and C is the nanoparticle volume fraction.

$$\Delta T\,LMTD = 38.92 - 19.86\,A + 23.22\,B + 23.3\,C + 3.062\,A^2 - 23.99\,B^2 - 171\,C^2 + 4.53\,AB + 3.4\,AC - 12.3\,BC \quad (4)$$

Figure 5a shows the response normality test TLMTD, also called the normal probability plot ΔT_{LMTD}, which can determine whether the regression to be analyzed in the study is normally distributed or not. The points or data from research through CFD experiments approach the diagonal line which states that the residual value is normally

Table 4. Analysis of variance (ANOVA) result from ΔT_{LMTD}

Source	DF	Contribution (%)	F-Value	P-Value
Model (R-*squared*)	3	87.25	25.1	0.00
Nanofluid flow	1	14.33	12.37	0.00
Volume fraction	1	0.09	0.08	0.005
Number of tube	1	72.83	62.84	0.78
Error	11	12.75		
Total	14	100		

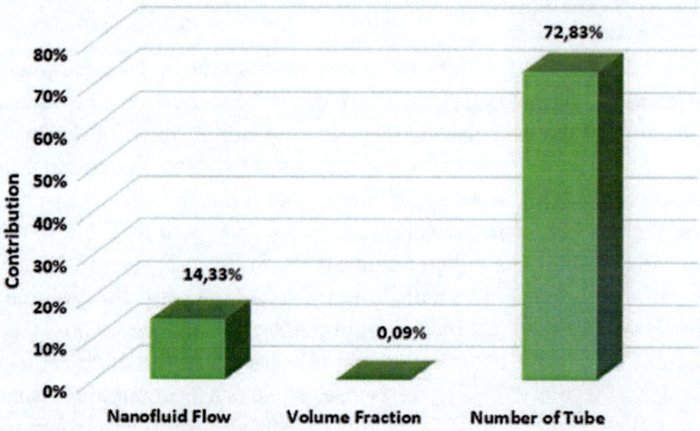

Fig. 3. Percentage contribution of parameter factor to ΔT_{LMTD}

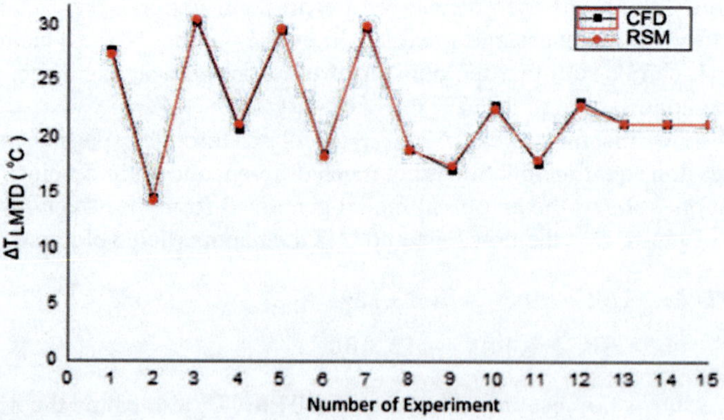

Fig. 4. Comparison of the value of ΔT_{LMTD} through CFD method and regression equation from RSM

distributed. Figure 5b shows the resulting contour plot which consists of various color variations, each of which shows a range of response values. The dark green color on the contour plot shows the maximum condition and this color range will give an indication of the location of the optimum point for the response variable to ΔT_{LMTD} [20–22].

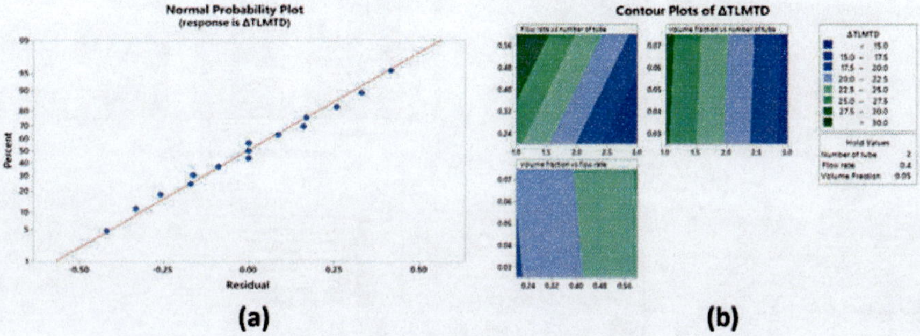

Fig. 5. **a** Normal probability plot and **b** Contour plot of ΔT LMTD

It is clearly seen in Fig. 6a that the value of ΔT_{LMTD} increases when there is an increase in the nanofluid flow rate and volume fraction of nanoparticles. However, it is inversely proportional to the variation in the number of tubes in the heat exchanger; if the number of tubes is increased, the value of ΔT_{LMTD} will decrease so that it cannot reach the optimum value, which according to Silaipillayarputhur and Khurshid [22] that more the contacts between fluids, the faster temperature equivalence occurs due to the black principle law [22]. Figure 6b shows that the overall heat transfer coefficient will be greater if there is an increase in the nanofluid flow rate and volume fraction of nanoparticles. While the variation in the number of tubes in the heat exchanger is inversely proportional because if the number of tubes is increased, the overall heat transfer coefficient value will decrease so that it cannot reach the optimum value where according to previous research conducted by Guo et al. [23] that the thermal resistance value is inversely proportional to the transfer coefficient through heat [23]. Figure 6c shows that the value of the heat transfer rate will be greater if there is an increase in the nanofluid flow rate and volume fraction of nanoparticles. If the variation in the number of tubes in the heat exchanger is increased, the value of the heat transfer rate will also increase according to the previous research conducted by Shahrul et al. [24] that more tubes increase the surface area of the tube so that the value reaches the optimum point in heat transfer [24].

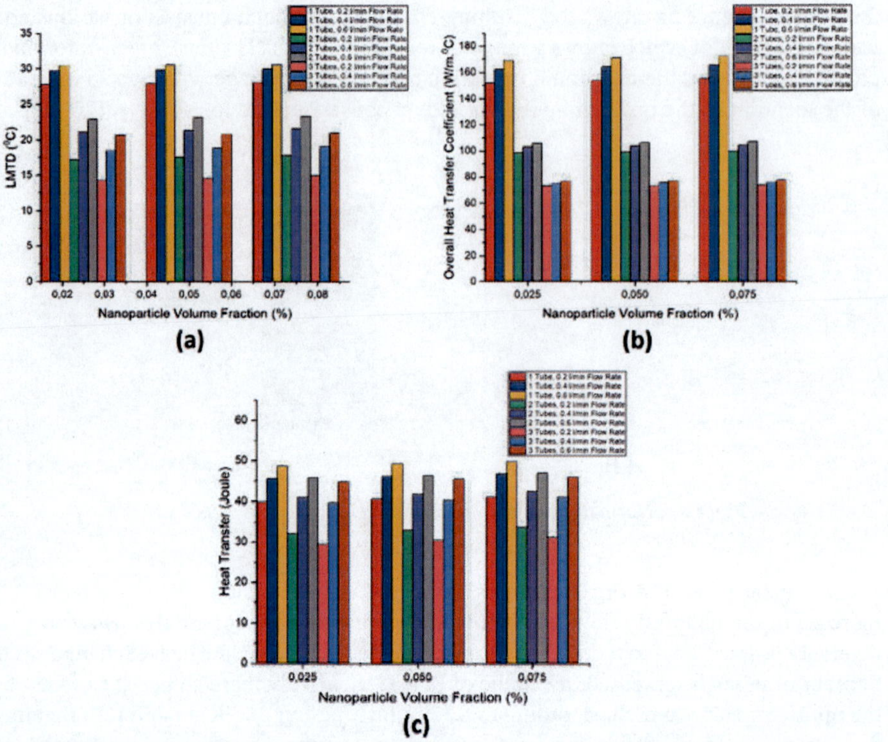

Fig. 6. Heat transfer characteristics for **a** ΔT_{LMTD}, **b** Overall heat transfer coefficient, and **c** Heat transfer with parameter factor

References

1. Permanasari AA, Kuncara BS, Puspitasari P, Sukarni S, Ginta TL, Irdianto W (2019) Convective heat transfer characteristics of TiO2-EG nanofluid as coolant fluid in heat exchanger. In: AIP Conference Proceedings, vol 2120, no. July. https://doi.org/10.1063/1.5115691
2. Nahumury FE (2009) Rekayasa Nanofluida Berbasis TiO2 Sebagai Media Pendingin Pada Sistem Penukar Kalor Rekayasa Nanofluida Berbasis TiO2 Sebagai Media. Universitas Indonesia
3. Sreelakshmy KR, Nair AS, Vidhya KM, Saranya TR, Nair SC (2014) An overview of recent nanofluid research. Int Res J Pharm 5(4): 239–243. https://doi.org/10.7897/2230-8407.050451
4. Hussein AM, Sharma KV, Bakar RA, Kadirgama K (2013) Heat transfer enhancement with nanofluids—A review. J Mech Eng Sci 4(June):452–461. https://doi.org/10.15282/jmes.4.2013.9.0042
5. Paper C (2014) CFD simulation of heat transfer in Cfd simulation of heat transfer in nanofluids containing graphene, no. April, pp 0–7
6. Amani M et al (2017a) 'Modeling and optimization of thermal conductivity and viscosity of MnFe2O4 nanofluid under magnetic field using an ANN. Sci Rep 7(1). https://doi.org/10.1038/s41598-017-17444-5

7. Senthilkumar R, Vaidyanathan S, Sivaraman B (2010) Thermal analysis of heat pipe using Taguchi method. Int J Eng Sci Technol 2(4):564–569
8. Kourkah FF (2017) Optimization of double pipe heat exchanger with 82 response surface methodology using nanofluid and twisted tape. Fluid Mech 3(3):20. https://doi.org/10.11648/j.fm.20170303.12
9. Han HZ, Li BX, Wu H, Shao W (2015) Multi-objective shape optimization of double pipe heat exchanger with inner corrugated tube using RSM method. Int J Therm Sci 90:173–186. https://doi.org/10.1016/j.ijthermalsci.2014.12.010
10. Yang LX, Wang F, Meng YF, Tang QH, Liu ZQ (2013) Fabrication and characterization of manganese ferrite nanospheres as a magnetic adsorbent of chromium. J Nanomater 2013. https://doi.org/10.1155/2013/293464
11. Pak BC, Cho YI (1998) Hydrodynamic and heat transfer study of dispersed fluids with submicron metallic oxide particles. Exp Heat Transf. https://doi.org/10.1080/08916159808946559
12. Shahrul IM, Mahbubul IM, Khaleduzzaman SS, Saidur R, Sabri MFM (2014) A comparative review on the specific heat of nanofluids for energy perspective. Renew Sustain Energy Rev 38:88–98. https://doi.org/10.1016/j.rser.2014.05.081
13. Bhanvase BA, Sarode MR, Putterwar LA, Abdullah KA, Deosarkar MP, Sonawane SH (2014) Intensification of convective heat transfer in water/ethylene glycol based nanofluids containing TiO_2 nanoparticles. Chem Eng Process Process Intensif 82:123–131. https://doi.org/10.1016/j.cep.2014.06.009
14. Senthilraja S, Vijayakumar K, Gangadevi R (2015) A comparative study on thermal conductivity of Al_2O_3/water, CuO/water and Al_2O_3 – CuO/water nanofluids. Dig J Nanomater Biostruct 10(4):1449–1458
15. Holdsworth SD, Simpson R (2016) Heat transfer. Food Eng Ser 17–88. https://doi.org/10.1007/978-3-319-24904-9_2
16. Holman JP (2010) Heat transfer Tenth Edition
17. Kleijnen JPC (2015) Response surface methodology. In: International series in operations research and management science
18. Fadakar Kourkah F (2017) Optimization of double pipe heat exchanger with response surface methodology using nanofluid and twisted tape. Fluid Mech 3(3):20. https://doi.org/10.11648/j.fm.20170303.12
19. Permanasari AA, Sukarni, Puspitasari P, Utama SB, Yaqin FA (2019) Experimental investigation and optimization of floating blade water wheel turbine performance using Taguchi method and analysis of variance (ANOVA). IOP Conf Ser Mater Sci Eng 515(1): 0–10. https://doi.org/10.1088/1757-899X/515/1/012086
20. Permanasari AA, Fadel F, Poppy P, Sukarni S (2020) Effect of additional Mnfe2O4 on a combination of Eg-water nanofluid and volumetric flowrate variations towards heat transfer in shell and tube heat exchanger system. Key Eng Mater 851 KEM, no. January 2018:38–46. https://doi.org/10.4028/www.scientific.net/KEM.851.38
21. Firlianda DA, Permanasari AA, Puspitasari P, Sukarni S(2019) Heat transfer enhancement using nanofluids (MnFe2O4-ethylene glycol) in mini heat exchanger shell and tube. AIP Conf Proc 2120, no. July. https://doi.org/10.1063/1.5115690
22. Silaipillayarputhur K, Khurshid H (2019) The design of shell and tube heat exchangers – A review. Int J Mech Prod Eng Res Dev 9(1):87–102. https://doi.org/10.24247/ijmperdfeb201910

23. Guo ZY et al (2010) Effectiveness-thermal resistance method for heat exchanger design and analysis. Int J Heat Mass Transf Elsevier Ltd 53(13–14):2877–2884. https://doi.org/10.1016/j.ijheatmasstransfer.2010.02.008
24. Shahrul IM et al (2016) Experimental investigation on Al2O3-W, SiO2-W and ZnO-W nanofluids and their application in a shell and tube heat exchanger. Int J Heat Mass Transf Elsevier Ltd 97:547–558. https://doi.org/10.1016/j.ijheatmasstransfer.2016.02.016

Check for updates

Numerical Simulation of the Effect of SA/PA on Combustion and Flow Characteristics in a Front-Rear Burner 400 MWe Boiler

Dimitra Meidina Kusnadi(✉) and Wawan Aries Widodo

Department of Mechanical Engineering, Institut Teknologi Sepuluh Nopember, Surabaya 60111, Indonesia
wawanaries@me.its.ac.id

Abstract. This paper will discuss computational fluid dynamics (CFD) modeling based on a 400 MWe boiler with the once-through tower type which is operated using low-rank coal to determine the effect of secondary air and primary air (SA/PA) ratios on boiler performance. This model was made with confidence after conducting an independence test and validation data on the boiler design. Then the boiler is simulated with SA/PA ratio conditions with different values, namely, 4.89, 3.87, and 3.15. The results show that the recommended SA/PA ratio, in this case 4.89 and 3.15, indicates a lower temperature value and lower flue gas flow velocity, which is also related to the ideal gas. The heat is too high which can potentially cause local heating and slagging in the heat exchanger area. However, low temperatures also indicate a decrease in boiler efficiency. It can be concluded that the SA/PA ratio based on recommendation can produce a temperature that is not too high which is good enough for avoiding slagging, fouling, and local heating but can reduce the efficiency of the boiler. where the low temperature is caused by the increasing number of primary air used can reduce the temperature in the combustion chamber.

Keywords: SA/PA ratio · Combustion · Computational fluid dynamics

1 Introduction

Coal is one of the main fuels that have a low calorific value more abundantly and has been distributed throughout the world [1]. When compared with medium calorific value coal, low-rank coal has several advantages including high reactivity, high volatility, low operating costs, and low pollutant impurities formed [2]. However, some types of low-rank coal have high moisture and ash content which can significantly limit their application in power plants [2]. From the above shortcomings, the moisture content is considered the most detrimental to the efficiency of power plants, where transportation costs are higher and spontaneous combustion occurs during storage [3, 4]. Given the negative impact on combustion behavior by moisture content in low-rank coal, then there is a pre-drying technology that is used to reduce the water content in coal. And many studies have been conducted to investigate the increase in power generation efficiency

I. Tolj et al. (eds.), *Recent Advances in Mechanical Engineering*, Lecture Notes in Mechanical Engineering, https://doi.org/10.1007/978-981-19-0867-5_35

when the pre-drying technique is used and how the pre-drying coal burns [4–6]. From several studies on the effect of primary air ratio on combustion, Zixiang Li et al. said that increasing PAR has a large effect on combustion performance, the overall combustion process worsens with increasing PAR [7]. The amount of primary air and secondary air is considered as a determinant that affects the performance of a boiler with low-rank coal because it affects the mixing process and the combustion behavior of coal particles in the main combustion area. The combustion behavior and the rate of degradation of the boiler performance are obvious when the PAR is increased to 0.425 [8].

In this paper, there are operating variations of the SA/PA ratio that will be investigated, to find out how the performance of boilers with low-rank coal operated under the design SA/PA ratio or operating conditions. To achieve this goal, a 3D CFD model was made based on a 400 MWe boiler-type once-through tower with a front-rear burner model. After the model is validated, it is used to study the combustion behavior and boiler performance at different SA/PA ratio conditions. The first case is based on the minimum value of primary air velocity at Babcocks and Wilcox so that the maximum SA/PA ratio is 4.89. The second case is based on the median value of the first and third cases, which is the median value of the recommendation, and the minimum SA/PA ratio of Babcocks and Wilcox, which is 3.87. And the third case is based on current operating conditions and is very close to Babcocks and Wilcox's recommendation, which is 3.15. The results show that the difference in SA/PA ratio has a significant effect on the distribution of flue gas produced, both in terms of combustion, heat transfer, and flow.

2 Numerical Scheme and Methodology

2.1 Boiler Operation and Configuration

The simulation domain is based on a 400 MWe once-through tower boiler with a front-rear burner type. As shown in Fig. 1a, the boiler is 65 m high with a horizontal cross section of 18 m × 12.5 m on the furnace side. Inside the boiler, there are several heat exchanger panels including a water wall tube on the furnace wall, primary superheater (PSH), secondary superheater (SSH), reheater (RH), and economizer. To simplify the modeling of the heat exchanger, the water wall is considered a wall that absorbs heat and the other is modeled as a heat exchanger in FLUENT 18.1. There are five lines of nozzles for inlet coal and air to burn, each line has seven nozzles. The coal inlet is united with the primary air, while the secondary air is outside the primary air envelope. The details can be seen in Fig. 1b. During the simulation process, the composition of the coal used is identical to the operating data at the power plant. More details with proximate and ultimate analysis can be seen in Table 1.

2.2 Boundary Condition and Simulation Scenario

In CFD simulation, the boundary condition is very important to simulate the initial condition of the boiler. There are several boundary conditions used, including mass flow inlet for primary air and coal in the coal pipe, mass flow inlet for secondary air, and pressure outlet for flue gas flow out of the boiler. And in the simulation, there is one row of nozzle burner that is disabled, namely, elevation B in Fig. 1a.

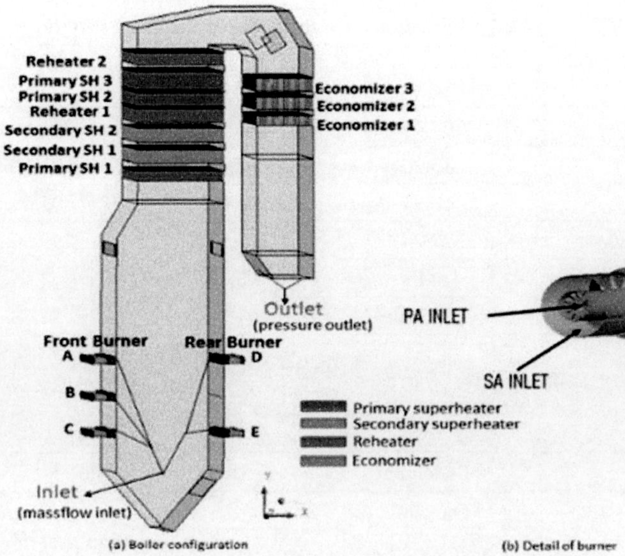

Fig. 1. Schematic of the whole boiler and details of the burner nozzle

Table 1. Properties of the coal used in the simulation

Proximate analysis (wt%,ar)				Ultimate analysis (wt%,daf)					LHVar (KCal/kg)
Fixed carbon	volatile	Ash	moisture	C	H	O	N	S	
33.84	31.85	4	30.31	76.6	5.56	16.5	1	0.37	4800

To determine the effect of the SA/PA ratio, three case variations were used in this simulation. As stated in Table 2, case 1 is the maximum value of SA/PA calculated based on the minimum velocity limit of primary air in the coal pipe, case 2 is the middle value of case 1 and case 3, and case 3 is the lowest SA/PA based on the operating data obtained. very close to Babcocks And Wilcox's recommendation.

Processing is a set process in commercial CFD software to simulate the combustion process that occurs in the boiler, here are some settings on commercial CFD software that will be used in this study, one of which is modeling (Table 3).

3 Result and Discussion

In the following sections, the distribution of combustion temperature, flue gas flow rate, and CO mass fraction will be used to analyze boiler performance in different cases.

Table 2. Detailed information on velocity burner and case variations

Case	1	2	3
Total air feed (kg/s)	394.15	394.15	394.15
Velocity PA (m/s)	15.20	18.40	21.60
PA flow total (kg/s)	66.91	81.00	95.09
SA flow total (kg/s)	327.24	313.15	299.06
SA/PA	4.89	3.87	3.15
Coal feed (kg/s)	51.38	51.38	51.38

Table 3. Model used in CFD simulation

Model	Statement	Rationale
Energy equation	On	Includes convection and radiation heat transfer
Viscous (turbulence model)	On (species transport) K-ε standard	Can solve the problem of heating, bouyancy, compressibility, and combustion
Heat exchanger	On	Design data PLTU, NTU model
Species	Species transport	Because the chemical species injected have a chemical reaction (combustion)
Discrete phase	On	Define injection for coal inlet

3.1 Flue Gas Temperature Distribution

Figure 2 depicts the contours of the temperature distribution in the vertical section (Z = 0). We can see in the figures that the temperature distribution pattern is the same for all cases in the cone hopper, where the temperature is very low because very few coal particles are burned inside.

The thing that is of concern is the furnace area to the primary superheater 1, where the SA/PA ratios of 4.89 and 3.15 produce a lower temperature value than the SA/PA ratio of 3.87. the SA/PA ratios 4.89 and 3.15 almost have the same temperature distribution in the backpass area, but differ in the furnace area where the SA/PA 4.89 tends to be higher than 3.15. As for what is shown from the temperature contour above, at SA/PA 4.89 and 3.87 flue gas with high temperature leads to the left side or commonly called the front burner area. With this uneven temperature distribution, there is a tendency for several problems to occur, including slagging and fouling in the ranks of the heat exchanger and local heating.

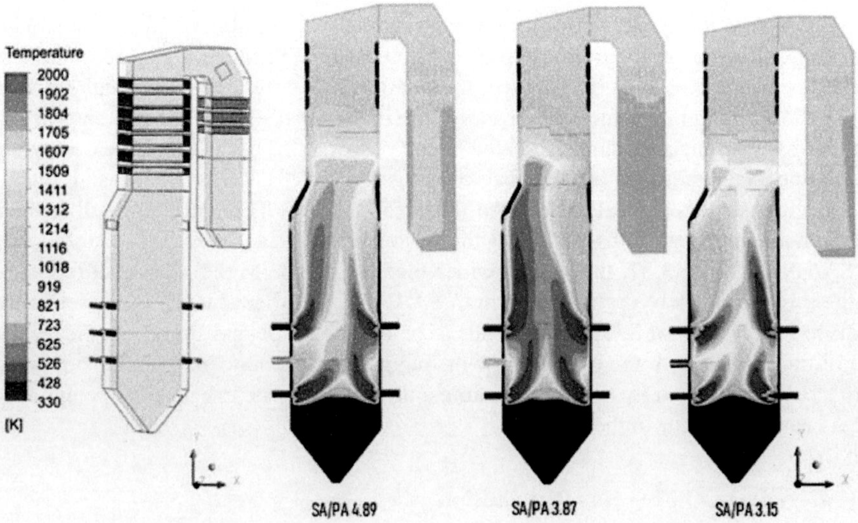

Fig. 2. Fluegas temperature distribution at vertical cross section (z = 0)

3.2 Flue Gas Velocity Distribution

Figure 3 depicts the velocity distribution contour on the vertical section (Z = 0). The figure shows that from the three cases the velocity contours show almost the same pattern, the lowest velocity is at the bottom of the boiler (cone hopper) and the boiler outlet with a value of 4–6 m/s.

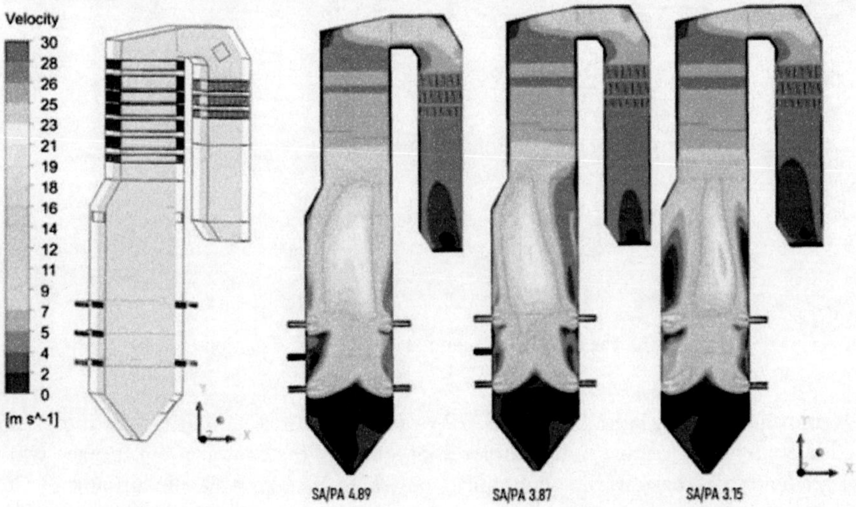

Fig. 3. Fluegas velocity distribution at vertical cross section (z = 0)

The high velocity is at the combustion air inlet with a value of 25–30 m/s. When it becomes flue gas in the furnace velocity decreases in the range of 14–18 m/s. When the temperature is high in the furnace, the flue gas flow rate also shows high numbers around 21–22 m/s in the furnace area after the highest burner level before entering the heat exchanger range, which is indicated by the yellow to the light orange area. However, as the flue gas passes through the heat exchanger velocity decreases because it hits the heat exchanger tubes and is related to the ideal gas law. There are some differences between the three variations, including the yellow area with a value of 21–23 m/s, where in SA/PA 4.89 and 3.87, the area is wider than SA/PA 3.15. the second difference is in the direction, where the yellow area in SA/PA 4.89 and 3.87 tends to the left of the furnace wall, while for SA/PA 3.15 tends to be in the middle and slightly to the right of the furnace wall. With the uneven flow of flue gas and the tendency to one side of the furnace wall followed by high temperatures, things like this can cause the potential for local heating to occur in the boiler.

3.3 CO_2 and CO Flue Gas Distribution

In the combustion process, there are several chemical compounds that will be produced, one of which is carbon dioxide (CO_2). The result of the combustion reaction that produces CO_2 is an indication of the perfect combustion reaction of carbon that reacts with oxygen. And here is the CO_2 contour contained in the XY plane boiler (Fig. 4).

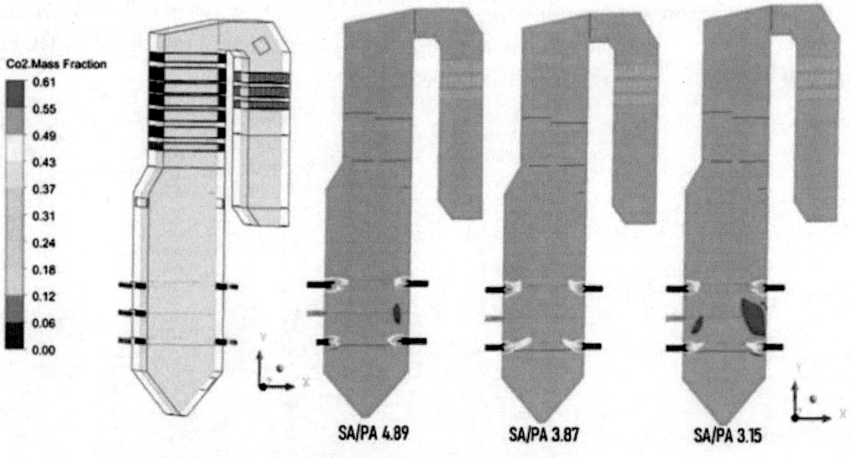

Fig. 4. Fluegas CO_2 distribution at vertical cross section (z = 0)

From the three variations of the SA/PA ratio, there is a similarity that the carbon dioxide content is the most in the burner inlet area and the center of the furnace which will gradually decrease to the economizer outlet. In SA/PA 4.89, the amount of CO_2 is found on the side of the level 2 burner (burner B) to the middle of the furnace and decreases and stabilizes around 0.5 in the entire boiler room. In SA/PA 3.87, the CO_2 content does not show a high value and tends to be stable throughout the boiler room, indicated by the orange color with a value of 0.5. And in case (c), there is a lot of CO_2 content in burner level 2 (burner B) which is higher up to 0.6. This can indicate that a certain SA/PA value based on recommendations in SA/PA 4.89 and 3.15 indicates more complete combustion.

In the combustion process, there are several chemical compounds that will be produced, one of which is carbon monoxide (CO). The result of the combustion reaction that produces CO is an indication that a combustion reaction is not yet complete. This can cause several losses including being a pollutant to the environment if released into the air, and also incomplete combustion where these compounds can produce heat if completely burned. Therefore, this CO compound is included as losses in boiler efficiency. And the following is a picture of the distribution of CO exhaust gas in the boiler room in Fig. 5.

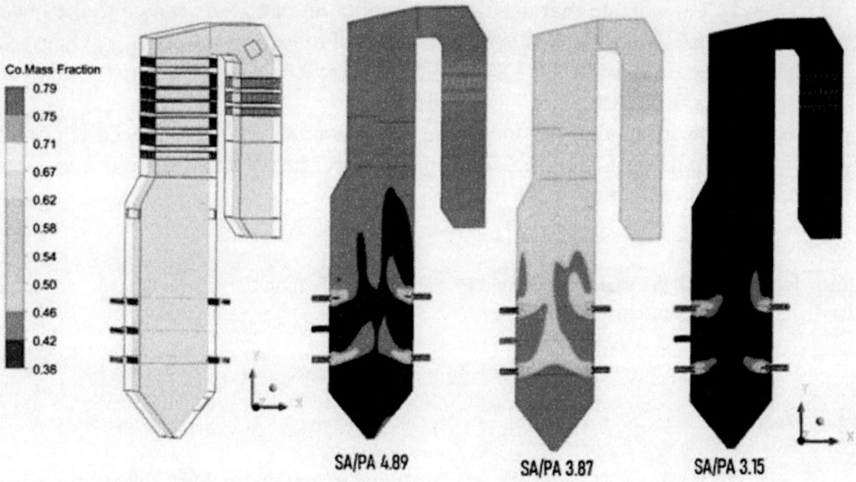

Fig. 5. Fluegas CO distribution at vertical cross section (z = 0)

For the three variations of SA/PA, the amount of CO contained in the inlet area of the combustion air and fuel alone is shown in red and yellow with mass fraction values of 0.67–0.79. There is a significant difference between the three variations in the overall boiler room. The CO content from the largest to the smallest was in SA/PA 3.87, 4.89, and 3.15, respectively. In SA/PA 3.87, the CO mass fraction in the boiler room ranges from 0.42 to 0.5, while in SA/PA 4.87, it ranges from 0.36 to 0.46 and for SA/PA 3.15, it is 0.36–0.42.

4 Conclusions

In this paper, the CFD model is based on a 400 MWe boiler which has been validated using design data. Then it is used to simulate three cases of SA/PA ratio with a value of 4.89, 3.87, and 3.15, respectively. From the above results, several conclusions can be drawn:

1. The heat transfer characteristics indicated by the temperature distribution show that the best to the worst heat absorption in the furnace area, respectively, are SA/PA 3.15, 4.89, and 3.89. Poor heat absorption in the furnace area results in high flue gas temperatures which can heat the heat exchanger ranks, especially in the primary superheater and secondary superheater, so that there is a local heating area that can cause considerable slagging and fouling. The potential for local heating and the formation of slagging and fouling from the highest to the lowest occurred, respectively, at SA/PA 3.87, 4.89, and 3.15 from how much temperature the boiler has.
2. The flow characteristics indicated by the velocity distribution illustrate that the highest to lowest velocity in the flue gas flow is SA/PA 3.87, 4.89, and 3.15. The greater the temperature in the boiler, the greater the velocity in the boiler room, this is related to the ideal gas.
3. Combustion characteristics indicated by the distribution of exhaust gases in the form of CO and CO_2 indicate that the highest production of CO_2 is shown to the lowest occurring at SA/PA 3.15, 4.89, and 3.89. The highest and lowest CO exhaust gas production occurred in SA/PA 3.87, 4.89, and 3.15. It can show the most combustion characteristics of SA/PA 3.15, 4.89, and 3.89.
4. There are indications that the lower the flue gas temperature produced is good to avoid the formation of slagging, fouling, and local heating, but there is a tendency to decrease the efficiency of the boiler.

Acknowledgements. We thank the related PLTU and ITS Techno Sains who have provided data and assisted in this research.

References

1. Si CD XS, Wu JJ, Wang Y, Zhang YX (2015) Drying of low-rank coals: a review of fluidized bed technologies. Dry Technol 33:277–287
2. You C, Wang HM (2014) Experimental investigation into the spontaneous ignition behavior of upgraded coal products. Energy Fuels:2267–2271
3. Defu C, Gao XZ, Man CB, Hu SJ, Xu XY (2012) Theoretical and experimental study on spontaneous ignition of lignite during the drying process in a packed bed, pp 6876–6887
4. H J, Rao ZH, Zhao YM, Huang CL, Duan CL (2015) Recent developments in drying and degring for low rank coals. Prog Energy Combust Sci 46:1–1
5. M A, Osman H, Jangam SV, Lease JD (2011) Drying of Low-Rank Coal (LRC)—A Review of Recent Patents and Innovations. Dry Technol, 29:1763–1783
6. B C, Nikolopoulos N, Violidakis I, Karampinis E, Agraniotis M, Grammelis P et al (2015) Report on comparison among current industrial scale lignite drying technologies (A critical review of current technologies). Fuel 155:86–114
7. Li Z, Miao Z, Zhou Y, Wen S, Li J (2018) Influence of increased primary air ratio on boiler performance in a 660 MW brown coal boiler. Energy 152:804–817. https://doi.org/10.1016/j.energy.(2018)
8. Li Z, Miao Z, Shen X, Li J (2018) Prevention of boiler performance degradation under large primary air ratio scenario in a 660 MW brown coal boiler. Energy 155:474–483. https://doi.org/10.1016/j.energy.(2018)

Combining Correlation Technique with Exhaustive Search Feature Selection Method for Rotating Machinery Fault Diagnosis

Mochammad Solichin[1,2], Yoojeong Noh[1(✉)], and Young-Jin Kang[3]

[1] School of Mechanical Engineering, Pusan National University, Busan, South Korea
[2] Department of Mechanical Engineering, Institut Teknologi Sepuluh Nopember, Surabaya 60111, Indonesia
[3] Research Institute of Mechanical Technology, Pusan National University, Busan, South Korea

Abstract. The feature selection in classification tasks as a need for diagnosing a fault in rotating machinery exceedingly plays a very notable role in the machine learning framework. Consequently, this study proposes a combination of a correlation technique with exhaustive search as a feature selection method for diagnosing rotating machinery faults. In any case, this method can be called as a hybrid method because it combines between the correlation technique as a filter method and an exhaustive search as a wrapper method. The correlation of each feature with the target of labeling normal and abnormal information is measured by Pearson's correlation matric, in which abnormal conditions indicate a failure in the rotating machinery. The top five best correlations at this stage are taken as the selected feature in the filtering stage. Based on these features selected, each feature is combined and its performance is considered through the training of a classification model. The combination of features with the highest accuracy is the final selected feature subset. Finally, the proposed method is able to successfully demonstrate the diagnosis of rotating machinery faults with normal and abnormal classification.

Keywords: Correlation technique · Exhaustive search · Feature selection · Rotating machinery · Fault · Diagnosis · Classifications

1 Introduction

Rotating machinery diagnosis is now required to be able to work in a sophisticated manner so that a problem that occurs can be quickly resolved. Rotating machinery is widely applied in modern industry with a very important role [1]. While one of the accelerations of technological developments in the field of diagnosis is the application of machine learning methods. Machine learning has several tasks that can be implemented. In the case of diagnosing rotating machinery, the task that can be used is the construction of a classification model such as diagnosing failures that can be classified into normal and abnormal rotating machinery conditions.

I. Tolj et al. (eds.), *Recent Advances in Mechanical Engineering*, Lecture Notes in Mechanical Engineering, https://doi.org/10.1007/978-981-19-0867-5_36

The need for effectiveness and efficiency to build a classification model with machine learning is determined by the accuracy in selecting features. The features are input data that will be trained in the classification model development process in the form of data extracted from raw data or data transformation. Redundant features which are commonly called unimportant features, will only make learning chaotic and do not have a good impact on the construction of the classification model.

Several selection methods are available such as filter, wrapper, hybrid, and embedded methods. Each method has advantages and disadvantages. The filter method can quickly select features based on certain criteria, for example, proximity based on feature correlation with a very strong output target classification to identify relevant variable relationships [2], or statistically use the chi-square method that has been developed to reduce the number of features [3]. Although the filter method has the ability that does not take much time, the results of the selected features do not involve the training process in the classification model. So, this method does not guarantee high accuracy performance after the feature is trained, then produces a classifier. The wrapper method is a combination of features that involves developing a classification model to obtain the best features based on the highest accuracy of the classification performance [4]. Thus, this method is more effective than the filter method, which is expected to have higher computations. An example of a wrapper method is an exhaustive search. It is capable of finding great feature combinations as it iterates sequentially according to a set number of features [1]. Another feature method is the hybrid method. The hybrid method takes advantage of the filter method, which can recommend a good ranking of features and combine it with a wrapper method to get the best combination of features according to the classification model training. If feature selection at the filter stage can be obtained properly and quickly, the hybrid method will be an effective and efficient feature selection method. The latter feature selection method is embedded. Embedded is a feature selection method that is similar to a hybrid which both utilizes filter and wrapper methods. The difference is that embedded learning is built by building its own feature selection [5].

Based on existing feature selection methods such as the affordable mentioned, the hybrid method by combining filter and wrapper methods has many advantages. Irrelevant features can be reduced from the filter method, while the combination of feature subsets can be optimized using wrappers. Two methods that, if both can be optimized, will generate accurate and efficient classifiers.

On the other hand, several classifier algorithms can be used to train and test classification models such as support vector machines (SVM), k-nearest neighbors (KNN), and multi-layer perceptron (MLP). The three classifiers have different characteristics in the feature data learning process. SVM is supervised learning that utilizes the kernel. SVM is widely used to diagnose failures in mechanical components [6]. KNN is widely used in the case of pattern recognition, which has the power to always perform accurately. Meanwhile, MLP has great power for complex problems in industrial processes [7].

To sum up, this study proposes combining correlation techniques with exhaustive search as a hybrid feature selection method for diagnosing failures in rotating machinery. Some of the highlights of this study are proposing a feature selection method by combining correlation techniques and exhaustive search to build a classification model for diagnosing rotating machinery failures and verifying the proposed method by running several classifiers such as SVM, KNN, and MLP including cross-validation for generalizing the proposed method to a variety of classifiers.

2 Data Collection

Raw data for this study were conducted using data from the National Aeronautics and Space Administration (NASA). The data repository used is a bearing dataset which is experimental data generated by the Center for Intelligent Maintenance Systems (IMS), University of Cincinnati, http://ti.arc.nasa.gov/project/prognostic-data-repository [8]. The format of the available data is in the form of text which has then been extracted by the researcher into numeric data.

The failure occurred in the bearing, which is part of a series of rotating machinery. Experiments were carried out by involving the motor as a driver, shaft, and bearing. The type of bearing failure that has occurred is a roller element defect that occurs after going through 100 million revolutions. The data obtained from the experimental results are in the form of a vibration response in the form of acceleration which is measured using an accelerometer. The data has 20,480 points with a sampling rate of 20 kHz for one measurement. While the total measurement consists of 2155 times over 5 weeks which are recorded every 10 min. Figure 1 displays data from the first week to the last week, with the length of the data being 44,134,400. The vibration response in the form of acceleration, which is plotted continuously from the first measurement until the failure happened. The data represent normal and stable conditions that started from the beginning of the measured data until the 1434th measurement of the total number 2,155 measurement data. So, the total abnormal data is 471 times the measurement data.

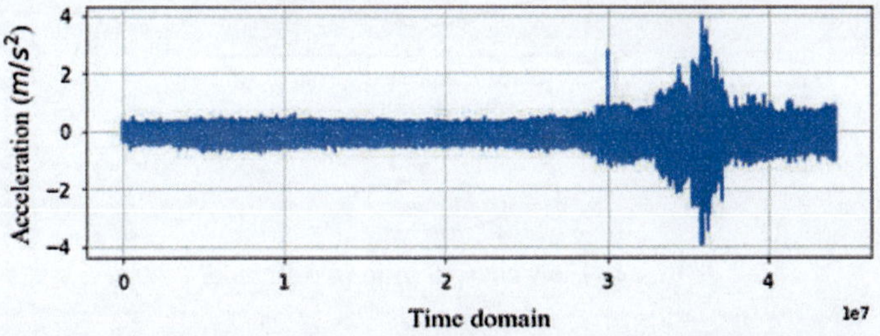

Fig. 1. Acceleration response with length of data points during 5 weeks.

3 Proposed Method

This study proposes a combination of correlation techniques to reduce the dimensions of the features generated from the transformation and extraction of raw data with exhaustive search. Data transformation is a process to add a domain that was originally only a time domain, but in this study, the domain was expanded by applying a fast Fourier transform to obtain a frequency domain. Meanwhile, the extraction is done based on both the time domain and frequency domain statistically. So, as an overview of the proposed method, it is shown in Fig. 2.

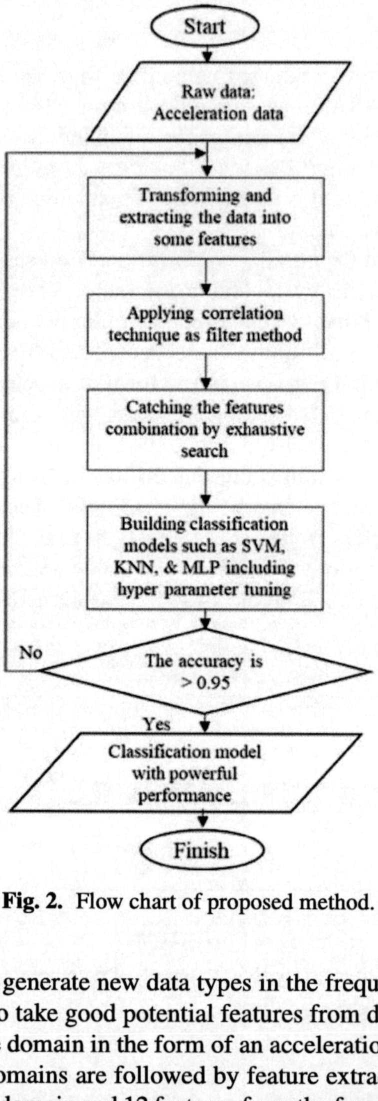

Fig. 2. Flow chart of proposed method.

Data transformations generate new data types in the frequency domain. This transformation is carried out to take good potential features from different domains because it is possible that the time domain in the form of an acceleration response does not have relevant features. Both domains are followed by feature extraction which will produce 12 features from the time domain and 12 features from the frequency domain. These features are absolute mean value (abs_mean), peak to peak value (peak_m), kurtosis (kur.), skewness (skew.), mean value (mean), standard deviation (std.), first–third quantiles (25%, 50%, and 75%), and maximum value (max.).

The available features are reduced to eliminate redundancy and reduce computation time for further processing. This stage is called the filtering stage. The strategy used for the process of reducing the number of features at this stage is using Pearson's correlation techniques. If a feature has an independent linear relationship with the target in the form of normal and abnormal classifications on rotating machinery, the correlation coefficient value is $+1$ or -1. However, the value of the correlation coefficient will be 0 if the feature has absolutely no independent linear relationship with its output target. In addition, this correlation technique can also be used to determine the relationship between features that can be presented in a correlation matrix. The high coefficient value between features and targets, namely, the best top 5, is chosen as an important feature which in addition is a deleted feature. The formula for the correlation coefficient, which is denoted by r, can be calculated using Eq. (1).

$$r = \frac{\sum(X_i - \overline{X})(Y_i - \overline{Y})}{\sqrt{\sum(X_i - \overline{X})^2}\sqrt{\sum(Y_i - \overline{Y})^2}} \tag{1}$$

where (X, Y) is a pair of variables with the ith observation. Based on the calculation of the correlation technique Fig. 3 displays the correlation coefficient matrix.

Figure 3 shows that the features with the strongest relationship with initial as output representing normal and abnormal bearings with the best top five are features max_F, mean_T, peak_m_F, 50%_T, and skew_F. Max_F which denotes the maximum value of the frequency domain, has the highest correlation coefficient value of 0.87. The frequency domain is the best alternative when the time domain cannot provide strong information to distinguish between normal and abnormal conditions in rotating machinery. The maximum value in the frequency domain will determine whether an unusual condition occurs in rotating machinery, in this case, a bearing failure so that the maximum value in the frequency domain is able to provide useful information to distinguish normal and abnormal data. Likewise, the best top five were determined by the researcher because they had a good deviation in the coefficient value. Therefore, it is determined that the best top five is the best feature in this data dimension reduction stage. The five selected features are then used to find a good combination of features with an exhaustive search. The iteration process of feature combinations that can be formulated to find out the number of combinations is denoted as H that can be formulated as Eq. (2).

$$H = \sum_{s=1}^{s} \frac{m!}{(m-s)!s!} \tag{2}$$

Simplified, the list of combined feature sets becomes Y_h as Eq. (3).

$$Y_h = \{y_j | j = 1, 2, 3, \ldots, m; y_j \in F\}, \tag{3}$$

where $m = (1, 2, \ldots, h)$, h is the number of selected features, s is the total subset, and F is the feature selected based on correlation technique.

The results of the running process of a combination of features that are integrated with the process of building a classification model for normal and abnormal rotating machinery are shown in Fig. 4.

Based on Fig. 4, the combination of 50%_T and skew_F features is the best subset feature that can be used as the final selected feature. Skewness in the frequency domain

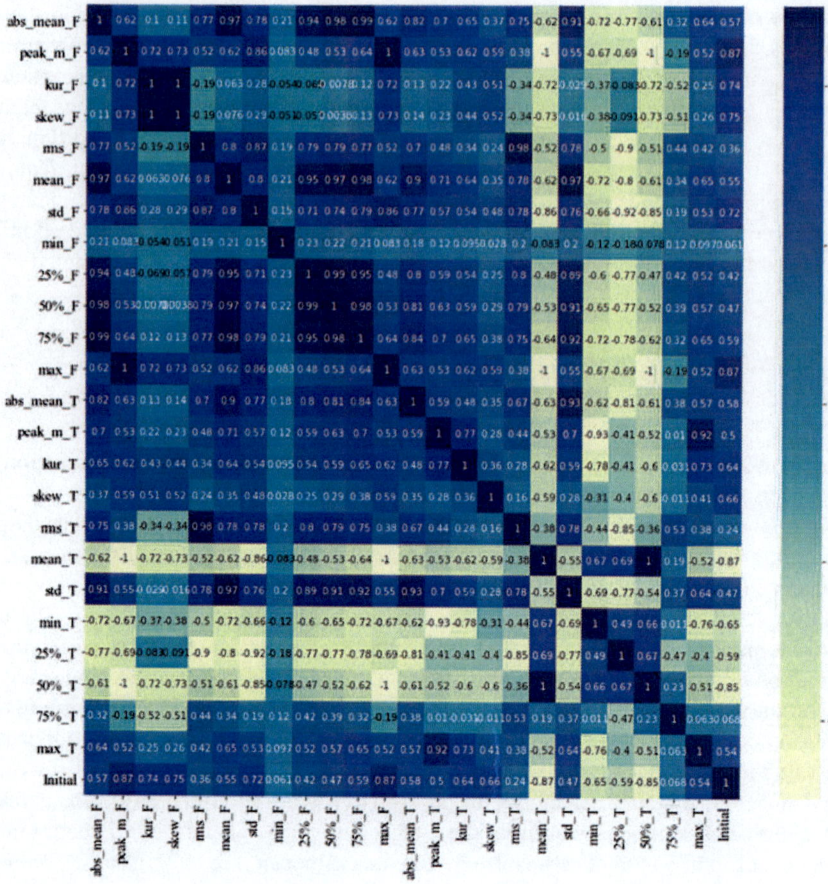

Fig. 3. Correlation coefficient matrix of all features and the initial as the target.

(skew_F) can show the degree of asymmetry of the probability density relative to the mean value between normal and abnormal rotating machinery. Meanwhile, 50% of the data in the time domain which is below the value point of the 50%_T feature also has good information to distinguish between normal and abnormal conditions. The combination of features derived from the frequency and time domains, where the second quantile data from the time domain and skewness from the frequency domain can present the highest accuracy to build a classification model.

To verify the proposed method, the selected subset features are applied using three different classifiers, namely, SVM, KNN, and MLP. In addition, validation is also carried out by considering cross-validation. The method used to validate is ten-fold cross-validation, in which the training data is split into ten parts to be tested with ten folds. In addition, to maximize the results of the model development, this study also performed a hyper-parameter optimization process using a grid search. Grid search is a tuning process by creating all possible combinations of hyper-parameters to find optimum values.

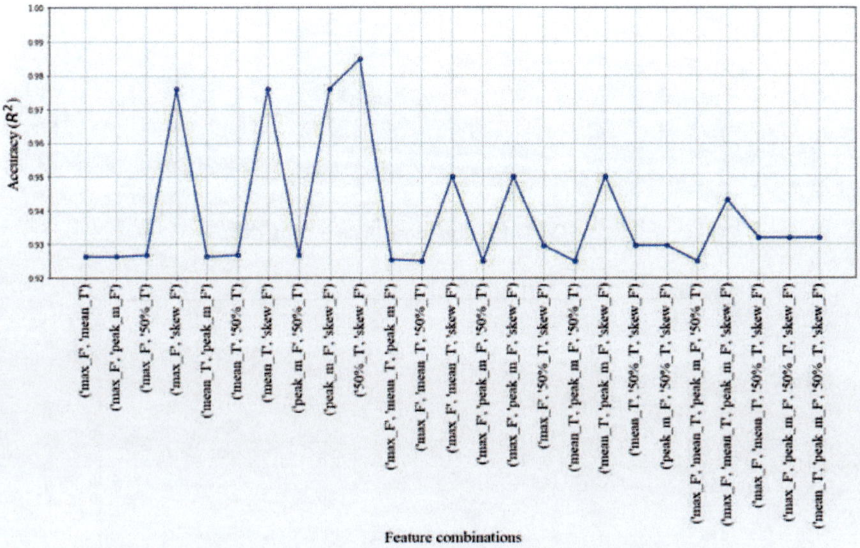

Fig. 4. Accuracy for all feature set combinations.

These are all processes to obtain the best and general classification model, with the final decision is the accuracy must be greater than 0.95.

4 Classification Results

The proposed method that has been presented is applied by using one of the datasets from NASA as described in Sect. 2. After the final feature set is obtained, Fig. 5 displays a two-dimensional visualization of the classification model built using SVM, KNN, and MLP, while Fig. 6 demonstrates ten-fold cross-validations. All classification models have satisfactory performance. The accuracy of the three models was 0.99 in testing. Meanwhile, in terms of efficiency, the execution time required for each training with hyper-parameter tuning up to test the data is 5.13 s, 100.06 s, and 60 s, respectively, SVM, KNN, and MLP. In addition, the results of the ten-fold cross-validation show that the model is not overfitting and robust because the model still has an accuracy distribution around the accuracy value of the test data. Thus, the proposed method for selecting features can work very satisfactorily, although there is still a slight distribution of accuracy in the cross-validation of the SVM and KNN models and an outlier in the KNN model.

5 Conclusions

The proposed method to select the most relevant features in building a classification model as a failure diagnosis of rotating machinery performs successfully which is satisfied in terms of effectiveness and efficiency. Three classification models built with

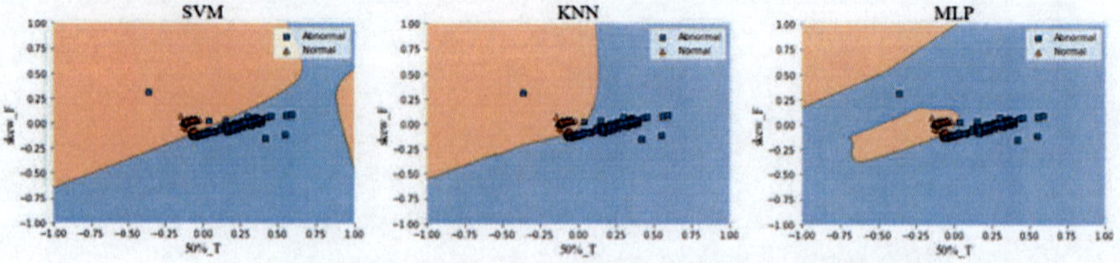

Fig. 5. Classification models on testing data.

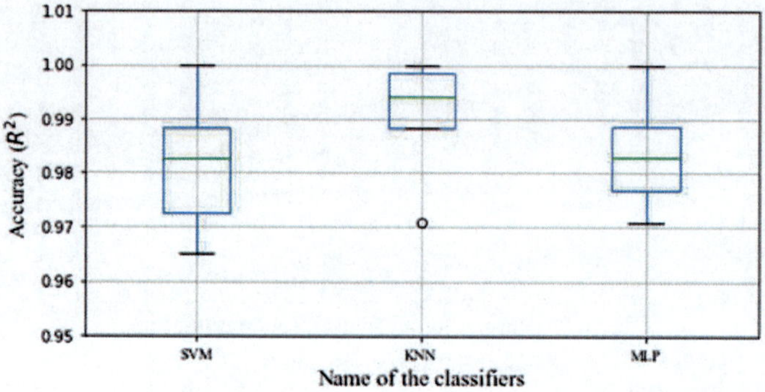

Fig. 6. 10-Fold cross-validation.

selected features and ten-fold cross-validation prove that the proposed method can be applied to several classification models while maintaining not overfitting and also robustness. In addition, from the computational aspect, the use of the features proposed by this method does not require an expensively time-consuming. This study was only trained and tested using one type of dataset, so further studies are still needed to apply this proposed method with several datasets to create a proposed method that can generalize to all rotating machinery fault conditions.

Acknowledgements. This work was supported by the National Research Foundation of Korea (NRF) grant funded by the Korea Government (MSIT) (No. 2020R1A5A8018822).

References

1. Mochammad S, Kang YJ, Noh Y, Park S, Ahn B (2021) Stable hybrid feature selection method for compressor fault diagnosis. IEEE Access 9:97415–97429
2. Chandra E, Karthikeyan E (2012) A feature selection algorithm using correlation based method. Algorithms Comput Tech 6(3):385–394
3. Said B, Abdellah M, Mohammed A, Mohamed K (2020) Feature selection using an improved Chi-square for Arabic text classification. King Saud Univ Comput Inf Sci 32:225–231
4. Aydemira O, Ergünb E (2019) A robust and subject-specific sequential forward search method for effective channel selection in brain computer interfaces. Neurosci Methods 313:60–67
5. Maldonado S, Lopez J (2018) Dealing with high-dimensional class-imbalanced datasets: embedded feature selection for SVM classification. Appl Soft Comput 67:94–105

6. Songyot N (2018) High-dimensional hybrid feature selection using interaction information-guided search. Knowl-Based Syst 145:59–66
7. Kai S, Xiuliang W, Jingyu X, Ma F (2019) Development of a new multi-layer perceptron based soft sensor for SO2 emissions in power plant. Process Control 84:182–191
8. Lee J, Qiu H, Yu G, Lin J, Rexnord Technical Services (2007) IMS, University of Cincinnati. "Bearing Data Set", NASA Ames Prognostics Data Repository (http://ti.arc.nasa.gov/project/prognostic-data-repository), NASA Ames Research Center, Moffett Field, CA

Trajectory Planning of Cable-Driven Parallel Robot with Mobile Cranes

Made Bhaswara Wiranugeraha, Hor Tan, and Latifah Nurahmi(✉)

Department of Mechanical Engineering, Institut Teknologi Sepuluh Nopember, Surabaya 60111, Indonesia
latifah.nurahmi@me.its.ac.id

Abstract. Cable-Driven Parallel Robot (CDPR) with mobile crane can be an alternative solution for helping evacuation process in post-disaster sites. This paper aims to carry out the trajectory planning of CDPR with mobile crane and hence the reconfiguration of mobile crane can be executed to avoid instability. This instability occurs due to high cable tension. Trajectory was generated based on fifth-degree polynomial that provides position, velocity, and acceleration profiles of the end-effector. Then, the end-effector moves along the trajectory while the mobile crane adjusting its length. The prototype of CDPR with mobile crane was built to verify the theoretical results. Reconfiguration of CDPR is also proposed to lift higher payload capacity while maintaining stability of the mobile crane.

Keywords: Cable-driven parallel robot · Mobile crane · Trajectory · Reconfiguration · Optimization

1 Introduction

Cable-Driven Parallel Robot (CDPR) with mobile crane can be a novel solution to accelerate the evacuation process in the post-disaster cites. CDPR is a type of parallel manipulator where the rigid links are replaced by cables. CDPR has been developed for many purposes, namely, material handling, telescopic, sport camera, search-and-rescue, and medical activities. FASTKIT [1] has been developed for providing a low cost and accomplished robotic solution for logistics. Another developed prototype of CDPR is called MoPICK [2] for performing multiple manipulation tasks in a constrained environment. Another application of CDPR has been proposed for carrying heavy loads in construction [3]. This robot consists of mobile crane, cable driving devices, platform, cable, and booms. The global path planning of CPRMC was performed with the grid-base artificial potential field method.

Trajectory planning should be carefully defined to obtain significant performance of the robot, especially in high-speed operations. Trajectory planning of CDPR can be generated in vertical oscillation and horizontal circular path by considering static, dynamic, and kinematic behaviors [4]. Tension distribution, velocity, and cable length errors are the parameters that should be investigated to see if the trajectory can be fulfilled while maintaining the requirements of the robot. Another research has been carried out

I. Tolj et al. (eds.), *Recent Advances in Mechanical Engineering*, Lecture Notes in Mechanical Engineering, https://doi.org/10.1007/978-981-19-0867-5_37

regarding dynamic trajectory generation of reconfigurable suspended cable robot [5]. Additional reconfigurability ensure positive cable tension throughout the trajectory. The reconfigurability is important to be considered so that the robot can easily adapt to different work environments. New type of CDPR by using three mobile bases with reconfigurable crane was introduced in [6]. The static equilibrium is evaluated in the reconfiguration to determine the cable tension and static feasible workspace of the end-effector. It is noted that cable tension has a significant effect on the stability of mobile cranes.

This paper will present the trajectory planning of CDPR with three mobile cranes for post-disaster sites. Mathematical analysis of trajectory and reconfiguration method of CDPR are discussed. The reconfiguration of mobile crane is determined through optimization based on pareto optimal solution. The experimental setup is provided to demonstrate and verify the theoretical results.

2 Geometric Analysis

CDPR developed in this paper is composed of three mobile bases that have global coordinate (O) and axes (X, Y, Z) located at the ground center. Each mobile base has local coordinate $(O_i)(i = 1, 2, 3)$ of distance ρ_i from point O_i to point O. Geometric model and prototype of the robot are shown in Fig. 1.

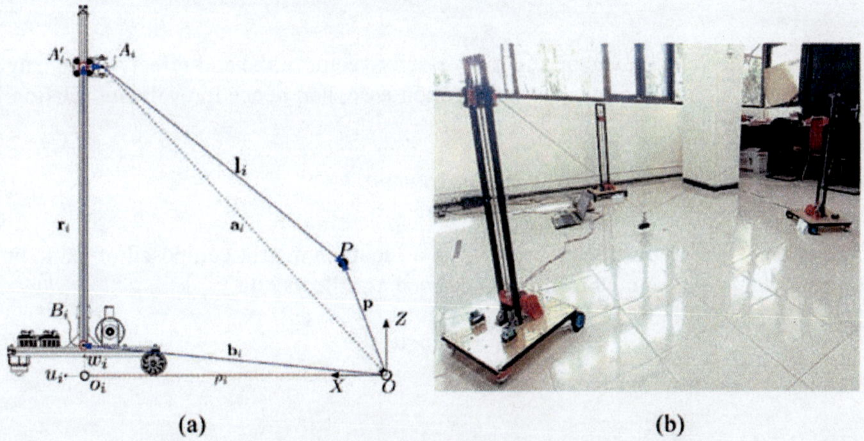

Fig. 1. **a** Geometric model of CDPR **b** Prototype of CDPR

r_i denotes crane length connecting from point B_i to point $A_i{}'$. The end-effector is considered as a point mass P vector $\mathbf{p} = [xyz]^T$ with respect to global coordinate. The exit point A_i is located on the tip of the pulley and connected to point P by one cable.

The concept of reconfiguration is introduced in this paper by extending and retracting the length r_i. Via reconfiguration, position of pulley will be changed to modulate the cable tension, hence maintaining the mobile crane stability.

3 Kinematic Analysis

Based on Fig. 1, the geometric relations of each mobile crane can be determined in the global coordinate as follows:

$$\mathbf{l}_i = \mathbf{a}_i - \mathbf{p};\ i = 1, 2, 3 \tag{1}$$

where $\mathbf{l}_i$ is cable vector from point P to A_i and $\mathbf{a}_i$ is position vector of point A_i. The cable length l_i thus can be determined as follows:

$$l_i = \|\mathbf{l}_i\| = \sqrt{(\mathbf{a}_i - \mathbf{p})^T(\mathbf{a}_i - \mathbf{p})} \tag{2}$$

Cable unit vector $\mathbf{u}_i$ can be computed from Eqs. (1)–(2) as follows:

$$\mathbf{u}_i = \frac{\mathbf{l}_i}{\|\mathbf{l}_i\|} \tag{3}$$

By deriving Eq. (1) with respect to time, the relations of input–output velocities can be derived. For non-reconfiguration case where length r_i is fixed, the velocity relation is given as follows:

$$\dot{\mathbf{l}}_i = \mathbf{J}\dot{\mathbf{p}} \tag{4}$$

where $\dot{\mathbf{l}}_i$ is cable velocity vector and $\dot{\mathbf{p}}$ is velocity vector of the end-effector. The length r_i is changing with time in the reconfiguration case, and hence the velocity relation is given by

$$\dot{\mathbf{l}}_i = \mathbf{J}(\dot{\mathbf{p}} - \dot{\mathbf{r}}_i) \tag{5}$$

where $\dot{\mathbf{r}}_i$ denotes crane velocity vector. $\mathbf{J}$ is a Jacobian matrix composed of cable unit vector $\mathbf{u}_i$, which can be mathematically defined as follows:

$$\mathbf{J} = [\mathbf{u}_1\mathbf{u}_2\mathbf{u}_3]^{\mathrm{T}} \tag{6}$$

4 Trajectory Planning

In this paper, trajectory is generated based on the fifth-degree polynomial for pick-and-place operations, as shown in Fig. 2. It consists of two symmetrical segments of 400 mm distance (w) and 100 mm height (h). For each segment, there are two points having three initial conditions with respect to position, velocity, and acceleration. This trajectory lays on the XZ-plane.

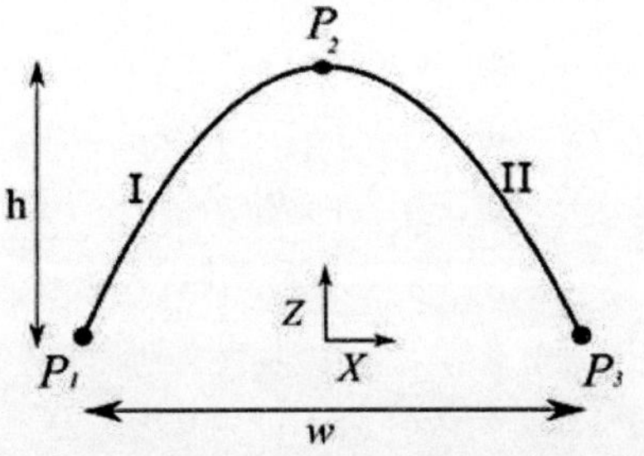

Fig. 2. Geometric path for polynomial trajectory

$$\begin{cases} p_{1x}(t) = b_{1x}t^5 + c_{1x}t^4 + d_{1x}t^3 + e_{1x}t^2 + f_{1x}t + g_{1x} \\ \dot{p}_{1x}(t) = 5b_{1x}t^4 + 4c_{1x}t^3 + 3d_{1x}t^2 + 2e_{1x}t + f_{1x} \quad \text{for t} \in [t_1, t_2] \\ \ddot{p}_{1x}(t) = 20b_{1x}t^3 + 12c_{1x}t^2 + 6d_{1x}t^2 + 2e_{1x} \end{cases} \tag{7}$$

with boundary conditions

$$\begin{cases} p_{1x}(t_1) = -0.2\ \dot{p}_{1x}(t_1) = 0\ \ddot{p}_{1x}(t_1) = 0 \\ p_{1x}(t_2) = 0\ \dot{p}_{1x}(t_2) = 0.0372\ \dddot{p}_{1x}(t_2) = 0 \end{cases} \tag{8}$$

$$\begin{cases} p_{1z}(t) = b_{1z}t^5 + c_{1z}t^4 + d_{1z}t^3 + e_{1z}t^2 + f_{1z}t + g_{1z} \\ \dot{p}_{1z}(t) = 5b_{1z}t^4 + 4c_{1z}t^3 + 3d_{1z}t^2 + 2e_{1z}t + f_{1z} \quad \text{for t} \in [t_1, t_2] \\ \ddot{p}_{1z}(t) = 20b_{1z}t^3 + 12c_{1z}t^2 + 6d_{1z}t^2 + 2e_{1z} \end{cases} \tag{9}$$

with boundary conditions

$$\begin{cases} p_{1z}(t_1) = 0.15\ \dot{p}_{1z}(t_1) = 0\ \ddot{p}_{1z}(t_1) = 0 \\ p_{1z}(t_2) = 0.25\ \dot{p}_{1x}(t_2) = 0\ \ddot{p}_{1z}(t_2) = 0 \end{cases} \tag{10}$$

p_{1x} and p_{1z} represent position of point P in X- and Z-axes first segment of trajectory. Position vector of point P in second segment is determined as follows:

$$\begin{cases} p_{2x}(t) = b_{1x}t^5 + c_{1x}t^4 + d_{1x}t^3 + e_{1x}t^2 + f_{1x}t + g_{1x} \\ \dot{p}_{2x}(t) = 5b_{2x}t^4 + 4c_{2x}t^3 + 3d_{2x}t^2 + 2e_{2x}t + f_{2x} \quad \text{for t} \in [t_2, t_3] \\ \ddot{p}_{2x}(t) = 20b_{2x}t^3 + 12c_{2x}t^2 + 6d_{2x}t^2 + 2e_{2x} \end{cases} \tag{11}$$

with boundary conditions

$$\begin{cases} p_{2x}(t_2) = 0\ \dot{p}_{2x}(t_2) = 0.0372\ \ddot{p}_2(t_2) = 0 \\ p_{2x}(t_3) = 0.2\ \dot{p}_{2x}(t_3) = 0\ \ddot{p}_2(t_3) = 0 \end{cases} \tag{12}$$

$$\begin{cases} p_{2z}(t) = b_{2z}t^5 + c_{2z}t^4 + d_2t^3 + e_{2z}t^2 + f_2t + g_2 \\ \dot{p}_{2z}(t) = 5b_{2z}t^4 + 4c_{2z}t^3 + 3d_{2z}t^2 + 2e_2t + f_2 \quad \text{for t} \in [t_2, t_3] \\ \ddot{p}_{2z}(t) = 20b_{2z}t^3 + 12c_{2z}t^2 + 6d_{2z}t^2 + 2e_2 \end{cases} \tag{13}$$

with boundary conditions

$$\begin{cases} p_{2z}(t_2) = 0.25 \; \dot{p}_{2z}(t_2) = 0 \; \ddot{p}_{2z}(t_2) = 0 \\ p_{2z}(t_3) = 0.15 \; \dot{p}_{2z}(t_3) = 0 \; \ddot{p}_2(t_3) = 0 \end{cases} \tag{14}$$

The results of trajectory planning are shown in Fig. 3 with respect to position, velocity, and acceleration profiles. The end-effector moves from point P_1 of coordinate (−0.2,0,0.15) at t_1= 5 s to point P_2 of coordinate (0,0,0.25) at t_2= 15 s and continues to point P_3 of coordinate (0.2,0,0.15) at t_3= 25 s. The maximum velocity achieved by the end-effector is 0.31 m/s in X-axis and 0.17 m/s in Z-axis.

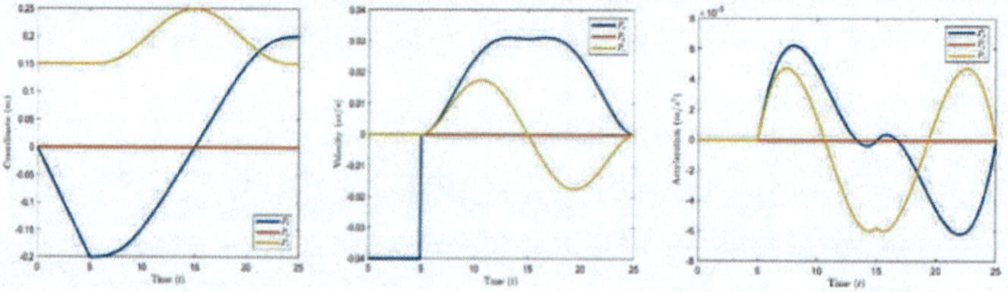

(a) (b) (c)

Fig. 3. **a** Position profile, **b** Velocity profile, and **c** Acceleration profile of trajectory CDPR

5 Results and Discussions

5.1 Non-Reconfiguration Scheme

Non-reconfiguration scheme is applied to CDPR when the end-effector moves along the generated trajectory. Fixed position of crane is determined for this case at 0.53 m (r_i = 0.53 m). During the task, the end-effector carries 1 kg payload. Figure 4 shows the comparison of theoretical and experimental results. The variation of cable length is expressed to represent position of end-effector.

Theoretical and experimental results of the variation of cable length are compared in Fig. 4. Motor encoder that connects to the pulley is measured to get experimental results. The maximum error of cable length in mobile crane 1 is $\varepsilon l_1 = 4 \times 10^{-3}$, whereas the maximum errors of cable length in mobile crane 2 and mobile crane 3 are εl_2=3.7 $\times 10^{-3}$ and $\varepsilon l_3 = 2 \times 10^{-3}$.

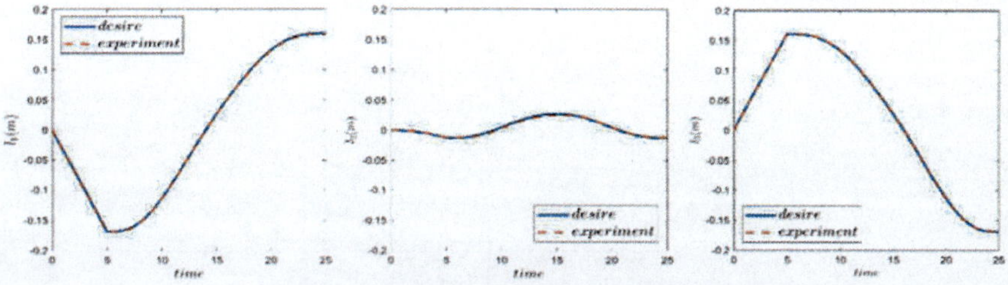

(a) Variation of Cable Length (l_i)

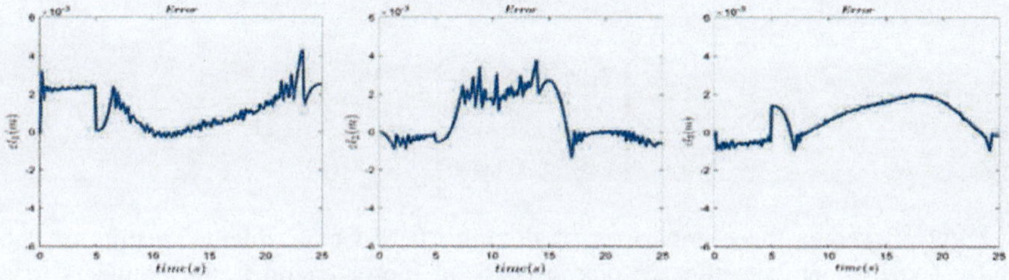

(b) Variation of Cable Length Error (εl_i)

Fig. 4. Theoretical and experimental results of non-reconfiguration scheme

5.2 Reconfiguration Scheme

Reconfiguration of crane length is obtained by using optimization method based on the pareto optimal solution. Two objective functions are defined, minimizing cable tension and cable velocity. Constraints are determined to maintain, i.e., positive value of cable tension, maximum and minimum boundaries of cable velocity, and to avoid mobile crane instability. Optimization calculations are expressed as follows (Table 1).

Table 1. Parameter definition of the optimization

Description	Symbol	Value
Cable tension [N]	τ_i	
Maximum position crane [m]	$\overline{r_i}$	0.83
Minimum position crane [m]	$\underline{r_i}$	0.53
Maximum velocity crane [m/s]	$\overline{\dot{r}_i}$	−0.015
Minimum velocity crane [m/s]	$\underline{\dot{r}_i}$	0.015
Zero moment point	d_i^u	

$$
\begin{gathered}
\textit{minimize}: \sqrt{\sum_{i=1}^{3} (\dot{l}_i)^2} \\
\textit{minimize}: \sqrt{\sum_{i=1}^{3} (\tau_i)^2} \\
\textit{over } r_i \in [\underline{r_i}, \overline{r_i}] \\
\textit{Subject to} \begin{cases} \underline{\dot{r}_i} \leq \dot{r}_i \leq \overline{\dot{r}_i} \\ \tau_i \geq 1N \\ c^u_{fri} \leq d^u_i \leq c^u_{rri} \end{cases} \quad \forall i \in [1, 2, 3]
\end{gathered}
\tag{15}
$$

Figure 5 shows the comparison of theoretical and experimental results of cable length. Experimental results are measured from motor encoder that connects to the pulley. In reconfiguration scheme, the robot can lift up to 1.3 kg instead of 1 kg in non-reconfiguration. The error has occurred because the motor velocity that connects to the pulley is limited. The error increases also due to the changing direction of the motor.

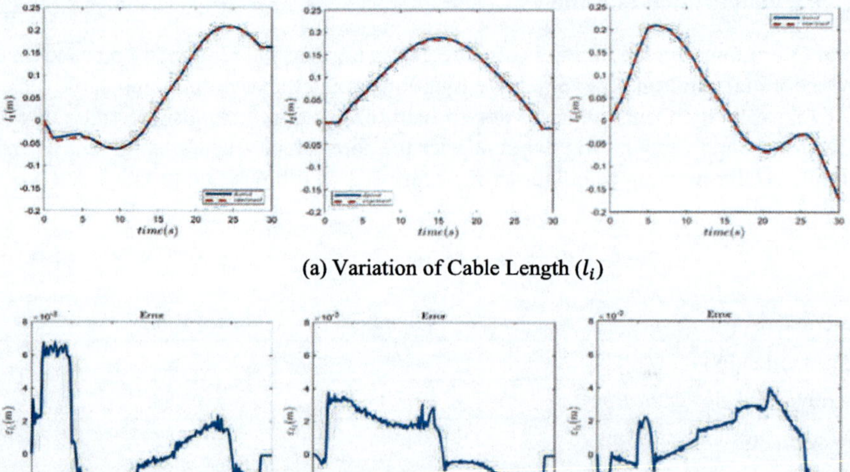

(a) Variation of Cable Length (l_i)

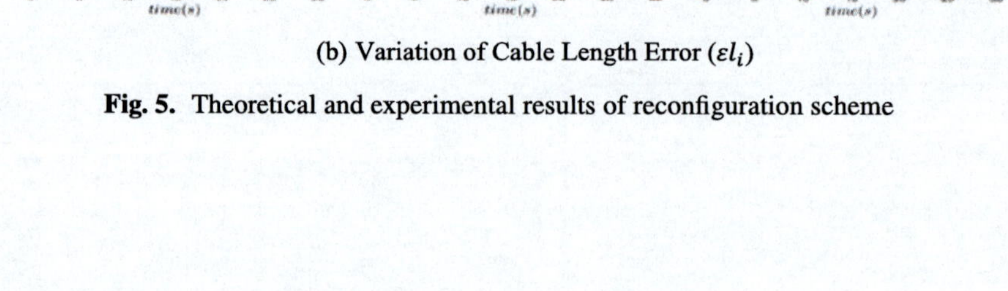

(b) Variation of Cable Length Error (εl_i)

Fig. 5. Theoretical and experimental results of reconfiguration scheme

6 Conclusion

This paper presented the trajectory planning of Cable-Driven Parallel Robot (CDPR) with mobile crane. Trajectory was generated in two segments by using fifth-order polynomial. A notion of reconfiguration was proposed in this paper by extending and retracting the crane lengths. While non-reconfiguration scheme defines the fixed crane length. The end-effector was tested to execute the generated trajectory while performing reconfiguration and non-reconfiguration schemes. It is noted that the error of experimental results occurred for both reconfiguration and non-reconfiguration schemes. This condition has caused the limitation of motor velocity that connects to the cable pulley and crane. The cable length error appeared because of the changing direction of the motor's rotation.

References

1. Pedemonte N, Rasheed T, Marquez-Gamez D, Long P, Hocquard É, Babin F, Fouché C, Caverot G, Girin A, Caro SF (2020) A mobile cable-driven parallel robot for logistics. In: Advances in robotics research: from lab to market. Springer, pp 141–163
2. Rasheed T, Long P, Roos AS, Caro S (2019) Optimization based trajectory planning of mobile cable-driven parallel robots. In: International conference on intelligent robots and systems (IROS) 2019. IEEE, pp 6788–6793
3. Zi B, Lin J, Qian S (2015) Localization, obstacle avoidance planning and control of a cooperative cable parallel robot for multiple mobile cranes. Robot Comput Integr Manuf 34:105–123
4. Syamlan AT, Nurahmi L, Tamara MN, Pramujati B (2020) Dynamic trajectory planning of reconfigurable suspended cable robot. Int J Dyn Contr 8(3):887–897. Springer
5. Syamlan AT, Nurahmi L, Pramujati B, Tamara MN (2019) Dynamic trajectory generation of suspended cable-driven parallel robot. In: AIP conference proceedings. AIP Publishing LLC, Vol 2187(1), pp 050006
6. Tan H, Nurahmi L, Pramujati B, Caro S (2020) On the reconfiguration of cable-driven parallel robots with multiple mobile cranes. In 5th international conference on robotics and automation engineering (ICRAE), pp 126–130

Experimental Study on a Side-By-Side Configuration of Two Cylinders Helically Covered by Triple Rods

Rudi Walujo Prastianto(✉) and Bagas Pumbarino

Department of Ocean Engineering, Institut Teknologi Sepuluh Nopember, Surabaya 60111, Indonesia

rudiwp@oe.its.ac.id

Abstract. Many investigations had been done both numerically and experimentally, on the characteristics of multiple-arranged cylinders especially in terms of the induced fluid forces and responses caused by external fluid flows acting on them. This paper discusses the investigation result on induced drag of two cylinders helically wrapped around by triple rods with certain gap in air flow fields. The model cylinders were installed in the test section of a wind tunnel as a side-by-side arrangement and treated as roll supports at ends of the cylinders. The model was executed at subcritical regime of incoming uniform air flow acting perpendicular to the cylinders that related to Reynolds number of 3.9×10^4 and evaluated for different gaps between the cylinder of 1.75D, 2.0D, and 2.50D, respectively, where D is diameter of the bare cylinder. Load cells were used for measuring the drag occurring on the cylinders. For all gap variety evaluated results demonstrated that greater coefficient of drag (C_D) happened in the two cylinders which was caused by more prominent interference between them. The C_D increased greater than other gaps for the cylinder's gap of 2.0D that was well confirmed against others relevant works. For this side-by-side configuration, the C_D of the lower and upper cylinders significantly increased in contrast with the single cylinder case by 55.8% and 55.5%, respectively.

Keywords: Experimental study · Drag · Side-by-side arrangement · Triple rods

1 Introduction

When offshore structural components such as mooring lines, risers, conductors, and marine pipelines are exposed to water currents, vortex shedding often occurs behind the components at specific water velocity. The vortex shedding causes continuous excitation force to act on the structure, which will cause structural vibration. Under certain conditions, the vortex-induced vibration (VIV) is usually attributed to this vibration phenomenon.

As continued efforts several investigations had been done for reducing induced fluid forces and the VIV on cylinders. Classically, Zdravkovich [1] reviewed helical rods as an effective suppression device to reduce VIV which is then followed by many researchers

I. Tolj et al. (eds.), *Recent Advances in Mechanical Engineering*, Lecture Notes in Mechanical Engineering, https://doi.org/10.1007/978-981-19-0867-5_38

in recent decade. Baarholm et al. [2] explained implementation to the data from VIV experiments of a vertical riser equipped with suppressing devices such as fairings and strakes into a theoretical VIV model. Gao et al. [3, 4] experimentally investigated a flexible riser in two current conditions which are uniform and linearly distributed sheared flows. The results showed that between the two strake parameters of the height and pitch, the first parameter has larger effect to suppress the VIV when helically installed within the riser. Meanwhile, the drag which occurred on the riser would be larger when the two parameters increased.

A new model of device for suppressing VIV and drag acting on a cylinder has been investigated through numerical and experimental works. The device consists of three rods helically attached within a cylinder with certain gap [5–7]. Through their numerical work the device could significantly reduce the drag force by 50.42% for Reynolds number (Re) of 10^5 compared to the case of a bare cylinder [6]. In addition, Prastianto et al. [5] experimentally at lower Re of 2.36×10^4 proved that the drag force could be reduced by 50% by applying the proposed VIV suppression device. By using the same device model and method of [5], Arianti et al. [7] obtained result that the cross-flow vibration of the cylinder could be suppressed 49.28% by utilizing the device for the case of Re = 3.5×10^4.

Nevertheless, the characteristic of multiple cylinders under action of fluid flows will be totally unique than that occurred in the case of the single cylinder in terms such as their responses, induced drag force, and fluid stream pattern. Previously, Prastianto, et al. [8–10] experimentally studied the VIV of bare cantilever cylinders including the cases of several variation on the cylinder arrangements that consist of two and three cylinders. In these cases, hydrodynamic forces, dynamic of the cylinder, and wake interference effects of the cylinders were evaluated.

Verma and Govardhan [11] numerically studied side-by-side configurations of two bare cylinders. Induced fluid forces were evaluated at various gaps between the cylinders (T) of 1.5D, 2.0D, 3.0D, and 4.0D, respectively. For Re = 200 if the cylinder's gap (T) increased, the fluid forces decreased. But a phenomenon called as "*flopping phenomenon*" occurred when T = 2.0D which is the result showing that fluid forces acting on the cylinders increased. Meanwhile, at higher Re of 6×10^4, Pang et al. [12] have also numerically worked on the same configuration of two cylinders. They examined the various gaps between the cylinders of 1.1D ~ 7.0D. The result demonstrated that at single body regime of T < 1.1D—1.2D the C_D were so high, which is then followed by decreasing the C_D if the T increased. But for the case of T in the range of 1.5D ~ 2.0D the C_D increased. Alam et al. [13] conducted experimental study on the characteristics of aerodynamic of a side-by-side configuration which consisted of two cylinders and tested in a uniform stream for Re = 5.5×10^4. The results examined wake frequencies around and fluid forces acting on the cylinder, and switching phenomena in such cylinder's configuration. Meanwhile, Prastianto et al. [14, 15] experimentally examined the drag on two cylinders in a tandem configuration. The results for the cylinder's gap of 1.75D at Re = 4.2×10^4 showed that maximum effects of wake interference were experienced by the rear cylinder. This effect resulted in decreasing the C_D by 93.6% in contrasted with the case of a single cylinder.

This current paper presents interference effects on induced drag of two cylinders arranged in side-by-side configuration under uniform wind flow. In this study, both cylinders are helically covered by triple rods in a way that created a small gap between the rods and cylinders.

2 Method of Experiment

Partly the model proportions utilized in current investigation were taken on from [5, 7]. Aluminum tubes of 5.08 cm in diameter (D) and length (L) of 48 cm were used for the model cylinders. The cylinder is helically wrapped around by three rods for one pitch with a certain gap. Each rod is made of stainless steel with a diameter (d) of about 0.317 cm and arranged to make a gap ratio (d/D) of 0.0625. Circular end plates with larger diameter are attached at both ends of the cylinder to keep the flow still uniform along the length of the cylinder without getting affected by end-effect of the cylinder. The complete model of each cylinder can be seen in Fig. 1.

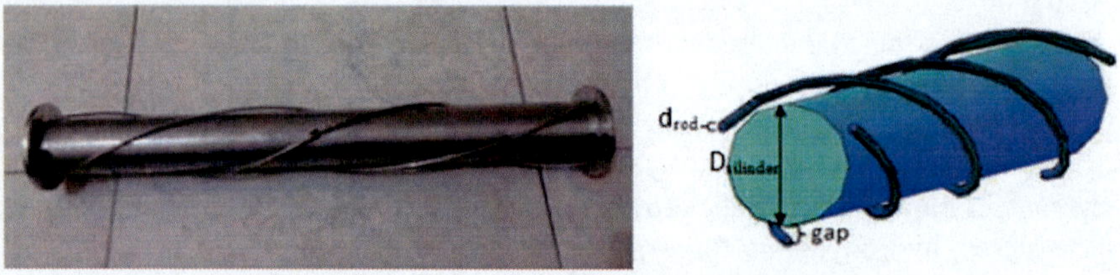

Fig. 1. Model of the cylinder with its nomenclature.

The experimental work was performed at the LSWT (low-speed wind tunnel) research center of BBTA3-Agency for the Assessment and Application of Technology, Indonesia. The test segment size of the tunnel is 0.5 m × 0.5 m × 1.25 m. A maximum limit of air stream velocity that can be generated is 45 m/s. In the test segment, two cylinders were placed at horizontal position as upper and lower cylinders to turn into a side-by-side configuration, and oppositely facing to the air stream. Each end of the model cylinder was treated as roll supports by using special frame equipped by load cells for measuring the drag and lift. Figure 2 shows the experimental setup. Measurement of the drag and lift which occurred on the upper and lower cylinders was conducted at Re of 3.9×10^4 for each variation of the cylinder's gap (T) which are 1.75D, 2.0D, and 2.5D, respectively. Calculation of the Re is based on the diameter of the bare cylinder. For more detail about the experiment, it can be directed to [14, 15].

3 Results and Discussion

Firstly, C_D measurement for the case of bare cylinder was performed in order to guaranty the accuracy of the test setting. Then the result was compared to other relevant works [12] to validate the measured C_D values as presented in Fig. 3. The approval demonstrated

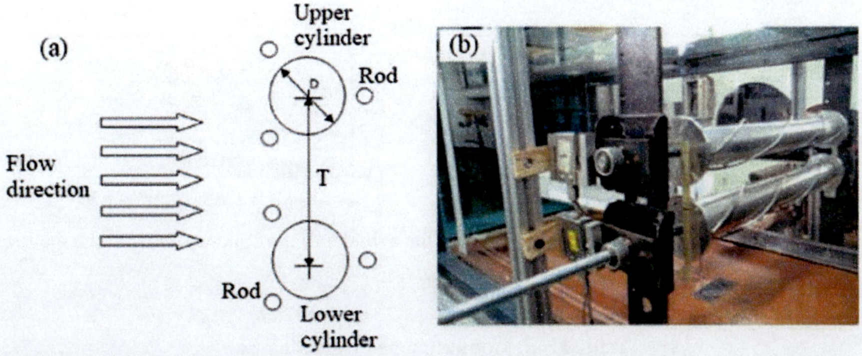

Fig. 2. **a** Side view of the model configuration, **b** position of the model in the test section.

decent results for the range of Re $= 3.65 \times 10^4 \sim 5.3 \times 10^4$. It implies that the experiment devices were in acceptable condition and the test setting could be acknowledged. So that the model and setting could be used for further analysis of the current study case in that limit of Re.

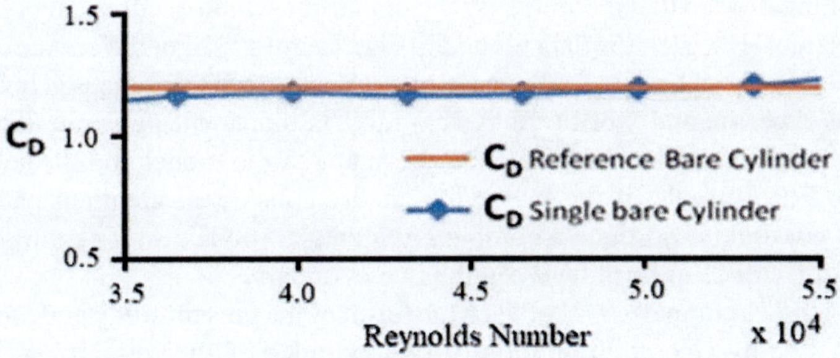

Fig. 3. Validation of the experiment for a single cylinder case.

Results of the drag measurement (in terms of C_D) are depicted in Fig. 4a, b for the two cylinders in a side-by-side configuration at three different gaps between the two cylinders (T) as well as for the case of a single cylinder covered by triple helical rods. The C_D for both the upper and lower cylinders are shown in Fig. 4a, b, respectively.

In general, the C_D for the upper and lower cylinders were bigger compared to the case of a single cylinder, for both bare or with triple helical rod conditions. The greatest value of around 1.65 happens at a proportion of the gap (T/D) of 2.0 yet at (T/D) = 1.75 and 2.5, the C_D decreased. This outcome shows that on every case of the three tried gap variations there is as yet significant interaction between the two cylinders because of the air current at Re $= 3.9 \times 10^4$ that caused C_D to drastically increase than the case of single cylinder one. Then at further T/D of 2.0 ~ 2.50 we can see that interaction of the two cylinders is gradually debilitated as shown by diminishing the C_D. Subsequently, it could be anticipated that each cylinder will generally act like the case of a single cylinder. Figure 4 presents the C_D of 1.072 as also shown in Table 1.

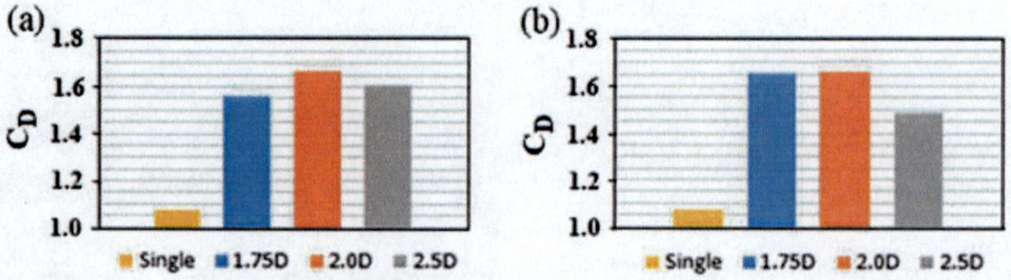

Fig. 4. A comparison based on the gap between the cylinder: **a** the upper cylinder and **b** the lower cylinder.

Table 1. C_D for the single cylinder cases.

Present work	Pang et al. [12]	Alam et al. [13]
1.180 (single, bare)	1.206	1.160
1.072 (single, helix)		

For the case of side-by-side arrangement, position of the cylinder will be symmetrical to the incoming flow so that theoretically the drag force which occurred in both cylinders will be the same. However, the data plotted in Fig. 4 shows a slight difference in the C_D of the upper and lower cylinders for the case of 1.75D and 2.5D distances. This is because naturally in experimental work on physical models there will be some uncertainty in some of its parameters, such as related to the shape of the model and the setting of the experiment. For example, if the two cylinders are considered identical, in practice, it will not be possible to produce a completely identical model. At a certain level, there will exist a difference in form between the two cylinders.

Figure 5 shows comparisons between the result of the present study and other relevant works on a side-by-side configuration of two cylinders. Other researchers performed experimental and numerical works to obtain C_D of bare cylinders at Re $= 5.5 \times 10^4$ for gap ratio T/D of 1.1 ~ 7.0 [12, 13]. The C_D of the present study have greater values than that the smooth single cylinder case. However, the trend curve will have the same shape as the T/D value increases. For all three curves C_D from the small of T/D gradually increase as T/D is increasing until it reached peaks at T/D of around 2.0, except the curve of Alam et al. [13] that the peak slightly shifted at greater T/D of about 2.2. Subsequently, the C_D gradually decreased if the gap ratio incrementally increased.

Therefore, the pattern of C_D as a function of the gap ratio T/D in the current study has a comparable form like the investigation of Pang et al. [12] and Alam et al. [13]. However, the C_D in current study case is bigger than the C_D for the case of the bare cylinder in various gap ratios as shown previously in Table 1 and Fig. 4. This could be due to the existence of the rod that modifies the flow as well as the gap between the cylinders becomes closer that caused interference effect more pronounced than the bare case. Also, it should be noted that different Re in this comparison could slightly affect the results even still in the same subcritical regime of Re.

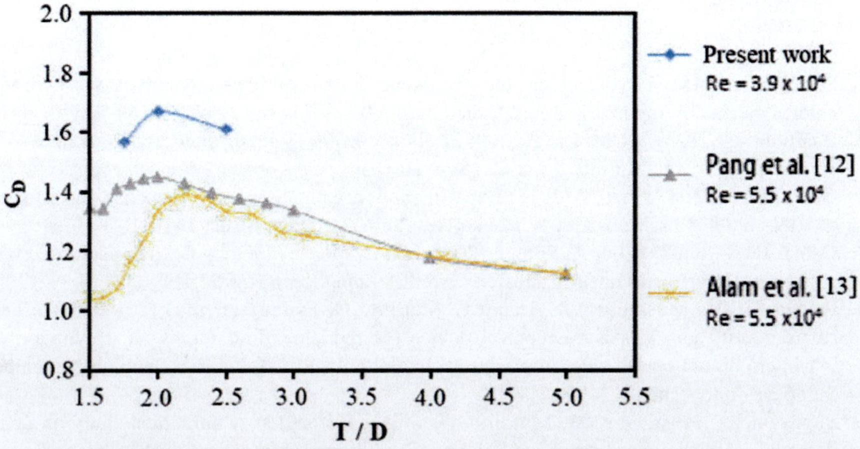

Fig. 5. A comparison of the C_D among other relevant works.

4 Conclusions

The result of the experiment on side-by-side configuration consisted of two cylinders which demonstrated a good comparison contrasted with some applicable past works with comparable pattern of the C_D based on the gap between the two cylinders. The results of the experiment at $\mathrm{Re} = 3.9 \times 10^4$ proved that greater C_D happened in the two cylinders for all gap variations evaluated because of solid obstruction between them. There is an increase on the C_D at the cylinder's gap of 2.0D that is greater than for other gaps. For this side-by-side configuration, the C_D of the lower and upper cylinders significantly increased which contrasted with the single cylinder condition by 55.8% and 55.5%, respectively. In this way, it tends to be anticipated that when the cylinder's gap increased then obstruction effect between the two cylinders gradually will decrease until reaching a certain distance that no more interference effect exists and each cylinder will act like a solitary cylinder.

Acknowledgements. The authors wish to appreciate Directorate General of Higher Education (DIKTI)–Ministry of Research, Technology and Higher Education, Republic of Indonesia for the financial support through a research scheme grant of "Penelitian Unggulan Perguruan Tinggi (PUPT)". All significant supports from Institute for Research and Community Services (LPPM)–Institut Teknologi Sepuluh Nopember (ITS) Surabaya and National Laboratory for Aerodynamics, Aeroelastic and Aeroacoustics Technology, Agency for the Assessment and Application of Technology (BPPT), Indonesia, are greatly appreciated.

References

1. Zdravkovich MM (1981) Review and classification of various aerodynamics and hydrodynamics means for suppressing vortex shedding. J Wind Eng Ind Aerodyn 145–189
2. Baarholm GS, Larsen CM, Lie H (2005) Reduction of VIV using suppression devices—An empirical approach. Marine Struct 18:489–510
3. Gao Y, Fu S, Ma L, Chen Y (2014) Experimental investigation of the response performance of VIV on a flexible riser with helical strakes. Ships Offshore Struct 1–16
4. Gao Y, Fu S, Ren T, Xiong Y, Song L (2015) VIV response of a long flexible riser fitted with strakes in uniform and linearly sheared currents. Appl Ocean Res 52:102–114
5. Prastianto RW, Musthofa AZ, Arianti E, Murdjito H, Fariduzzaman S (2014) Triple helical rods with gap as passive control device for reducing fluid forces on a cylinder. In: 9th International conference on marine technology. Institut Teknologi Sepuluh Nopember, Surabaya-Indonesia, pp 24–25
6. Sugiwanto A, Prastianto RW, Murdjito, Djatmiko, EB (2013) A numerical study on cylinders with passive control device of helical rods with gap for reducing vortex-induced vibration. In: International conference on sustainable infrastructure and built environment. Institut Teknologi Bandung, Bandung-Indonesia, pp 357–368
7. Arianti E, Prastianto RW, Djatmiko EB, Handayanu, Murdjito, Fariduzzaman (2014) Studi eksperimental pengaruh penambahan Helical Rods Ber-gap pada silinder rigid tertumpu fleksibel terhadap respon geraknya. In: Prosiding Seminar Pascasarjana XIV. Institut Teknologi Sepuluh Nopember, Surabaya-Indonesia, pp 246–251. (In Indonesia language).
8. Prastianto RW, Otsuka K, Ikeda Y (2008) Dynamics of two flexible hanging-off circular cylinders in staggered configurations. In: Proceedings of the eighth ISOPE Pacific/Asia offshore mechanics symposium, ISOPE, Bangkok-Thailand
9. Prastianto RW, Otsuka K, Ikeda Y (2008) Experimental study on two flexible hanging-off circular cylinders undergoing wake interference. In: OCEANS '08 MTS/IEEE Kobe Techno Ocean'08—Voyage toward the future OTO'08, Kobe, Japan
10. Prastianto RW, Otsuka K, Ikeda Y (2009) Hydrodynamic forces on multiple hanging-off circular cylinders in uniform flows. Int J Offshore Polar Eng 19(2):108–114
11. Verma PL, Govardhan M (2011) Flow behind bluff bodies in side-by-side arrangement. J Eng Sci Technol 745–768
12. Pang JH, Zong Z, Zou L, Wang Z (2016) Numerical simulation of the flow around two side-by-side circular cylinders by IVCBC vortex method. J Ocean Eng 119:86–100
13. Alam MM, Moriya M, Sakamoto H (2003) Aerodynamic characteristic of two side by-side circular cylinders and application of wavelet analysis on the switching phenomenon. J Fluids Struct 18(3):325–346
14. Prastianto RW, Dwipayana KH, Syahroni N, Pumbarino B (2018) Experimental study on interference effects of two tandem cylinders wrapped around by triple helical rods with gap on induced drag, In: IOP Conference series: earth and environmental science, vol 135, pp 1–6, paper number: 012025
15. Prastianto RW, Syahroni N, Murdjito (2016) Kajian Evaluasi Kinerja Piranti Inovasi Baru Pereduksi Getaran akibat Vorteks pada Komponen Struktur Anjungan Lepas Pantai dalam Konfigurasi Jamak. Penelitian Unggulan Perguruan Tinggi (PUPT) – Tahun ke-2, LPPM – Institut Teknologi Sepuluh Nopember, Surabaya-Indonesia. (in Indonesia Language)

Multi-performance Optimization in End Milling of GFRP Composites Using Backpropagation Neural Network and Differential Evolution Algorithm

M. Khoirul Effendi[1(✉)], Bobby O. P. Soepangkat[1], Dinny Harnany[1], and Rachmadi Norcahyo[2]

[1] Department of Mechanical Engineering, Institut Teknologi Sepuluh Nopember, Surabaya 60111, Indonesia
khoirul_effendi@me.its.ac.id

[2] Department of Mechanical and Industrial Engineering, Universitas Gadjah Mada, Yogyakarta, Indonesia

Abstract. An integrated approach has been applied to predict and optimize multi-performance characteristics, namely, cutting force (CF) and surface roughness (SR), in the end-milling process of glass fiber-reinforced polymer (GFRP) composites. The experiments were performed by varying spindle speed, feeding speed, and depth of cut. The quality characteristics of cutting force and surface roughness were the smaller, the better. Full factorial design $3 \times 3 \times 3$ was used as the design of experiments. Backpropagation neural network (BPNN) was used to model the end-milling experiment and also to determine the objective function. This objective function will be modified into a fitness function optimized by using a differential evolution algorithm (DEA) to find the combination of drilling parameters' levels that yield minimum cutting force and surface roughness simultaneously. As a result, the minimum cutting force can reduce the energy consumption, and the end-milling process can be performed with higher energy efficiency. Based on BPNN-DEA, the depth of cut of 2 mm, the spindle speed of 4797.5 rpm, and the feeding speed of 579.7 mm/min can simultaneously minimize the cutting force and surface roughness in the end milling of GFRP.

Keywords: Glass fiber-reinforced polymer · End milling · Backpropagation neural network · Differential evolution algorithm · Energy consumption

1 Introduction

Glass fiber-reinforced polymer (GFRP) composites are used in quite a number of industrial sectors, such as aerospace, marine, aviation, automotive, electronics, and many others. It has extraordinary material properties such as high specific stiffness, high fracture toughness, high strength-to-weight ratio, good impact resistance, and lightweight. The products produced by GFRP composites generally acquire their shape during the

I. Tolj et al. (eds.), *Recent Advances in Mechanical Engineering*, Lecture Notes in Mechanical Engineering, https://doi.org/10.1007/978-981-19-0867-5_39

cycle of the curing process. Hence, machining processes are required to obtain the dimension's accuracy during the assembly of the final product [1]. One of the most commonly used composite material removal operations is end milling. This process is considerably applied in a number of manufacturing industries, including the automotive and aerospace areas, where quality is considered as a determining factor in the manufacture of dies, slots, and keyways.

GFRP composites consist of glass as a filler and polymer as a matrix. This type of composite is usually inhomogeneous and has anisotropic properties as well as abrasive behavior. Hence, its end-milling process would be substantially affected by the catastrophic nature of the characteristics of the composites [2]. During the end-milling process, GFRP may fail because of the emergence of unoptimized machining forces. The quality of the workpiece surface has a vital function in the performance of end milling because a good quality of the end-milled surface would improve creep life, corrosion resistance, and fatigue strength [3]. Saikumar and Shunmugam [4] have proposed an optimization comparison in the milling process using DEA and response surface method (RSM). The optimized parameters are surface roughness, while the input parameters are cutting speed, feeding speed, and depth of cut. It was concluded from the experimental result that the performance of DEA was better for predicting the surface roughness compared to the RSM method. Yildiz [5] has applied hybrid differential evolution to optimize the milling process compared to other algorithms (i.e., particle swarm optimization, ant colony algorithm, immune algorithm, hybrid immune algorithm, genetic algorithm, and feasible direction method). The result shows that the performance of hybrid differential evolution has a significant improvement compared to other algorithms [5]. The cutting force, delamination, and surface roughness are standard optimized parameter output in the end-milling process for composite material. Robbany et al. [6] used a combination of BPNN and genetic algorithm (GA) to solve multi-objective optimization in end milling for composite fiber-reinforced polymer (CFRP) material. The experiment aimed to find machining parameters with small delamination and low cutting force.

This study attempts to obtain the minimum cutting force and surface roughness simultaneously in the end milling of GFRP composites. BPNN is applied in modeling, determining objective function, and predicting cutting force and surface roughness. An optimization method is utilized to simultaneously produce the combination of end-milling parameters' levels with minimum cutting force and surface roughness.

2 Research Methodology

2.1 Backpropagation Neural Network (BPNN)

Artificial Neural Network (ANN) is a popular method that is used in end milling to predict the surface roughness of the Ti-6AI-4V alloy material [7], the damage factor in GFRP [8], and tool breakage during operation [9]. BPNN is one type of ANN used to model the relationship between parameter input and response or target. BPNN consists of three layers, namely, input layers, hidden layers, and output layers. All data are then used to update the weight on the hidden layer through a series of processes (epochs) to minimize mean squared error (MSE) between the target and the output predicted by

BPNN. The general flowchart of BPNN to find a network with the best MSE is displayed in Fig. 1.

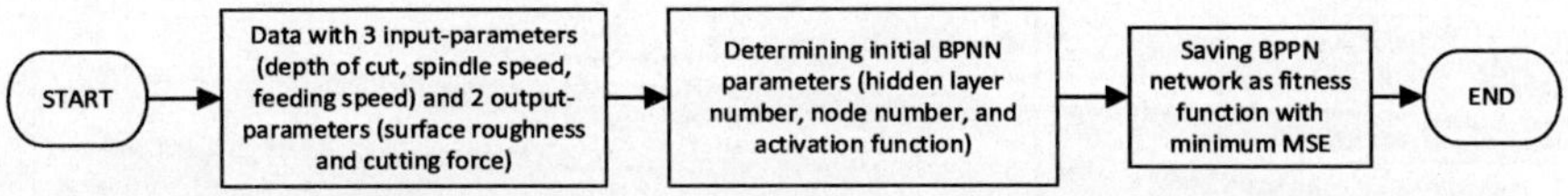

Fig. 1. The BPPN flowchart for producing a fitness function with minimum MSE.

2.2 Differential Evolution Algorithm (DEA)

DEA owns a unique capability because it can optimize a problem by using an iterative approach in attempting to improve a candidate solution concerning a given quality measure. Optimization of a problem is conducted by (1) maintaining a population of candidate solutions, (2) developing new prospective solutions by incorporating existing ones conforming to its simple formulae, and (3) retaining any candidate solution which has the best score or fitness on the solve optimization problem. In simple way, DE will go over every solution. If the solution matches the criterion (e.g., meets minimum score), it will be included in the candidate solutions list. The new solutions might be obtained by performing simple math operations on solution candidates.

DEA is generally applied for multidimensional real-valued functions, and it also does not utilize the gradient of the problem being optimized. Hence, DEA does not need the optimization problem to be differentiable, unlike the classical optimization methods such as quasi-Newton and gradient descent methods. Further, DEA can also be utilized on optimization problems that are not even continuous, change over time, noisy, etc. [10]. The general flowchart of the DEA algorithm can be seen in Fig. 2.

3 Experiment and Result

3.1 Experimental Work

This study used a GFRP produced by using the hand lay-up technique. The type of glass fiber utilized in this study was woven fiber, which consisted of 18 layers. Layering is performed with an orientation angle of 0 degrees. The applied resin was epoxy. The specimen dimensions were 60 mm (length), 30 mm (width), and 3.7 mm (thickness). Hartford S-Plus 10 vertical milling was used to perform the slotting milling process without coolant. The experiment utilized a YG solid end-milling carbide tool (type K-2 EMC 54,060) with four flutes and a 6 mm diameter. The number of flutes was 4. F_x, F_y, and F_z were feed, cutting, and thrust forces, respectively, measured using Kistler dynamometer 9272. Table 1 presents the levels of the end-milling process parameters. The design experiments applied were a completely randomized factorial design with two replications.

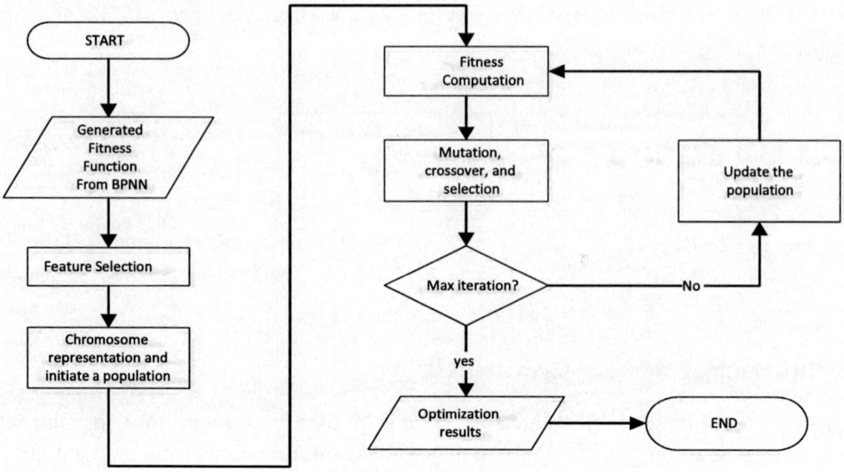

Fig. 2. The DEA flowchart for obtaining the optimization results.

Table 1. End-milling process parameters.

End-milling variables	Unit	Levels		
		1	2	3
Depth of cut (a)	mm	1	1.5	2
Feeding speed (V_f)	mm/min	500	750	1000
Spindle speed (n)	rpm	3000	4000	5000

3.2 Experiment Results

The following (Figs. 3 and 4) show the measured cutting force and surface roughness with three replications as the results of the combination of depth of cut (a), spindle speed (n), and feeding speed (V_f).

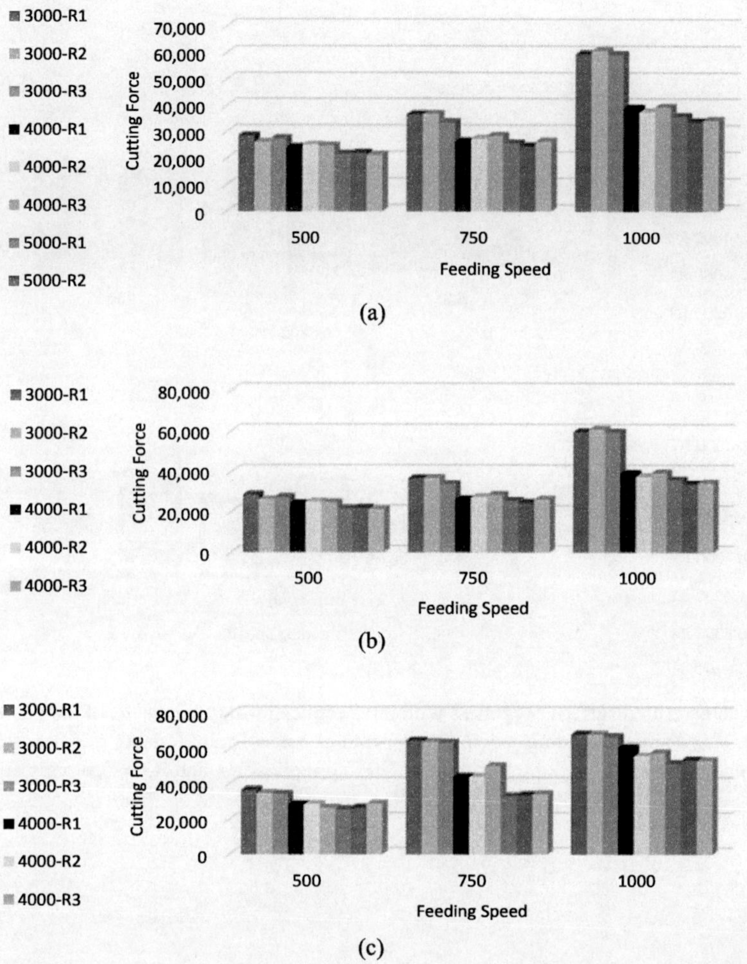

Fig. 3. The measured cutting force with three replications as the results of the combination of spindle speed (n), and feeding speed with **a** depth of cut = 1 mm, **b** depth of cut = 1.5 mm, **c** depth of cut = 2 mm [R1 = 1st replication, R2 = 2nd replication, and R3 = 3rd replication].

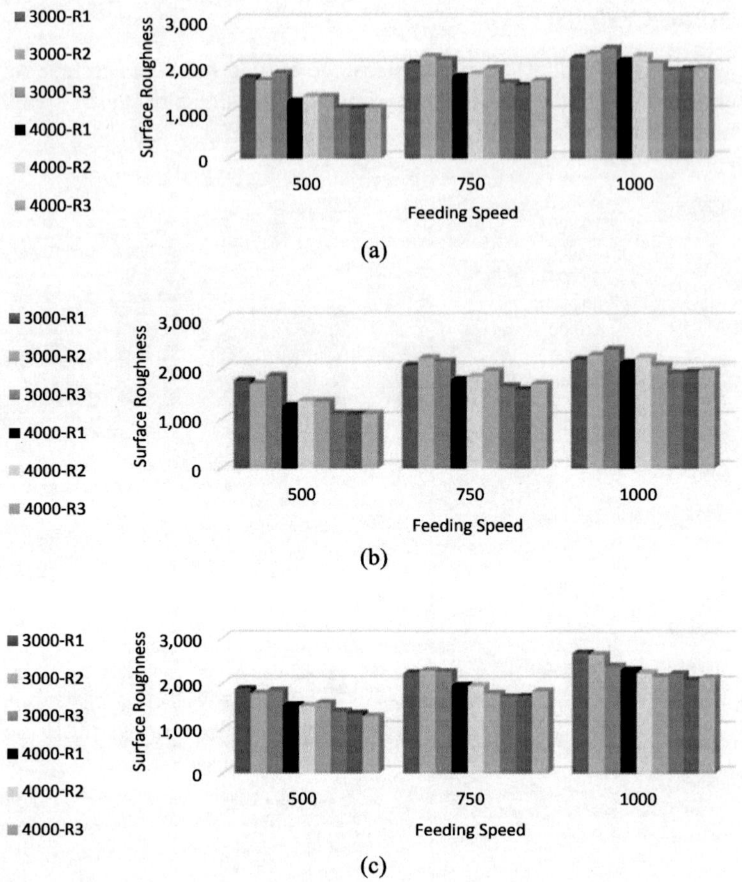

Fig. 4. The measured surface roughness with three replications as the results of the combination of spindle speed (n) and feeding speed with **a** depth of cut = 1 mm, **b** depth of cut = 1.5 mm, **c** depth of cut = 2 mm [R1 = 1st replication, R2 = 2nd replication, and R3 = 3rd replication].

4 Multi-objective Optimization

4.1 BPNN Modeling

The BPNN tuning parameters in modeling the relationship between parameter input and target are presented in Table 2.

Table 2. BPNN tuning parameters.

No	Parameter	Values
1	Ratio data for training, testing, and validation	70%:15%:15%
2	Variation of activation function	*Harlim, Hardlims, Satlin, logsig, tansig,* and *purelin*
3	Number of hidden layers	1–10 layers
4	Number of nodes each hidden layer	1–10 nodes
5	Stopping criterion	Max epoch (10,000 epochs)
6	Parameter for selecting best BPNN network	mean squared error (MSE)
7	Training algorithm	Levenberg–Marquardt

Information of the most refined BPNN's network can be described as follows: the number of hidden layers is seven, the number of neurons in each hidden layer is six, the activation function of the hidden layers is *tansig, and the* MSE is 0.0084123. Figure 5 shows the most refined BPNN's network topology with the minimum MSE. Next, the correlation coefficient which BPNN produced is 0.98865 for training, 0.98765 for testing, 0.98017 for validation, and 0.98706 for all. Since all of the correlation coefficient values are close to one, the BPNN output or predicted result is in excellent conformity with the experimental values.

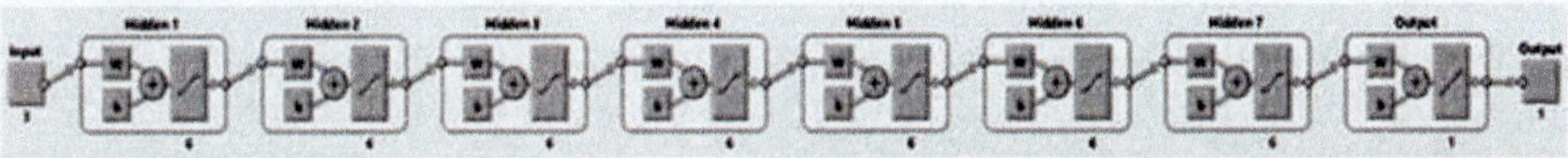

Fig. 5. The most refined BPNN's network topology with the minimum MSE.

The optimum responses could be obtained by setting the maximum iteration at 100, the number of solution candidates at 100, the scaling factor for permutation in the range 0–10, cross-over probability at 0.02, and the decision variable in the interval 0–2. Figure 6 shows the optimization result using the DEA method for 100 iterations, where the optimum value is −0.7092.

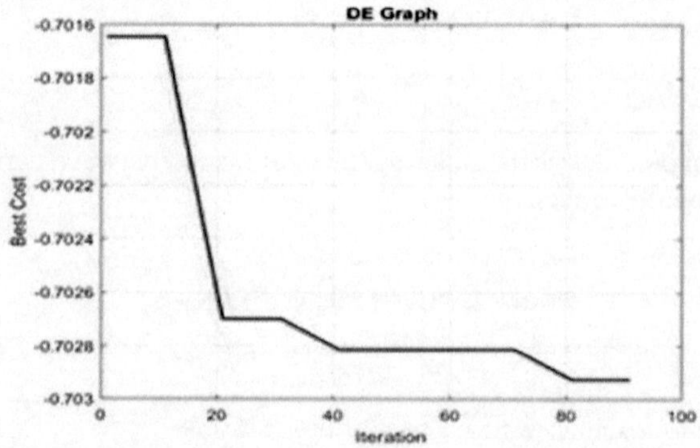

Fig. 6. The result of optimization using DE method for 100 iterations.

5 Result and Discussion

The optimum prediction results using DEA are then converted into a normal range through de-normalization. The optimum responses and end milling's input parameters after de-normalization are shown in Table 3. Furthermore, based on this table, the predictions of DEA for CF and SR are smaller than the experimental results. From the experimental results, the smallest value of CF is 15110 N, and the smallest value of SR is 1109 μm, where DEA's prediction for CF is 5719.8 N and 834.7 μm for SR. Therefore, based on the DEA's prediction, applying the obtained levels of the end-milling parameters can reduce the energy consumption in the end-milling process of GFRP composites.

Table 3. The levels of end-milling process parameters for obtaining minimum cutting force and surface roughness.

Depth of cut (mm)	Spindle speed (rpm)	Feeding speed (mm/min)	Cutting force (N)	Surface roughness (μm)
2	4797.5	579.7	5719.8	834.7

6 Conclusions

The following are the conclusions drawn from this study:

- The optimum BPNN topology of cutting force and surface roughness can be achieved using seven hidden layers, six neurons in each hidden layer, and a tansig activation function.
- The minimum value of cutting force and surface roughness in end milling of GFRP composites have been obtained by applying a combination of BPNN-DEA. The

depth of cut of 2 mm, the spindle speed of 4797.5 rpm, and the feeding speed of 579.7 mm/min can simultaneously minimize the cutting force and surface roughness.

References

1. PrasanthISNVR, Ravishankar DV, Hussain MM, Badiganti CM, Sharma VK, Pathak S (2018) Investigations on performance characteristics of GFRP composites in milling. Int J Adv Manuf Technol 99(5):1351–1360
2. AzmiAI, Lin RJT, Bhattacharyya D (2013) Machinability study of glass fibre-reinforced polymer composites during end milling. Int J Adv Manuf Technol 64(1):247–261
3. PrashanthISNVR, Ravi Shankar DV, Manzoor Hussain M, Chandra Mouli B (2018) Critical analysis in milling of GFRP composites by various end mill tools. Mater Today: Proc 5(6):14607–14617, (Part 2)
4. SaikumarS, Shunmugam MS (2006) Parameter selection based on surface finish in high-speed end-milling using differential evolution. Mater Manuf Process 21(4):341–347
5. YildizAR (2013) A new hybrid differential evolution algorithm for the selection of optimal machining parameters in milling operations. Appl Soft Comput 13(3):1561–1566
6. Robbany F, Pramujati B, Suhardjono, Effendi MK, Soepangkat BO, Norcahyo R (2019) Multi response prediction of cutting force and delamination in carbon fiber reinforced polymer using backpropagation neural network-genetic algorithm. AIP Conf Proc 2114(1):030012
7. ZhangS, Li JF (2011) Prediction of surface roughness using back-propagation neural network in end milling Ti-6Al-4V alloy. Adv Mater Res 325:418–423
8. ErkanÖ, Işık B, Çiçek A, Kara F (2013) Prediction of damage factor in end milling of glass fibre reinforced plastic composites using artificial neural network. Appl Compos Mater 20(4):517–536
9. HuangPB, Ma C-C, Kuo C-H (2015) A PNN self-learning tool breakage detection system in end milling operations. Appl Soft Comput 37:114–124
10. RoccaP, Oliveri G, Massa A (2011) Differential evolution as applied to electromagnetics. IEEE Antennas Propag Mag 53(1):38–49

Multi-objective Optimization Using Backpropagation Neural Network and Teaching–Learning-Based-Optimization Method in Surface Grinding Under Dry and Minimum Quantity Lubrication Conditions (MQL)

Dinny Harnany[1], M. Khoirul Effendi[1(✉)], H. C. Kis Agustin[1], Bobby O. P. Soepangkat[1], Sampurno[1], and Rachmadi Norcahyo[2]

[1] Department of Mechanical Engineering, Institut Teknologi Sepuluh Nopember, Surabaya 60111, Indonesia
khoirul_effendi@me.its.ac.id

[2] Department of Mechanical and Industrial Engineering, Universitas Gadjah Mada, Yogyakarta, Indonesia

Abstract. The present work focuses on the performance modeling of surface grinding to attain an optimum parameter setting for the minimum coefficient of friction and surface roughness. The experimental data were collected during the surface grinding process using dry conditions and minimum quantity lubrication (MQL) as a clean technology lubricant. The usage material was SKD 61 tool steel. The varied surface grinding parameters were the depth of cut and table speed, wherein each had three levels. The surface grinding operation was performed by using a full factorial design $2 \times 3 \times 3$. Backpropagation neural network (BPNN) was first applied to obtain the modeling of the surface grinding experiment, the objective function, the predictions of coefficient of friction, and surface roughness. The objective function is then modified into a fitness function. Finally, this fitness function is utilized in multi-objective optimization using the teaching–learning-based optimization (TLBO) method to attain the surface grinding parameters' levels that simultaneously produce a minimum coefficient of friction and surface roughness. Based on our experimental results, the combination of BPNN-TLBO can be applied to simultaneously minimize the coefficient of friction and surface roughness in the grinding of SKD 61 by implementing MQL and setting the feeding speed at 150 mm/s and the depth of cut at 0.01 mm. As the result, the minimum surface roughness is 0.376 μm, and the coefficient of friction is 0.333.

Keywords: Backpropagation neural network · Clean technology · Minimum quantity lubrication · Surface grinding · Teaching–learning-based optimization

I. Tolj et al. (eds.), *Recent Advances in Mechanical Engineering*, Lecture Notes in Mechanical Engineering, https://doi.org/10.1007/978-981-19-0867-5_40

1 Introduction

Surface grinding is different from the other metal machining processes such as drilling, milling, and turning since it needs more energy to remove a specific volume of material. In addition, there are many ploughing and rubbing actions since the grinding wheel possesses many abrasive grits or cutting edges [1]. Consequently, these phenomena cause a temperature rise in the grinding zone, which undermines the surface integrity of the ground surface in the form of grinding burn, uneven cracks, poor surface quality (geometrical and dimensional accuracies), induction of tensile residual stresses, and phase transformation [1, 2]. Effective lubrication and cooling can reduce the high grinding zone temperature. Nowadays, MQL or semi-dry machining has been widely applied as an effectual near-dry application in the grinding process. Furthermore, based on environmental preservation characteristics and conformity with the ISO 14000 standard this application is considered environmentally friendly. Surface roughness, specific grinding energy, and coefficient of friction are three criteria commonly used to evaluate the lubrication property of MQL during the surface grinding process are [3]. It has been reported that the outstanding lubrication property of the mixed vegetable oil resulted in lower values for both specific grinding energy and friction coefficient. The varied grinding parameters in some researches are the table speed and depth of cut. The coefficient of friction of grinding force ratio determines the magnitude of the interaction between the abrasive and grains workpiece surface. This coefficient has a close relationship with the quality of the workpiece surface and processing precision. The coefficient of friction is calculated as the ratio of tangential and normal forces. The machine component surface roughness has been considered as one of the key criteria in assessing the grinding process. A smaller surface roughness value indicates higher surface smoothness. Most of the assessment of the grinding process has included surface roughness as a key performance indicator.

Teaching–learning-based optimization (TLBO) is an algorithm that is inspired based on the effect of the influence of a teacher on the output of learners in a class. In this algorithm, the teacher puts his hard work and makes all the learners of a class educated. Then the learners interact with themselves to further modify and improve their gained knowledge. This algorithm requires only the common control parameters such as the population size and the number of generations and does not require any algorithm-specific control parameters. This method is one of the popular metaheuristic methods in optimization because it produces output with a high level of consistency, less computational effort. Furthermore, good performance was showed by this method to solve complex optimization problems since the limitations on benchmark functions, mechanical design constraints, and continuous non-linear numerical issues [4].

The present work attempts to obtain the minimum coefficient of friction and surface roughness simultaneously in the surface grinding process using dry and minimum quantity lubrication conditions for SKD 61 tool steel. BPNN is applied in modeling, determining objective function, and predicting coefficient of friction and surface roughness. In addition, the teaching–learning-based optimization (TLBO) method is utilized to produce the combination of surface grinding parameters' levels that simultaneously produce the minimum coefficient of friction and surface roughness.

2 Research Methodology

2.1 Backpropagation Neural Network (BPNN)

Simulated neural networks (SNN) are a subset of machine learning that is one of the standard optimization methods for predicting the grinding temperature of titanium matrix composite [5], predicting CNC grinding parameter [6], predicting grinding force [7], and monitoring tool condition [8]. Backpropagation neural network (BPNN) is a part of SNN for modeling the relationship between parameter input and target in surface grinding. BPNN consists of 3 layers, namely input layers, hidden layers, and output layers. All data is then used to update the weight on the hidden layer through a series of processes (epochs) to minimize MSE between the output predicted by BPNN and the target. The general flowchart of BPNN to find a network with the best MSE is shown in Fig. 1.

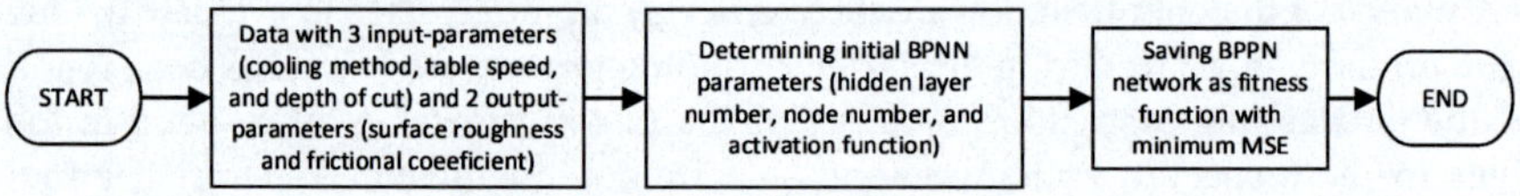

Fig. 1. The BPPN flowchart for producing a fitness function with minimum MSE.

2.2 Teaching–Learning-Based Optimization (TLBO)

There are many advantages of the TLBO method; several studies have been performed to compare the performance of this method with other techniques in the manufacturing process. For example, Pawar and Rao [9] used TLBO to perform the optimization problems in abrasive water jet machining, grinding, and milling. Furthermore, the TLBO's optimization result is compared with other methods such as simulated annealing, harmony search, genetic algorithm, artificial bee colony algorithm, and particle swarm optimization. In grinding, the author used the optimization model given by Wen et al. [10] to maximize production rate and minimize production cost by varying wheel speed, workpiece speed, and the depth and lead of dressing. Rao et al. [9] implemented the TLBO method in manufacturing processes such as drilling, turning, and grinding. In grinding, this method maximizes material removal rate (MRR) by varying feed rate, grit size, and depth of cut. TLBO's performance is then compared with other optimization methods such as quadratic programming, simulated annealing, harmony search, genetic algorithm, artificial bee colony algorithm, and particle swarm. From the comparison, it is found that the performance of TLBO is equal to the other optimization methods results. Tiwari et al. [11] proposed TLBO to optimize the manufacturing process such as drilling, turning, face milling, end milling, grinding, abrasive jet machining, electric discharge machining, electrochemical machining, laser beam machining, ultrasonic machining, and micromachining. The quality of the final product in surface grinding is influenced by cost and production time. Therefore, the authors used TLBO to minimize cost, surface roughness and maximize production rate by varying workpiece speed, wheel speed, lead of dressing, and depth of dressing. As a result, TLBO gives 90% improvement compared with the quadratic programming method and 10% improvement compared with the simulated annealing method, harmony search method, and artificial bee colony method (ABC). There are two phases of TLBO, the teacher phase, and the learner phase. Figure 2 shows the general flowchart of the TLBO algorithm.

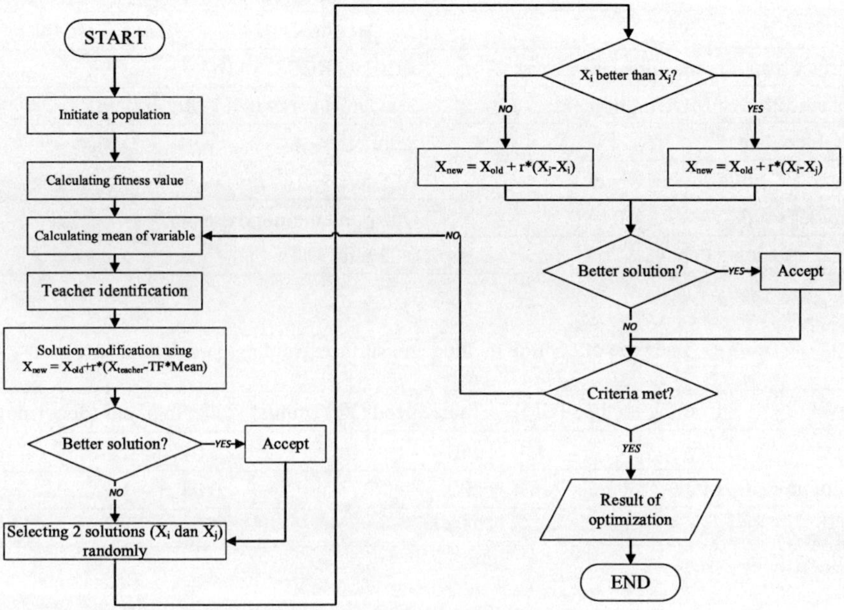

Fig. 2. The TLBO flowchart

3 Experiments and Results

3.1 Experimental Work

The material in this experiment was SKD 61, which hardness was 40 HRC. This material is frequently applied in several manufacturing areas, such as pressure dies, plastic injection mold, casting dies, stamping dies, punches, hot shear blades, and extrusion tools. The dimensions of the specimen were 40 mm (length), 10 mm (width), and 5 mm (thickness). The detail of the experimental surface grinding experiments is shown in Table 1. The type of cooling method and the surface grinding parameters' levels during the experiments are presented in Table 2. The design of experiments was a factorial design 2 × 3 × 3. The replication of the experiment was thrice. Hence, there were 36 experiments total. Randomization was applied during the investigation. The responses in this study were two forces (normal and tangential) and surface roughness, which measurements were performed after the fifth passes. Both forces were measured using Kistler dynamometer type 9272, while surface roughness was measured using Mitutoyo surftest SJ-310.

Table 1. The detail of the surface grinding experimental conditions

Parameters	Condition
Material of the specimen	SKD 61
Grinding process' type	Up cut plunge surface grinding
Grinding wheel's type	Al_2O_3(A46K)
Surface grinding machine	Krisbow KGS818AH
Specification of MQL equipment	SMC mist spray unit LMU 100-15
MQL coolant	Palm oil
Speed (V_s) of grinding wheel	27.5 m/s
Dresser's type	One point diamond dresser
Total dressing's depth	0.03 mm

Table 2. The type of cooling method and surface grinding parameters' levels

Level	Cooling method (CM)	Table speed (V_f) (mm/s)	Depth of cut (doc) (mm)
Low (1)	Dry	150	0.01
Medium (2)	MQL	200	0.03
High		250	0.06

3.2 Experimental Result

Figures 3 and 4 show the measured surface roughness and coefficient of friction resulted from the combination of table speed, depth of cut using dry and MQL cooling methods.

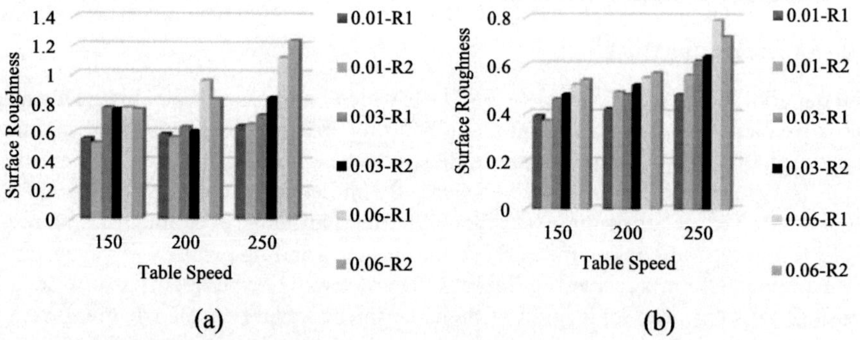

Fig. 3. Measured surface roughness as the results of the combination of table speed, and depth of cut for **a** dry method, **b** MQL method. [R1 = 1st replication and R2 = 2nd replication]

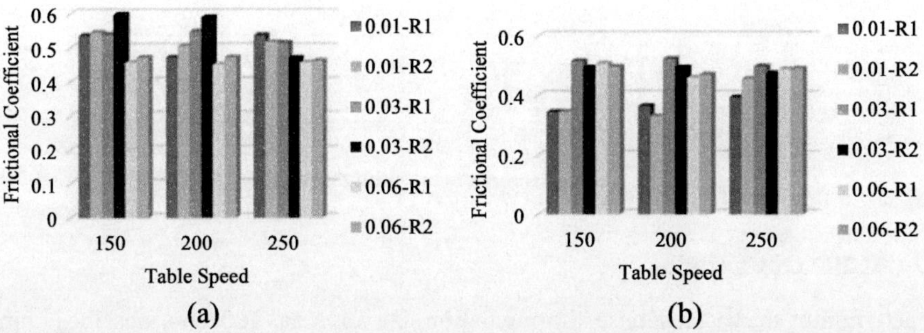

Fig. 4. Measured coefficient of friction as the results of the combination of table speed, and depth of cut for **a** dry method, **b** MQL method. [R1 = 1st replication and R2 = 2nd replication]

4 Modeling and Optimization

4.1 Backpropagation Neural Network (BPNN)

The selected BPNN parameters in modeling the relationship between parameter input and target are presented in Table 3.

Table 3. BPNN tunning parameters

No	Parameter	Values
1	Ratio data for training, testing and validation	70%:15%:15%
2	Variation of activation function	*Harlim, Hardlims, Satlin, logsig, tansig*, and *purelin*
3	Number of hidden layers	$1 \div 10$ layers
4	Number of nodes each hidden layer	$1 \div 10$ nodes
5	Stopping criterion	Max epoch (10,000 epochs)
6	Parameter for selecting best BPNN nework	Mean squared error (MSE)
7	Learning rate	0.001
8	Training algorithm	Levenberg–Marquardt

The best BPNN network possesses:

(1) three hidden layers,
(2) six neurons in each hidden layer,
(3) logsig activation function, and
(4) the MSE is 0.013259.

Figure 5 shows the best BPNN network based on the explained specification.

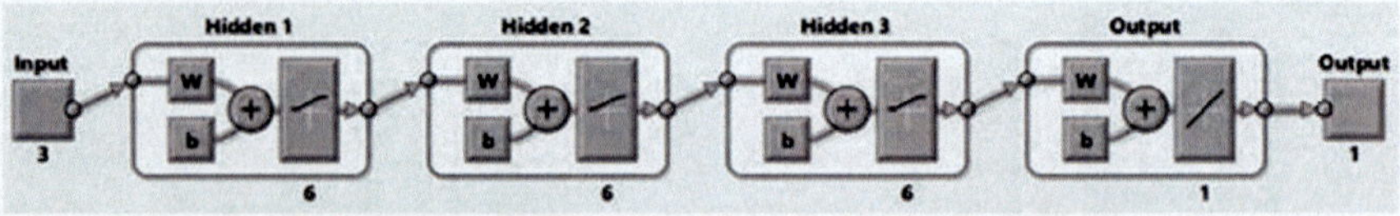

Fig. 5. The best BPNN's network with the smallest MSE

4.2 TLBO Optimization

In determining the best response, the population size was set as 100 iterations, the number of iterations was set as 100, teaching factor was selected in the range 0–10 randomly. Further, bounding method (min–max) was used in controlling violating variables, and minimum and maximum bound were set in the range 0–2. The result of the best response using the TLBO method for 100 iterations is shown in Fig. 6, where the optimum value is 0.1212.

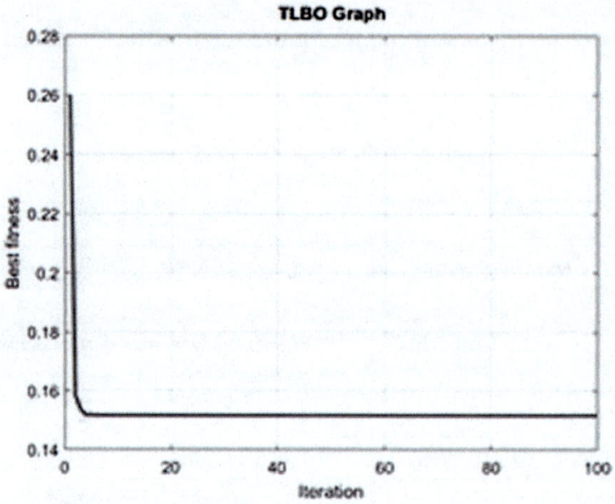

Fig. 6. The result optimization using TLBO method for 100 iterations.

5 Result and Discussions

The denormalization process is then used to transform the best prediction results using TLBO into the original limit value, and the result is shown in Table 4. The experimental data shows that the smallest surface roughness value of 0.37 μm has the coefficient of friction value of 0.3445. Furthermore, the smallest coefficient of friction value of 0.33 has a surface roughness value of 0.49 μm. On the other hand, the prediction of surface roughness and the coefficient of friction using BPNN-TLBO simultaneously is 0.376 μm and 0.333, respectively. Therefore, based on the comparison between the experimental data and TLBO's predictions, determining the parameter input of grinding or the levels of the grinding process parameters for producing the best responses can use the TLBO method.

Table 4. TLBO's prediction parameter process for obtaining minimum surface roughness and coefficient of friction in grinding

Cooling method	Feeding speed (mm/s)	Depth of cut (mm)	Surface roughness (Ra) (μm)	Coefficient of friction
2	150	0.01	0.376	0.333

6 Conclusions

The followings are the conclusions drawn from this study:

- The BPNN topology that can predict the coefficient of friction and surface roughness possesses seven hidden layers, six neurons in each hidden layer, and a tansig activation function.
- The combination of BPNN-TLBO can be applied to simultaneously minimize the coefficient of friction and surface roughness in the grinding of SKD 61. The minimum coefficient of friction and surface roughness can be obtained by applying MQL and setting the feeding speed at 150 mm/s and the depth of cut at 0.01 mm. The minimum surface roughness is 0.376 μm, and the coefficient of friction is 0.333.

References

1. Vashista M, Gaddam A, Paul S (2012) Study of surface integrity of ground bearing steel using Barkhausen noise technique. Int J Adv Manuf Technol 63(5):771–783
2. Shao Y, Fergani O, Ding Z, Li B, Liang SY (2015) Experimental investigation of residual stress in minimum quantity lubrication grinding of AISI 1018 steel. J Manuf Sci Eng 138(1)
3. Mia M, Gupta MK, Singh G, Królczyk G, Pimenov DY (2018) An approach to cleaner production for machining hardened steel using different cooling-lubrication conditions. J Clean Prod 187:1069–1081
4. Springer Homepage. https://www.springer.com/gp/book/9783319227313. Accessed 02 Sept 2021
5. Liu C, Ding W, Li Z, Yang C (2017) Prediction of high-speed grinding temperature of titanium matrix composites using BP neural network based on PSO algorithm. Int J Adv Manuf Technol 89(5):2277–2285
6. Šarić T, Šimunović G, Vukelić Đ, Šimunović K, Lujiće R (2018) Estimation of CNC grinding process parameters using different neural networks. Teh Vjesn 25(6):1770–1775
7. Sagepub Homepage. https://journals.sagepub.com/doi/abs/https://doi.org/10.1177/0954405418780166. Accessed 02 Sept 2021

8. Cheng C, Li J, Liu Y, Nie M, Wang W (2019) Deep convolutional neural network-based in-process tool condition monitoring in abrasive belt grinding. Comput Ind 106:1–13
9. Pawar PJ, Rao RV (2013) Parameter optimization of machining processes using teaching–learning-based optimization algorithm. Int J Adv Manuf Technol 67(5):995–1006
10. Wen XM, Tay AAO, Nee AYC (1992) Micro-computer-based optimization of the surface grinding process. J Mater Process Technol 29(1):75–90
11. Tiwari A, Pradhan MK (2017) Applications of TLBO algorithm on various manufacturing processes: a review. Mater Today Proc, Part A 4(2):1644–1652

Innovation Model to Improve BOP Transport Performance

Mohammad Kurniadi Rasyid(✉), Putu Mahayana Santika, and Erik Mubarik Romadhona

Department of Mechanical Engineering, Institut Teknologi Indonesia, Tangerang 15320, Indonesia
kurniadirasyid@gmail.com

Abstract. This study aims to design a blow out preventer (BOP) transport equipment. The design of this transport equipment must be as efficient as possible at a low cost. This designed equipment must be able to withstand the load without deforming or breaking. This process begins with the creation of a 3D geometry model which is then transformed into a finite element model. Three design models of the BOP transport skid were made using the Inventor 2014 software. After selecting the best model using pugh's concept selection, a computer-aided design (CAD) model was created, followed by the creation of finite element models. The FEM models with three different materials had been created and analyzed. The materials used were ASTM 53, ASTM A 36, and Aluminum 6063. The models made were then analyzed for deflections and stresses that occur due to the given load. The results showed that the ASTM A53 material and the selected dimensions could be used for the BOP transport equipment because it meets the required conditions.

Keywords: BOP · Finite element · Transport skid

1 Introduction

Although currently, the world of oil and gas is experiencing a decline in production, equipment and services in this field are still being developed [1]. Most of the equipment is made in the form of a modular design. The modular construction allows the equipment to be easily moved from one location to another, thereby further enhancing operational efficiency [2]. The background of this research is to find the most suitable design for transport equipment from blow out preventer (BOP) skid. The BOP skid is a steel frame used to accommodate the BOP equipment so that it can be transported. BOP itself is a tool that is useful for sealing and limiting the liquid that enters the borehole. There are many things that must be considered in designing a BOP skid such as strength, accuracy in use, warranty, etc. [3].

Until today, there are still few publications on Transport skid BOP that have been made and researched. Skid BOP ransport allows us to raise, lower and transport BOPs in an even easier way. This research will focus on finding the best material that can

I. Tolj et al. (eds.), *Recent Advances in Mechanical Engineering*, Lecture Notes in Mechanical Engineering, https://doi.org/10.1007/978-981-19-0867-5_41

withstand the stiffness of the Transport skid BOP and has a high degree of safety with the desired geometry. In the process of designing or making the Transport skid BOP design, the finite element method (FEM) is used to carry out various strength analyzes. The advantage of this method is very efficient and quite accurate. The right FEM software will provide solutions for an innovative research [4].

2 Basic Theory

The finite element method (FEM) is a numerical calculation method to solve engineering problems and has been widely used in the fields of mechanics, electromagnetics, and heat conduction [5, 6]. The use of the finite element method (FEM) to calculate the strength and stiffness of structures is growing rapidly, especially in the current era of digital computing. FEM is based on the calculation of the stiffness of the rods and the solution of the equilibrium equation of the system [7].

The use of Pugh's concept selection is based on the desire to find the best choice from several available options based on several established criteria [8, 9]. Pugh's concept is a systematic decision-making method and makes decisions more precise, functional, methodical, fast, and organized [10].

3 Research Method

The design process begins with collecting the input data for the BOP skid transport. The design process is then continued by conducting an initial design study and concept creation (see Fig. 1). There are 3 design models made, as shown in Fig. 2. After selecting the best model using pugh's concept selection, a computer-aided design (CAD) model was created, followed by the creation of a finite element model. The FEM model that has been created is then analyzed. If the results are as expected, a detailed drawing is made. The drawing will then be sent to the manufacturing section to be made.

Computer-aided analyze (CAA) begins with the creation of a 3D geometry model which is then transformed into a finite element model. The 3D computer model of the BOP transport skid was designed using the Inventor 2014 software. This 3D model was then analyzed using the finite element calculation tool in the software. Material loading is given to supports with a load of 10,000 lbs (4.5 tons). The reference standard used: (a)ASME II "Specification Material", (b) AWS D1.1 "Structural Welding Code—Steel", (c) API 4G dan API 4F and (d) API 8 A and API 8 B. The schematic of the forces on the steel skid and sling can be seen in Fig. 3.

Load that the plate supports is the weight of the BOP, namely, W = 10,000 lbs (4.5 ton), and the dimensions of the retaining/reinforcing pipe are 3.5". The bending stress occurs $\sigma_b = M_{\mathrm{bmax}}/w_b$ (N/mm^2). Compressive stress in skid structure $\sigma_c = F_c/A$. If the maximum stress that occurs is smaller than the allowable stress, the skid structure is considered strong to accept the load and is safe to use. The monitored results are the stress distribution and the deformation on the entire BOP skid transport support. The materials used are ASTM 53, ASTM A 36, and Aluminum 6063.

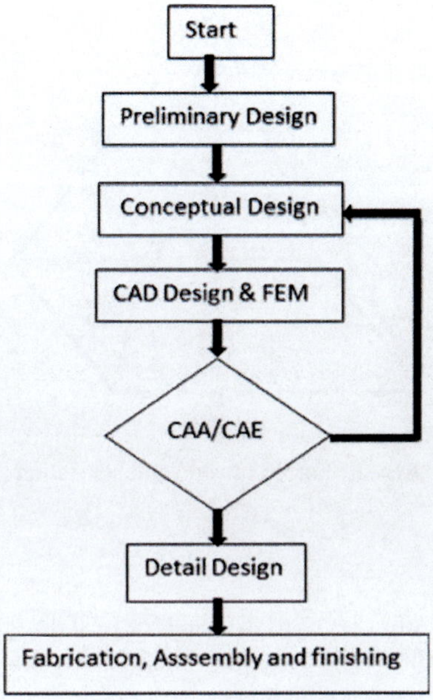

Fig. 1. Research design

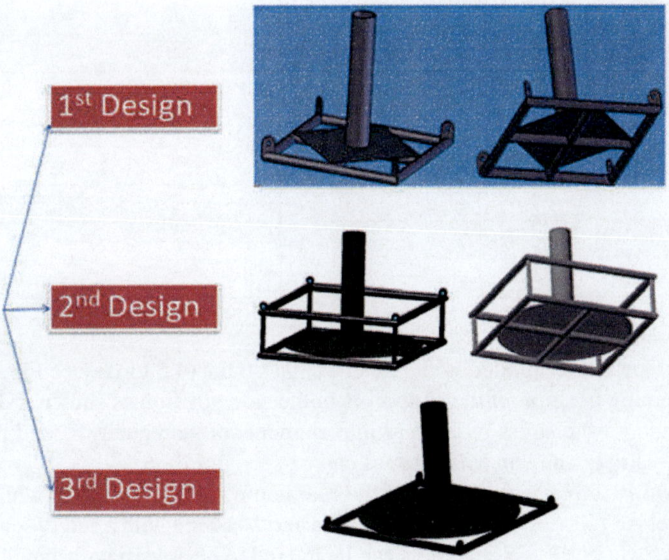

Fig. 2. Design models of the BOP transport skid

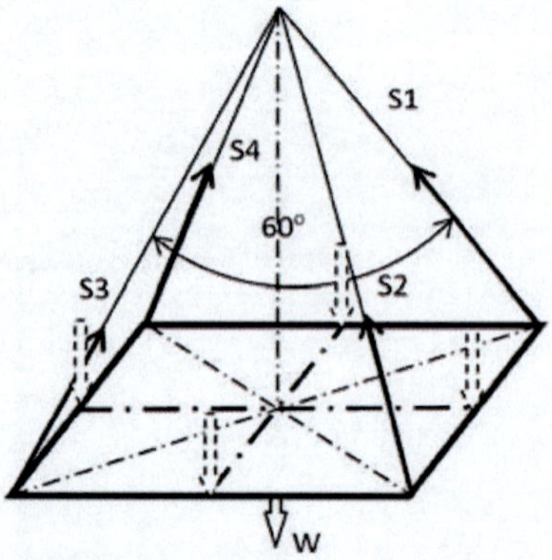

Fig. 3. Forces on skid and steel sling

4 Discussion

Pugh concept selection (Table 1) is used to select the best design model. After comparing the 3 design models using the Pughs concept, we decided to choose the first design model, because the score is the highest.

Table 1. Pugh concept-based selection

Criterion	Design 1	Design 2	Design 3
Easy to use	+	0	+
Light to use	+	0	+
Cost	+	0	+
Stable	+	0	+
Easy to manufacture	+	−	−
Safe to use	+	0	+
Final score	6	−1	4
Develop?	Yes	No	No

Material loading is applied with a load of 10,000 lbs (4.5 tons) (see Fig. 4).

The retaining position of the support is in the side position as shown in Fig. 5.

The results of the stress based on finite element software analysis and the location that occurs, namely, the von Mises stress can be seen in Fig. 6.

As shown in Table 2, the result of the maximum stress is 97 MPa while the yield stress of steel ASTM A36 and A53 is 248 MPa and 241 Mpa with a safety score showing a value of 2.5. So both steel materials can be judged to be able to withstand the load. Al 6063 was not chosen because of its low-security value, which is only 1.5. Material steel ASTM A53 is easier to get and to buy. While material steel A36 is a little cheaper than material A53. We recommend using material 2 because it is easier to obtain.

Fig. 4. Skid BOP model designed and the position of the load.

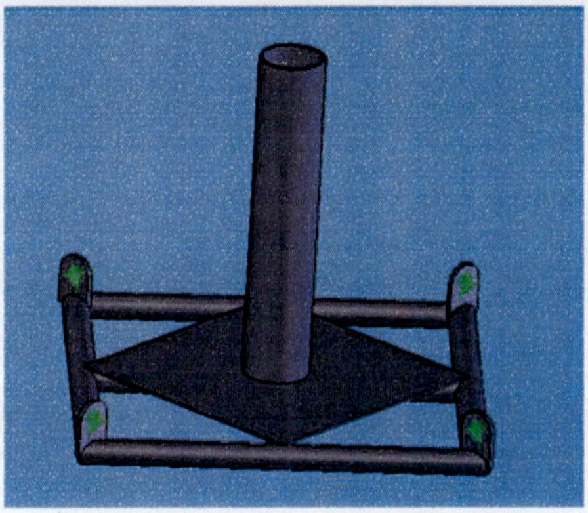

Fig. 5. Skid BOP model designed and the position of the constraints

Manual calculations are not much different from the results of the CAA software. Based on the calculation manually, the value for the bending stress in skid structure is 125.3 N/mm^2 and the compressive stress is 2.7 N/mm^2. The maximum stress in the skid retaining pipe is 116.2 N/mm^2. The skid reinforcement pipe is welded at four points. The shear stress in the reinforcing member is smaller than the allowable shear stress. Therefore the skid structure is considered strong enough to accept the load and is safe to use.

Meanwhile, the deformation that occurs can be seen in Fig. 7a. the maximum desired deformation is 1 mm, this is still greater than the recommended ideal value 0.1 mm/m [3]. The deformation of a structure made of steel A36 and A53 is 0.6 mm. These are still acceptable, while the deflection that occurs in the skid of aluminum is larger than 1 mm, so it is not recommended. The results of the design analysis using the software were continued with the manufacture of the BOP skid transport as shown in Fig. 7b.

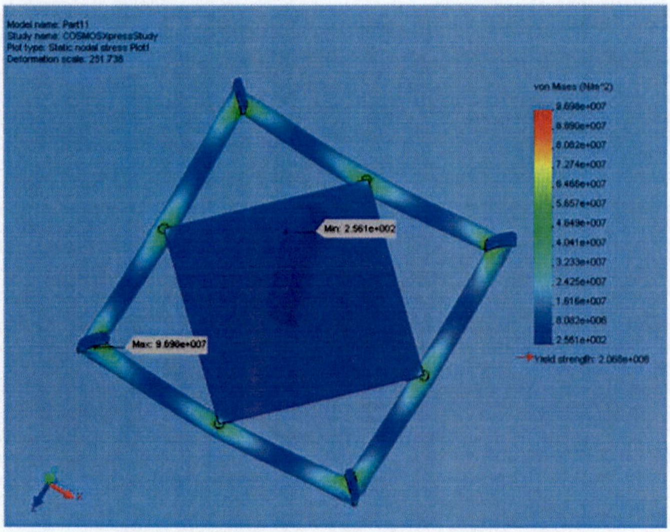

Fig. 6. Graph of von mises stress that occurs

Table 2. Computer-aided analyze (CAA) result

	Steel ASTM A36	Steel ASTM A53	Aluminum 6063
Yield stress (MPa)	248.2	241	145
Maximum von misses stress (MPa)	97	97	95.5
Safety	2.55	2.48	1.5
Deflection (mm)	0.6	0.6	1.74
Price/kg	US$ 0.55	US$ 0.652	US$ 2.0

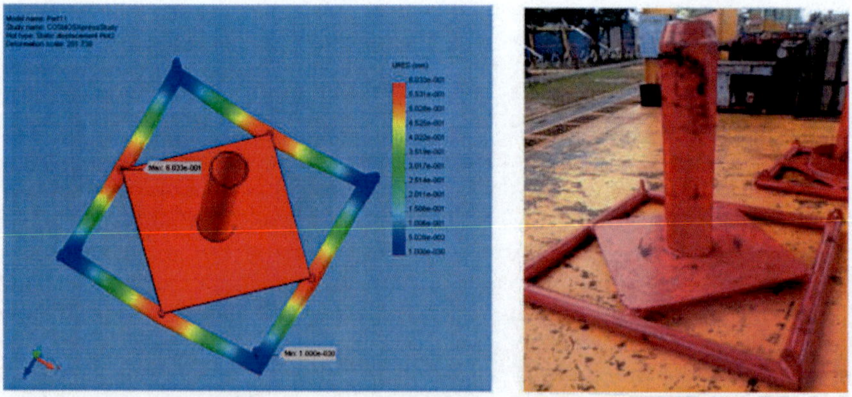

(a) (b)

Fig. 7. a Deformation that occurs, and **b** the completed BOP transport skid

5 Conclusion

1. BOP transport skid was made after the process of selecting the best design model. There were 3 design models made. After selecting the best model using pugh's concept selection, a computer-aided design (CAD) model was created, followed by the creation of finite element models. The FEM models with three different materials had been created and analyzed. The materials used were ASTM 53, ASTM A 36, and Aluminum 6063. After the results matched the desired criteria, a detailed drawing was made. The drawing was then sent to the manufacturing section to be made.
2. The maximum stress value was 97 Mpa. This value was far less than the yield stress of steel ASTM A53 and A36. So that these materials are safe to use for BOP skid transport. Steel ASTM A53 is easier to get and to buy. While material steel A36 is a little cheaper than material A53. We used material 2 because it is easier to obtain.
3. The maximum deformation for design model 1 was 0.6 mm which met the requirement.
4. By looking at the safety value and deformation that occurs, this design model 1 was safe and reliable. The design model can be used for BOP skid transport.

References

1. Munro N, Myers A, Lund C, Neate R (2019) Enhancing industry subsea well intervention response capabilities. In: SPE offshore Europe conference and exhibition. https://doi.org/10.2118/195757-ms
2. Duthie L, Otaibi A, Saiood H, Toal T (2019) Modular hydraulic workover unit enhances operational efficiency and cost savings for ESP replacement's.In: Society of petroleum engineers SPE/ICoTA well intervention conference and exhibition. https://doi.org/10.2118/194230-MS
3. Popa A, Scurtu IC (2018) Skid designing for offshore equipment transport. J Phys: Conf Ser 1122
4. Epelle E, Gerogiorgis D (2019) A review of technological advances and open challenges for oil and gas drilling systems engineering. AIChE J. https://doi.org/10.1002/aic.16842
5. Li R (2021) Research on application of finite element method in static analysis. J Phys: Conf Ser 1802(4). https://doi.org/10.1088/1742-6596/1802/4/042008
6. Hung N, Ngoc T (2004) Analysis of cracked plates and shells using "metis" finite element model. Finite Elements Anal Des 855–878
7. Bhattacharjee SS (20221) Finite element analysis of solids and structures, 1st edn. CRC Press. https://doi.org/10.1201/9781003027843
8. Thakker A, Jarvis J, Buggy M, Sahed A (2009) 3D-CAD conceptual design of the next-generation impulse turbine using the Pugh decision-matrix. Mater Des 30(7):2676–2684
9. Ulrich KT, Eppinger SD (2008) Product design and development, 4th edn. McGraw-Hill, New York
10. Andrade M, Valéria L, Valter E (2020) Application of the pugh method for the selection of structural adhesives in precision instruments: a case study. In: VI Simpósio Internacional de Inovação e Tecnologia, Blucher Engineering Proceedings, vol 7, pp 71–79. ISSN 2357-7592

Numerical Study of a Medium-Speed Train with the Influence of Wall on One Side

Vivien S. Djanali[1(✉)], Quraisy Amri[1], and Agus Windharto[2,3]

[1] Department of Mechanical Engineering, Institut Teknologi Sepuluh Nopember, Surabaya 60111, Indonesia
vivien_s@me.its.ac.id
[2] Department of Industrial Product Design, Institut Teknologi Sepuluh Nopember, Surabaya 60111, Indonesia
[3] Creative Center, Institut Teknologi Sepuluh Nopember, Surabaya 60111, Indonesia

Abstract. This study was conducted to determine the aerodynamic drag on the surface of the medium-speed train, especially when passing through a wall on one side of the railroad tracks. A steady, three-dimensional numerical study was carried out. The train was designed to operate on a narrow-gauge track. The turbulence model used was the k-ω SST. The type of meshing used was a combination of hexahedral and tetrahedral cells. The simulation was performed at velocities of 120 km/h (33.3 m/s) and 160 km/h (44.4 m/s), which correspond to the Reynolds numbers based on the height of the train of 7.76×10^6 and 10.34×10^6, respectively. There were two variations of the distance of the wall to the center of the railway tracks, i.e., 2.35 m (0.69H) and 6 m (1.76H), with H is the height of train. The results obtained showed that a train with the influence of a wall at a distance of 2.35 m has a higher drag coefficient than that with a wall at a distance of 6 m. The closer the distance of the wall increased the drag experienced by the train. At a wall distance of 2.35 m, there was a difference in the value of the pressure coefficient on the sides of train surface of the train. The resulting pressure coefficient on the surface close to the wall was smaller than the open airside.

Keywords: Medium-speed train · Aerodynamic · Pressure coefficient · Drag coefficient

1 Introduction

Railway is one mode of mass transport which is increasingly high in demand. In developing countries, the use of high-speed trains has many aspects to consider, such as high infrastructure cost. The use of medium-speed train is one of the alternatives in which the use of high-speed train is not possible. One of the advantages is that it will allow the use of existing infrastructure. In many developing countries, such as in Indonesia, the railway built was a narrow-gauge track. Therefore, consideration should be taken into account in terms of the aerodynamic performance of the train and safety issue. The clearance area applied at the existing infrastructure was designed for the train to operate at lower speed. In this study, we analyzed the influence of a wall on one side close to the

I. Tolj et al. (eds.), *Recent Advances in Mechanical Engineering*, Lecture Notes in Mechanical Engineering, https://doi.org/10.1007/978-981-19-0867-5_42

train designed to operate at medium speed. The wall was a representation of a structure near the train, such as air or noise barrier. A passing train induces slipstream velocity (gust of wind) to the surrounding [1] and wave of pressure-pulsation to a structure [2]. On the other hand, the presence of nearby structures, such as barrier or wall buildings reduces aerodynamics performance of the train. The evaluation of the aerodynamic performance of the train itself can be done by experimental or numerical with full-scale and scaled-model approaches. Experiments of full-scale and scaled-model in wind-tunnel provide different drag coefficients, and thus the correlation for extrapolating the scaled-model to the full scale is also important [3]. The test requirement and procedure for train assessment on open track is documented in [4].

Previous studies have been performed on train aerodynamic performance near a wall structure. The empirical approach for calculating the wind load due to passing train near a wall or other structure has been documented in [5]. Rocchi et al. [6] and Xiong et al. [7] performed full-scale experiment on high-speed train passing through wall or barrier. Hu et al. [8] conducted numerical simulation of high-speed train passing through wall barrier located on both sides of the train. Yang et al. [2] carried out experiment and simulation of high-speed train passing under a overhead bridge. Khayrullina et al. [9] performed a numerical study to determine the effect of wind gusts caused by trains in the platform area. wind exceeds the hazard limit. Li et al. [10] conducted a numerical study to analyze the pressure on the air barrier and found out that the wind pressure on the inside was greater than the outside on the air barrier. Vittozi et al. [11] carried out experiments to analyze the fluid dynamics interactions between noise barriers. This study found that the distribution of maximum pressure along the height of the barrier was not constant with the maximum pressure located at the bottom area.

Studies on medium-speed train had also been performed. Mario et al. [12] conducted simulation and experiment of the medium-speed train. Utomo et al. [13] performed simulation for train ranging from 100 up to 350 km/h. Since the length of its mask of car is relatively short, the medium-speed train has drag coefficient which is higher than that of the high-speed train. The drag coefficient of the medium-speed train can be as high as 0.647 for single car [12].

Maleki et al. [14] compared various turbulence model on a freight container to the wind-tunnel experiment. The steady Reynolds Averaged Navier-Stokes (RANS) model showed similar result to the unsteady RANS and Reynolds Stress Model (RSM) for a single freight car simulation. Nevertheless, the discrepancy of the drag coefficient of these method to the experiment was quite high of around 12.5%. Wang et al. [15] compared the unsteady RANS model to a higher accurate scheme for the high-speed train simulation. It was found that the unsteady RANS was unable to accurately predict the slipstream of the high-speed train, particularly in predicting the trailing vortices. However, the difference of the drag coefficient by comparing the models was small, of maximum 4%, and suggested that the unsteady RANS model can be used in which we want to evaluate the drag coefficient.

This study aims to analyze the drag coefficient of the full-scale train with the presence of a barrier wall on one side of the train, as well as to analyze the pressure induced by the train to the wall. The simulation was performed at velocities of 120 km/h (33.3 m/s) and 160 km/h (44 m/s), which correspond to the Reynolds numbers of 7.76×106 and 10.34×106, respectively. The distance of the wall was varied between 2.35 and 6 m.

2 Numerical Method

Steady, three-dimensional simulations at a full-scale single train were performed using ANSYS FLUENT software. The configuration of the train and wall is shown in Fig. 1. The train design has a height of 3.4 m (H), width of 3 m (0.88H), and length of 20.5 m (6.03H). The wall was located along the passage close to the track, at distances from the wall to the center of the railway track of 2.35 m (0.69H) and 6 m (1.76H). The height of the wall was equal to H and the thickness is 0.3 m (0.088H). As shown in Fig. 1a, the reference axis was placed, such that $x = 0$ is located at the front end of the mask of car, $y = 0$ is at the ground, and the flow is to z-axis direction.

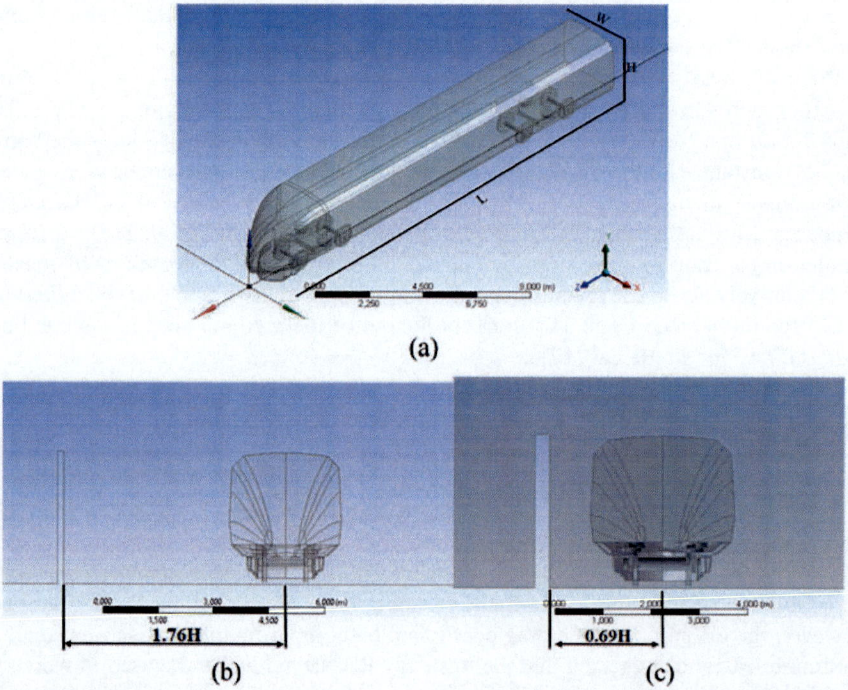

Fig. 1. Configuration of the train and wall. **a** Geometry of the train, **b** wall distance to the center of the railway track.

The computational domain is shown in Fig. 2. The boundary conditions used were velocity inlet at the inlet, pressure outlet at the outlet, symmetry for the top and side boundaries, solid wall for the train surface, and moving walls for the ground and wall surfaces. The turbulence model used was the k-ω SST model, with SIMPLE pressure–velocity coupling. The discretization schemes were second-order for pressure equations, and second-order upwind for all other equations. For convergence, the residual was kept below 10-4.

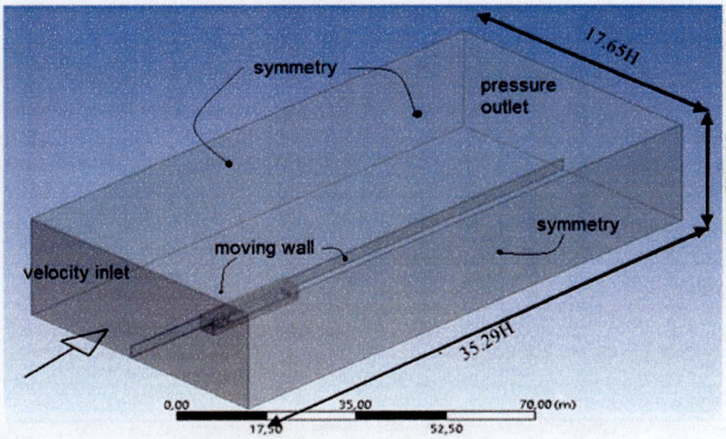

Fig. 2. Simulation domain and boundary conditions.

The mesh combined the tetrahedral and hexahedral cells. Inflation layer was used around the surface of the train, which represented the boundary layer region. The thickness of the first cell adjacent to the train surface was 0.2 mm. Grid independency test has been performed, and it was found that the mesh with a total number of about 5.5 million cells, as shown in Fig. 3, was sufficient with the difference of drag coefficient value less than 1%.

(a) (b)

Fig. 3. Meshing of the computational domain, **a** hybrid mesh, **b** enlarged view of the inflation layer adjacent to the train surface.

3 Results and Discussion

Figure 4 shows the dimensionless velocity contour on the x–z plane (horizontal plane) at height y/H = 0.53 from the ground. In general, we can see that the flow had lower velocity close to the mask of car of the train, close to the stagnation region. The flow further continued passing through the sides of the train, to the rear side of the train. Wake formed at the rear of the train body, showed by vortices in the wake region. At the wall distance to the center of the track s/H = 0.69, the contour of the air flow around the train formed was not symmetric on the left and right sides of the train surface. The contour of the air flow on the right side of the train tended to widen and extend toward the wall. While at wall distance of s/H = 1.76, the flow contours formed the same pattern for the left and right sides of the train surface. In addition, the normalized peak speed for wall distance s/H = 0.69 was greater than the distance s/H = 1.76 for each speed variation. For wall distance s/H = 0.69 had a maximum v/v_∞, which is 1.225 (for $Re_H = 7.76 \times 10^6$ and $Re_H = 10.34 \times 10^6$), while the distance s/H = 1.76 had a maximum v/v_∞ of 1.214 (for $Re_H = 7.76 \times 10^6$) and 1.215 (for $Re_H = 10.34 \times 10^6$). At different speeds with the same wall distance, the flow contours were similar. This shows that increasing the velocity did not cause a significant change to the resulting flow contour.

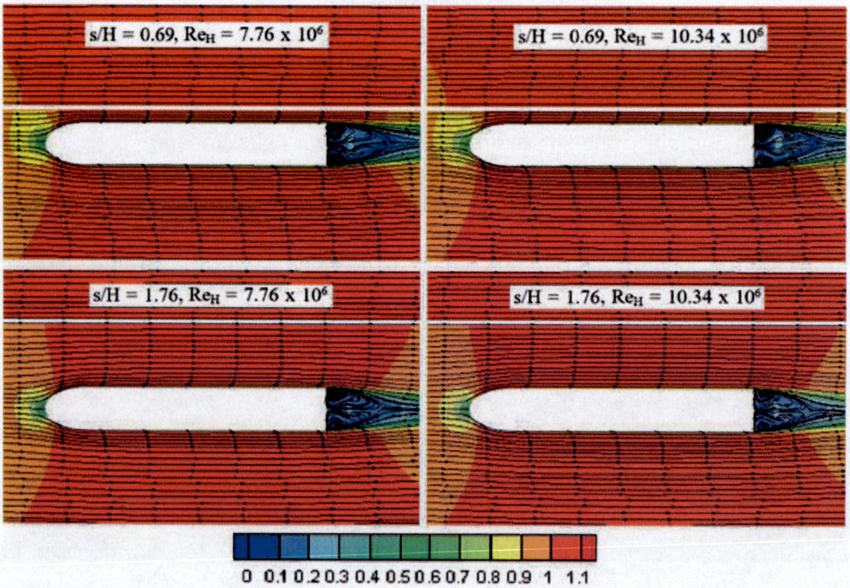

Fig. 4. Contour of flow and velocity streamline in the x–z plane, height y/H = 0.53.

The non-dimensional velocity contours at several planes on the transversal direction of the train are shown in Fig. 5. From the figures, we can see that the flow around the surfaces of the train was influenced by the presence of the wall. Three-dimensional flow was formed around the train and the wall, and this flow became more intense when the distance between the wall was closer. The underbody part of the train showed vortices in the ground clearance area. At the rear of the train, a wake area was formed which experiences strong turbulent flow and loss of energy.

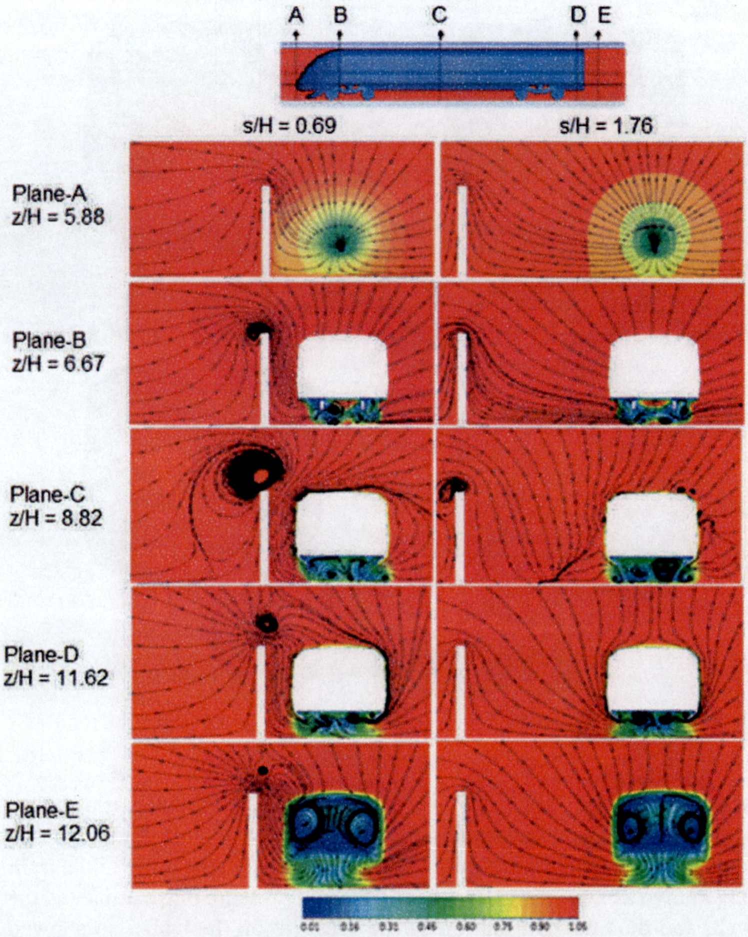

Fig. 5. Contours of dimensionless velocity in the several *x*–*y* planes at $Re_H = 10.34 \times 10^6$

The contours of pressure coefficients are shown in Figs. 6 and 7. Figure 6 shows the contour of the pressure coefficient on the surface of the train for each variation. It can be seen that with closer distance to the wall, the high pressure area around the stagnation region became wider. This suggested that the presence of wall induced slight blockage to the main flow. Figure 7 shows the contour of the pressure coefficient around the train on the *x*–*z* plane at y/H = 0.31, which was the plane at which the stagnation point occurred. It can be seen that pressure variations around the train were influenced by the wall. With closer gap to the wall, the pressure variations around the train increased and also impacted the wall surface.

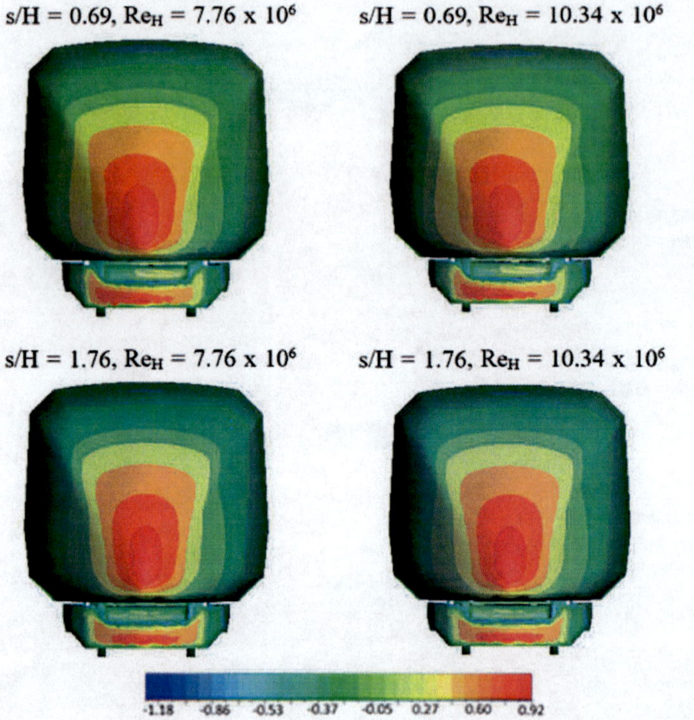

Fig. 6. Contour of the pressure coefficient on the surface of the train

Pressure distributions along the inner surface of the wall taken at a horizontal line at height y/H = 0.31 are shown in Fig. 8. The maximum and minimum pressure values occurred on the mask of car area, due to profile changes on the body of the train. This is commonly known as head pressure pulse. The head pressure pulse impacted the surface of the wall, and thus the distribution of the pressure on the wall also showed similar pattern. For wall with the distance of s/H = 1.76, the difference between the maximum and minimum values was below 120 Pa for all Reynolds numbers. However, for the wall with the distance s/H = 0.69, the difference between the maximum and minimum values was quite high up to 367 Pa and 655 Pa for Reynolds number of 7.76 $\times$ 10^6 and 10.34 $\times$ 10^6, respectively. This suggested that the wave of pressure and suction imposed on the wall and consideration should be taken whether this would be safe for the existing wall structure.

Table 1 shows the composition of the pressure and viscous drag coefficients to the total drag coefficients for varying wall distance and Reynolds number. In general we can see that the drag coefficient was larger under the influence of closer wall distance. The total drag coefficient of the closer wall distance of s/H = 0.69 was about 7–8% higher

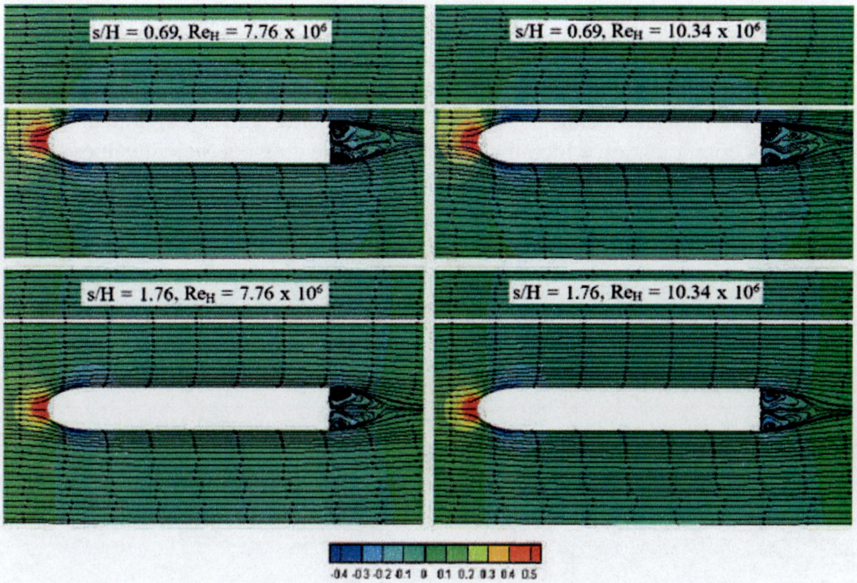

Fig. 7. Contour of the pressure coefficient in the *x–z* plane at y/H = 0.31.

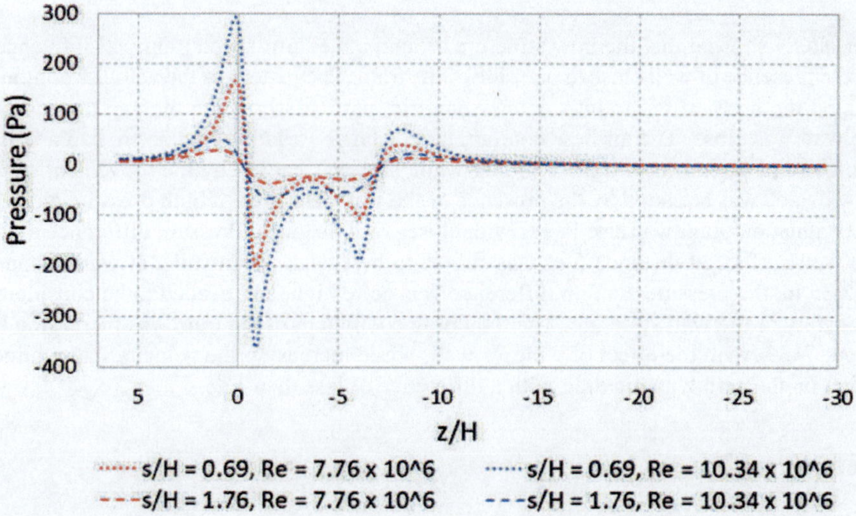

Fig. 8. Pressure distributions along the inner wall surface at height y/H = 0.31.

than that of s/H = 1.76. The increase in the total drag coefficient was contributed by the increase of pressure drag. Meanwhile, the changes in viscous drag were less significant of about only less than 1%. At ReH = 7.76 × 10^6, the difference between the drag coefficient value for distances s/H = 0.69 and s/H = 1.76 was 7.71%. Meanwhile, at

ReH = 10.34 × 10^6 the difference in drag coefficient values for distances s/H = 0.69 and s/H = 1.76 was 7.38%. It can be concluded that the difference in the distance between the walls of the building greatly affected the drag coefficient experienced by the surface of the train. On the other hand, the difference in velocities did not significantly affect the total drag coefficient, in which the difference of the drag coefficient values was less than 1%.

Table 1. Drag coefficient

s/H	Re_H	Pressure drag coefficient	Viscous drag coefficient	Total drag coefficient
0.69	7.76 × 10^6	0.298	0.044	0.342
	10.34 × 10^6	0.298	0.042	0.340
1.76	7.76 × 10^6	0.273	0.045	0.318
	10.34 × 10^6	0.274	0.043	0.317

4 Conclusions

This study showed that the flow structure around a medium-speed train was influenced by the presence of wall close to one side of the train. The pattern of the velocity contours around the body of the train was unsymmetric, particularly when the gap distance of the wall was close. The air flow contour, at a distance variation of 2.35 m, had a wider contour and extend toward the building wall. This was caused by the movement of air flow, which was confined by the presence of the wall. The area of high pressure close to the stagnation point was also larger with closer wall distance. Pressure difference along the wall surface at distance 6 m was shown to be saved. Meanwhile, at wall distance of 2.35 m, the pressure-suction difference was quite high, and needed to be considered for safety of the wall structure. The total drag coefficient of the train was increasing by about 7–8% with the effect of a closer wall, while increasing the velocity had a minor effect on the aerodynamic drag with a difference of less than 1%.

References

1. Sterling M, Baker CJ, Jordan SC, Johnson T (2008) A study of the slipstreams of high-speed passenger trains and freight trains. Proc Inst Mech Eng Part F: J Rail Rapid Transit 222(2):177–193
2. Yang N, Zheng XK, Zhang J, Law SS, Yang QS (2015) Experimental and numerical studies on aerodynamic loads on an overhead bridge due to passage of high-speed train. J Wind Eng Ind Aerodyn 140:19–33
3. Baker CJ, Brockie NJ (1991) Wind tunnel tests to obtain train aerodynamic drag coefficients: reynolds number and ground simulation effects. J Wind Eng Ind Aerodyn 38(1):23–28

4. Railway Applications—Aerodynamics Part 4: Requirements and test procedures for aerodynamics on open track, CEN EN 14067-4 (2013)
5. UIC (2013) Effect of the Slipstream of Passing Trains on Structures Adjacent to the Track, second ed. UIC 779-1
6. Rocchi D, Tomasini G, Schito P, Somaschini C (2018) Wind effects induced by high speed train pass-by in open air. J Wind Eng Ind Aerodyn 173:279–288
7. Xiong XH, Yang B, Wang KW, Liu TH, He Z, Zhu L (2020) Full-scale experiment of transient aerodynamic pressures acting on a bridge noise barrier induced by the passage of high-speed trains operating at 380–420 km/h. J Wind Eng Ind Aerodyn 204:104298
8. Hu HX, Lei WX, Zhang Y (2018) Complexity analysis on the aerodynamic performance of the mega high-speed train caused by the wind barrier on the embankment. In: Complexity of contruction mega infrastructure project, vol 2018, ID. 7130532
9. Khayrullina A, Blocken B, Janssen W, Straathof J (2015) CFD simulation of train aerodynamics: train-induced wind conditions at an underground railroad passenger platform. J Wind Eng Ind Aerodyn 139:100–110
10. Li HM, Xuan Y, Wang L, Li YL, Fang X, Shi GH (2013) Research on numerical simulation of high-speed railway noise barrier aerodynamic pressure. In: Applied mechanics and materials, vol 274. Trans Tech Publications Ltd, pp 45–48
11. Vittozzi A, Silvestri G, Genca L, Basili M (2017) Fluid dynamic interaction between train and noise barriers on High-Speed-Lines. Procedia Eng 199:290–295
12. Mario J, Halfina B, Bahtera D, Shalahuddin L, Windharto A (2020) Numerical and experimental analysis of drag force in medium speed train design. In: IOP conference series: materials science and engineering, vol 909, no 1. IOP Publishing, p 012031
13. Utomo T, Fajar B, Arpriyanto H (2018) CFD analysis on the aerodynamics characteristics of Jakarta-Bandung high speed train. In: MATEC web of conferences, vol 159. EDP Sciences, p 02038
14. Maleki S, Burton D, Thompson MC (2017) Assessment of various turbulence models (ELES, SAS, URANS and RANS) for predicting the aerodynamics of freight train container wagons. J Wind Eng Ind Aerodyn 170:68–80
15. Wang S, Bell JR, Burton D, Herbst AH, Sheridan J, Thompson MC (2017) The performance of different turbulence models (URANS, SAS and DES) for predicting high-speed train slipstream. J Wind Eng Ind Aerodyn 165:46–57

Check for updates

Facial Recognition-Based Automatic Door Security System Integrated with Internet of Things for Smart Home Actualization

Billy Indrawan(✉), Deanova Ghivari Alzamora, Muhammad Shiddiq Rahmatullah, and Alief Wikarta

Department of Mechanical Engineering, Institut Teknologi Sepuluh Nopember, Surabaya 60111, Indonesia
wikarta@me.its.ac.id

Abstract. Conventional door lock uses physical keys, which is a well-known technology but it also has its flaws, including the possibility of getting stolen, misplaced, lost, or even getting duplicated. The popular solution to that is to use biometrics data, but most of them require special hardware that needs direct contact. Conventional door lock also doesn't have any means of control and monitor for the owner, and when a crime happens it often doesn't have enough evidence for an investigation to proceed. Therefore, we designed a system that can replace physical keys, has multiple layers of automatic security system, gives access to control and monitor, and also helps with criminal investigations. From the research and experiment conducted, it is obtained that the door security system using face recognition can replace physical keys with a false rate as low as 0%. The system has multiple layers of automatic security system as a smart home actualization. The system provides access to control and monitor for the owner with an average time response of 0.553 s while also unaffected by the distance. The security system can also help crime investigations by providing evidence through images and videos taken by the system.

Keywords: Door lock · Face recognition · False acceptance rate (FAR) · False rejection rate (FRR) · Security System

1 Introduction

Conventional door lock uses physical keys to lock or unlock the door. It is the most natural way and everyone is acquainted with it. The physical key is a well-tested and well-known technology, but it also has its flaws [3]. The flaws include the possibility of getting stolen, misplaced, lost, or even getting duplicated so it can be troublesome and bad for the safety of your house. The popular solution regarding that problem is the use of biometrics data such as fingerprint, but there raises another problem because most of them require special hardware that needs direct contact [10]. Another flaw of a conventional door lock system is that it doesn't have any means of control and monitor for the owner, and this could be bad especially for people who live alone and often

I. Tolj et al. (eds.), *Recent Advances in Mechanical Engineering*, Lecture Notes in Mechanical Engineering, https://doi.org/10.1007/978-981-19-0867-5_43

away, and when a crime happens it often doesn't have enough evidence to proceed the investigations because still photography is useful in supplying significant bits and pieces with exact measurement of the site and evidence where the crime has occurred [5].

Several studies on automatic door security system have been conducted using methods such as RFID [8], password [4], QR-Code [7], Short Message Service (SMS) [1], fingerprint [2], smartphone application [3], and face recognition [6]. However, there are no studies that combine the use of face recognition and application. Therefore, our goal is to design a solution regarding that problem with an automatic door security system that can replace physical keys and solve its flaws, has multiple layers of automatic security, gives access for the owner to control and monitor, and also helps with the proceeding of crime investigations.

2 Designed System

The replacement of physical keys in our designed door security system is face recognition. Face recognition is a kind of a biometric data that is unique, since human beings are recognized by their unique facial characteristics [10] so there won't be any chance of losing and no way to duplicate. The advantage of using face recognition compared to other biometric data is that face recognition does not require any hardware to be touched. Furthermore, facial recognition is one of the fastest-growing fields, and it can be attributed to the increase of commercial interest including criminal investigations [10]. The algorithm of face recognition that is used in our system is Haar-Cascade using OpenCV.

The system designed also has multiple layers of security, one of them being a unique key of face recognition mentioned above, and the others include a motion sensor that can detect a movement, an alert when the face recognition system doesn't recognize the face scanned, a photo that the system will take, a video streaming that can be used to monitor, and an alarm that can be used as a last resort of security.

As a way to control and monitor, the system has a smartphone application, which acts as a Remote Control Unit (RCU) that can execute the system's security feature. A smartphone has now become a very common commodity and everyone carries it along with them [3], hence it is a good device to use as a means of control and monitor. The application is integrated with Internet of Things (IoT) so it can be accessed anywhere in a real time without any effect from the distance. Figure 1 shows the User Interface (UI) of the application.

The system mentioned above can possibly be helpful for the proceedings of crime investigations. The photo taken by the system and the live streaming videos can be used as an evidence when a crime took place.

3 System Architecture and Overview

The system mainly consists of Embedded Control Unit (ECU) and Remote Control Unit (RCU). Figure 2 shows the architecture of the system, and Fig. 3 shows how the overall system will work.

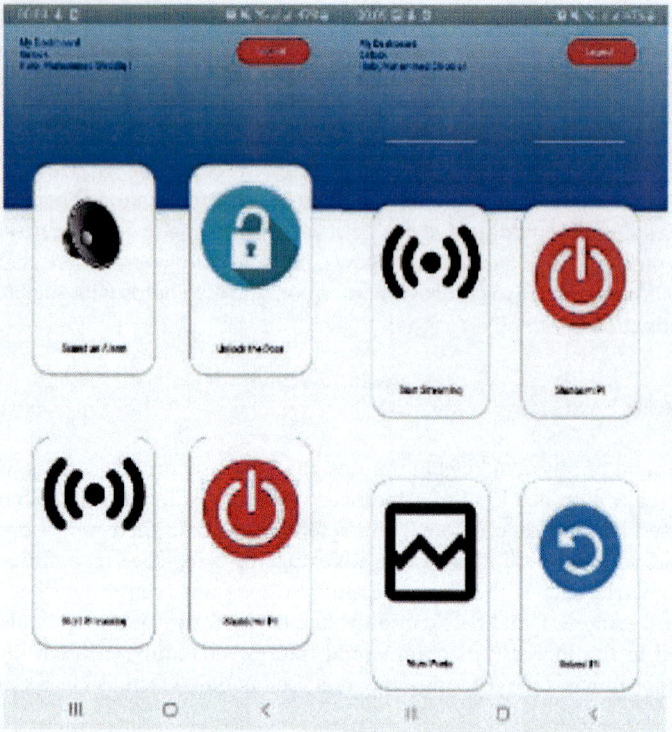

Fig. 1. User interface (UI) of the application

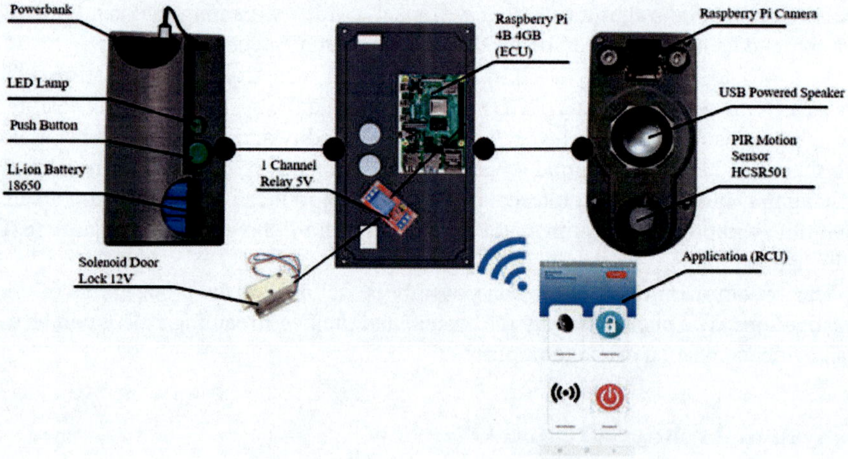

Fig. 2. System Architecture

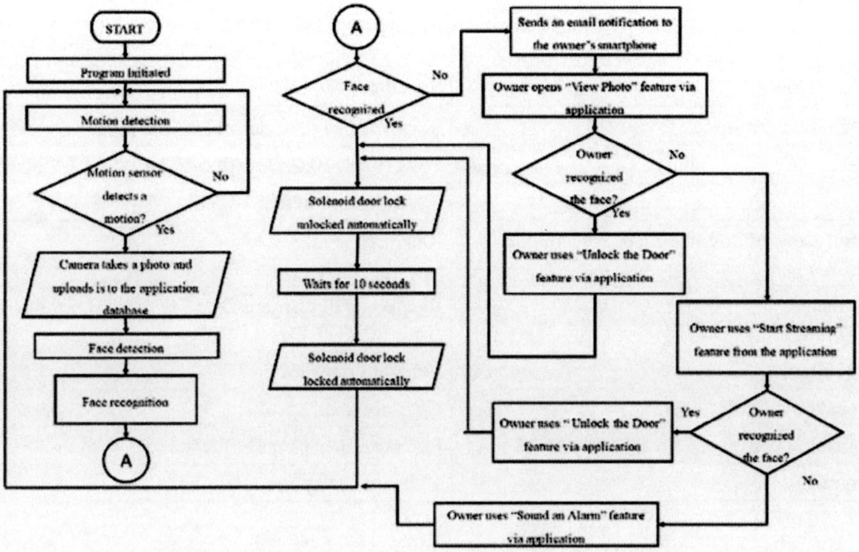

Fig. 3. Automatic door security system flowchart overview

3.1 Remote Control Unit (RCU)

The Remote Control Unit (RCU) is used as a way to provide an access to control the system remotely. In this system, RCU is a smartphone application integrated in a real time with the system and the database through Internet of Things (IoT).

The application has several features that can execute the system's functions. Features in the application include "Unlock the Door", which unlocks the solenoid door lock when it is executed, "Sound an Alarm", which sounds an alarm from the speaker when it is executed that is used to prevent unusual activities from an unknown figure such as preventing a crime, "View Photo", which views photos that are uploaded to the database when a motion sensor detects a sensor that is used to determine whether to open the door or to do a preventive measure, such as sounding an alarm, depending on the figure captured in the photo, and "Start Streaming" is used to view a live video captured by the camera. It can be used to monitor the outside of the door similar to "view photo", to decide whether or not a preventive measure is necessary.

3.2 Embedded Control Unit (ECU)

The Embedded Control Unit (ECU) is connected to the components and sensor through each of their respective ports so that it became one part of embedded systems with the specifications described in Table 1.

Table 1. Embedded system specification

Description	Specification
Main power source	Raspberry Pi 4 Adaptor 5 V 3A
Secondary power source	Powerbank 10,000 mAh 5 V 3A
Solenoid door lock power source	3 X Li-Ion batteries 18,650
Microcontroller/embedded Control unit (ECU)	Raspberry Pi 4B RAM 4 GB
Camera	Raspberry Pi Camera OV5647 1080 3.6 mm IR Auto IR-CUT Day
Door lock	Solenoid door lock 12 V
Speaker	USB powered speaker
Sensor	PIR motion sensor HCSR501
Switch	1 channel relay module 5 V

4 Method of Experiment

4.1 Face Detection

The face detection experiment is conducted to find out the effect of the variation of the dataset used (10, 25, and 50), IR mode (ON and OFF), and the uses of accessory (None, glasses, and mask) towards the False Acceptance Rate (FAR) and False Rejection Rate (FRR).

FAR is a percentage of a condition when a supposedly not recognized face is recognized by the system, and FRR is the opposite of that, a supposedly recognized face is not recognized by the system.

The experiment is conducted with the following steps. A dataset is gathered, it consists of a bunch of images to be recognized by the system. The dataset must follow the requirements set by ISO/IEC-19794-5 regarding biometric data. The dataset is then trained by extracting a certain part of the images, in this case the face. The test is conducted 100 times for each set of variables, with 50 for the supposedly recognized face and the other 50 for the supposedly not recognized face. The output of the test is False Acceptance Rate (FAR) and False Rejection Rate (FRR), which then will be compared with a result from an experiment conducted by Suwarno and Kevin [9]. The step is then repeated until each set of variables has been tested.

4.2 Application

The application experiment is conducted to find out whether or not the distance ($<$100 m, 1 km, $>$1 km) between the system and the Remote Control Unit (RCU) affects its response when a feature is executed through the smartphone application.

The test is conducted by executing a feature from the application for each distance variable. The output of the test is the time response of the system that is measured in seconds. The steps are then repeated until each variable is tested.

5 Experimental Result

5.1 Face Detection

The dataset is gathered compared to the 13 constraints as stated in ISO/IEC-19794-5 as shown in Table 2, and the example of an image that met the requirements in Fig. 4.

Table 2. Dataset requirements in ISO/IEC-19794-5 [9]

Attribute	Constraint
Posture	Control on deviation from frontal
Illumination	Uniformly illuminated with no shadow
Background	Plain light-colored
Eyes	Open and clearly visible
Glasses	No flash reflections, dark tint or heavy frames
Mouth	Closed and clearly visible
Head position	Placed in the center
Distance to the camera	Moderate head size
Color	Color neutral and no red eye
Exposure	Appropriate brightness
Focus	No out-of-focus and in good sharpness
Resolution	Width constraint of the head

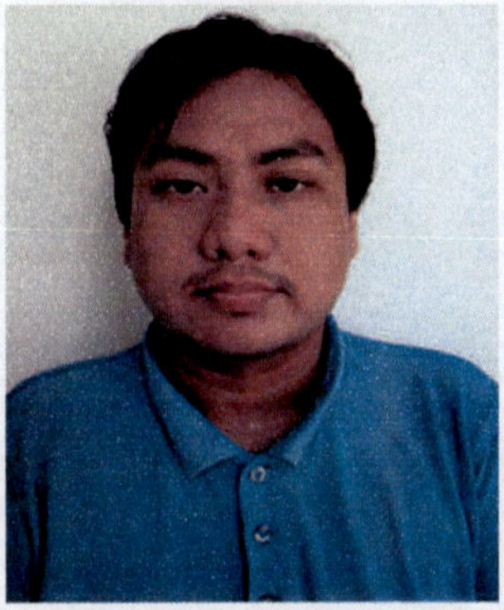

Fig. 4. Examples of an image used in the dataset

The experiment is then conducted with a variation of dataset, IR mode, and use of accessory as seen in Fig. 5. The result is then compared with test results conducted by Suwarno and Kevin as shown in Table 3 for IR mode OFF and Table 4 for IR mode ON.

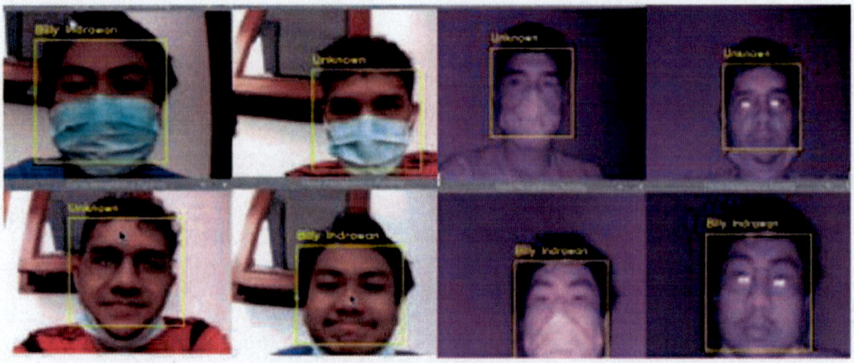

Fig. 5. Face detection experiment

Table 3. Face detection experiment result with IR mode OFF

	Dataset									Suwarno and Kevin's test results
	No accessory			Glasses			Mask			
	10	25	50	10	25	50	10	25	50	
FAR (%)	0	0	0	16	12	0	0	0	0	0.65
FRR (%)	0	0	0	0	0	0	14	6	0	34.42

Table 4. Face detection experiment result with IR mode ON

	Dataset									Suwarno and Kevin's test results
	No accessory			Glasses			Mask			
	10	25	50	10	25	50	10	25	50	
FAR (%)	8	0	0	4	4	0	12	4	0	0.65
FRR (%)	4	0	0	8	8	0	8	0	0	34.42

From the results, it can be implied that the amount of dataset used affects the value of FAR and FRR, with the increase of the dataset reduces the value until 0% in 50 dataset despite the use of IR camera and accessory. Compared to the result from Suwarno and Kevin's test [4], our system tends to have a higher false rate (precisely the FAR) when using 10 and 25 datasets, while when using 50 datasets, our system has a lower false rate with 0% on all sets of variables.

5.2 Application

The application is tested to find out the response of the system and its effect on distance with an example of experiment shown in Fig. 6.

Fig. 6. Application experiment on <100 m distance

From the result shown in Table 5, we can imply that the distance does not affect the response time of the system.

Table 5. Application experiment result

Distance	System response (s)
<100 m	0.56
1 km	0.6
>1 km	0.5
Average	0.553

6 Conclusion

This paper presents an automatic door security system that aims to replace physical keys while providing control access to the user as a means of smart home actualization. The use of face recognition proves to be able to replace physical key and its flaws, with a false acceptance rate (FAR) and false rejection rate (FRR) as low as 0% when using 50 images of dataset. The system has multiple layers of security that run automatically, which can be controlled and monitored by a smartphone application with an average response of 0.553 s while also being unaffected by distance. The security system can also be helpful for crime investigations by providing evidence through images and videos taken by the system.

References

1. Faroqi A, Fitriadi A, Adiningsih NU, Ramdhani MA (2018) Automatic door control system using SMS gateway base on arduino uno and ultrasonic sensor. Int J Eng Technol 7(3.4):122–126
2. Febriyanto E, Padeli P, Suprayogi D (2019) Prototype Sistem Smart Lock Door Dengan Timer dan Fingerprint Sebagai Alat Autentikasi Berbasis Arduino Uno Pada Ruangan. Jurnal Informatika 19(1):10–19
3. Gupte NN, Shelar RM (2013) Smart door locking system. Int J Eng Res Technol 2(11):2214–2217
4. Hlaing NNS, Lwin SS (2019) Electronic door lock using RFID and password based on arduino. Int J Trend Sci Res Dev 3(3):799–802
5. Rohatgi R, Kapoor AK (2014) Importance of still photography at scene of crime: a forensic versus judicial perspective. J Harmon Res Appl Sci 2(4):271–274
6. Saraf T, Shukla K, Balkhande H, Deshmukh A (2018) Automated door access control system using face recognition. Int Res J Eng Technol 5(4):3036–3040
7. Sholeh MB, Herlina A, Hasan F (2020) Design of automatic door lock control system library Nurul Jadid University based on arduino uno R3 and QR-code. Buletin Ilmiah Sarjana Teknik Elektro 2(2):91–102
8. Soni S, Soni R, Waoo AA (2021) RFID-based digital door locking system. Indian J Microprocess Microcontroller 1(2):17–21
9. Suwarno K (2020) Analysis of face recognition algorithm: dlib and OpenCV. J Inform Telecommun Eng 4(1):174–184
10. Timse P, Aggarwal P, Sinha P, Vora N (2014) Face recognition based door lock system using Opencv and C# with remote access and security features. Int J Eng Res Appl 4(4):52–57

Data-Driven LoRa Precision Agriculture as a Farm Monitoring Solution and Increasing Efficiency of Water Use and Consumption of NPK Fertilizer

Hafiz Rayhan Gunawan(✉), Wafiq Azhar Rizaldi, Nur Elisa Putri, Danial Farros Maulana, and Alief Wikarta

Institut Teknologi Sepuluh Nopember, Surabaya 60111, Indonesia
wikarta@me.its.ac.id

Abstract. The use of technology for agriculture is important where volatile agricultural growth conditions can be maximized by improving efficiency, one of which is engineering efficiency. By improving engineering efficiency, farmers are able to increase land productivity using minimum inputs and by utilizing technology. Some programs that have been offered there are still shortcomings that must be fixed. Therefore, it is necessary to make a tool which is able to help the efficiency of agriculture in Indonesia. One of the technologies that can help the effectiveness and increase of agricultural production of previous research is the concept of automatic monitoring and controller tools that use WiFi and Arduino systems but have not yet continued until application. In addition, the study has not yet reached the actual stage of application. Therefore, Data-Driven LoRa Precision Agriculture solution is needed that is able to monitor crops, control irrigation systems, and input fertilizer doses automatically. Research methods used in the manufacture of this tool are as follows: system identification and literature studies, model and system design, until tool testing. The results obtained from these tests include the temperature sensor, the value is 31–32°, the humidity sensor is 49%, while the distance test is 440 m, as well as the NPK sensor that works properly. The data sent from these sensors are obtained every 10 min the tool will only turn on for 2 s to send the data, then this tool can work for quite a long time because it does not consume a lot of battery power.

Keywords: Agriculture · LoRa · Sensor

1 Background

The use of technology for agriculture is important where volatile agricultural growth conditions can be maximized by improving efficiency, one of which is engineering efficiency. The ineffectiveness rate of agriculture in Indonesia itself reaches a range of 59–87%. By improving the efficiency of techniques, farmers are expected to be able to increase land productivity using minimum inputs and by utilizing technology.

I. Tolj et al. (eds.), *Recent Advances in Mechanical Engineering*, Lecture Notes in Mechanical Engineering, https://doi.org/10.1007/978-981-19-0867-5_44

The program that was once offered by itself is Supra Insus, and after this program, there is no longer a breakthrough in agricultural technology that is efficient enough. Supra Insus program itself recommends the use of fertilizers that are equally common between one plant and another plant. While each plant has different fertilizer needs. In addition, several factors that support the increase in crop productivity such as water control and fertilizer that are very important for rice [1]. Stagnant water on rice fields must be changed according to the age of rice, as well as the necessary nutrients of rice on the soil that can be increased by the use of fertilizer with the right dose so as not to pollute the environment. A breakthrough is needed to achieve a high level of technical effectiveness on land productivity that can study the use of fertilizers, water, pesticides, and other needs needed by plants [2].

Judging from the urgency above, it is necessary to create a tool through which the tool can help the efficiency of agriculture or agriculture in Indonesia. One of the technologies that can help the effectiveness and increase of agricultural production of previous research is the concept of automatic monitoring and controller tools that use RFID and Arduino systems but have not yet continued until application. In addition, the study has not yet reached the actual stage of application [3]. Therefore, Data-Driven Lora Precision Agriculture solution is needed that can monitor crops, control irrigation systems, and input fertilizer doses automatically. By using agricultural long-distance, farmers no longer need to go back and forth to rice fields or agricultural land to monitor. It is because by using this tool, crop development in the land can be monitored by mobile phones. In addition to having advantages in terms of distance technology, LoRa also has other advantages including only requires relatively low power, safe, and costs needed is also low [4].

The purpose of this research is to know the process of making and applying Data-Driven LoRa Precision Agriculture tools. Then knowing how to integrate the ESP8266 Node MCU microcontroller with a LoRa network and knowing how to integrate LoRa network with supporting sensors and users.

2 Methodology

The implementation methods used in this study are as follows, the first being system identification and literature studies. Secondly, the design of systems and models, the process of how the tool works begins by collecting data on the parameters needed, among others, soil NPK content, humidity, soil pH, and camera. The data are then temporarily entered on the LoRa module nodes 1 and 2 that will then be sent to the database via the LoRa network with the help of antennas. Once the data are accumulated and processed, it will be displayed on the smartphone application. From this application on the soil sensor tab, farmers can see the recommendations given by this monitoring tool so that the fertilizer used does not exceed the needs of plants. In addition, farmers can also monitor their land directly from smartphones. The third stage is the collection of tools and materials, the fourth is the manufacture of tools. The manufacturing process begins by collecting all the hardware needed ranging from microcontrollers, sensors, to plastic casings from the tool. Then the electronic components are assembled in such a way that it works and continued with microcontroller programming. After that, the tool was tested to contact the internet and created database software. If the database system gets data well from the field, the next step is to process the data to create an easy application. The fifth stage is tool testing. Testing is very important to do before field tests. The product is first conducted a simulated test of the system.

3 Results and Discussion

The results achieved in each method are as follows: the first is system identification and literature study. System identification includes brainstorming about how the system works and state of the art of LoRa technology for solving the farm monitoring problems. In this stage, we concluded how the device works that begins by sensors collecting the required parameter data, including water level, soil NPK content, humidity, and temperature. The data are then temporarily entered in the LoRa module nodes 1 and 2, which will then be sent to the database via the LoRa network with the help of an antenna. After the data are accumulated and processed, the data will be displayed on the web application. From this application, farmers can see the sensor parameters used such as NPK to optimize agricultural yields so that the fertilizer used does not exceed the plant's needs.

After the system is established, doing literature research is necessary to get the technology working properly such as the concept of Agriculture IOT using LoRa module SX1278 which uses hygrometer sensors, humidity and temperature. The next step is system and model design that obtained diagrams of how lora system works, electronic circuit diagrams, and flowchart coding diagrams. Here, we attach the progress of making the tools that we have done (Fig. 1).

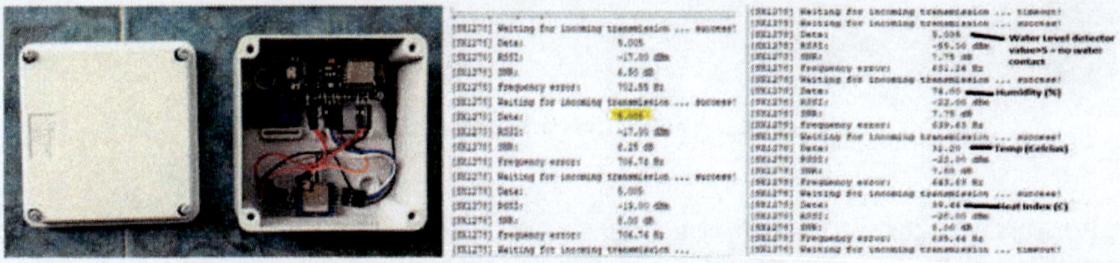

Fig. 1. Prototype and sensor results

After that is stage collection of tools and materials. This stage is carried out the purchase of tools and materials one of them are sensors, microcontroller, LoRa module, NPK sensor, watertight enclosure, etc. The next step is to do the process of designing tools using tools and materials that have been collected. The initial process is soldering, after soldering the next process is to combine or connect a WiFi module to several sensors. Each node has sensors according to the needs such as temperature, temperature, and water level. Then there is the coding process. The coding is using a Visual Studio Code application to give sets of instruction on the MCU board node. This process is done with the help of the open source github website as the main source code (Fig. 2).

After that, we produced a tool that can already receive data more than one node with a variety of different sensors, namely as follows temperature sensors, humidity, water

Fig. 2. Coding process

level, and others. Furthermore, initial testing in the field is directly to test the strength of the sensor and the success of the tool to receive data from the sensor while in the field. From this field test, it was obtained that the sensor can run well in the field or rice fields. Also, the measurable distance strength reaches 440 m. And lastly, web creation was done by creating html java script on the computer by using local host, then tested on the computer, and uploaded on the website. The webpage can be seen in the following pictures (Fig. 3).

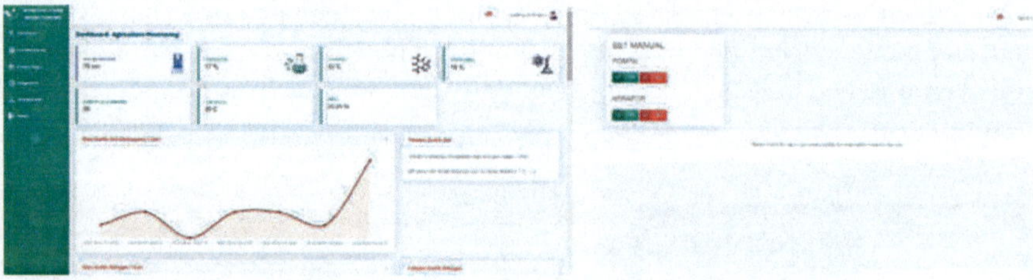

Fig. 3. Field trials and user interface

For distance testing, it is first testing in urban area, and then we test our sensor in rice field, as follows in Fig. 4.

The next step is the analysis of the initial results. In the analysis phase, data obtained from initial testing in the field are evaluated by validating the data using test equipment other to ensure the reliability of the data obtained. The result of temperature and humidity data can be seen in Table 1.

The data on this test were taken for an hour at a time interval of 10 min. While real data of temperature are taken using thermistor and humidity sensor sampling humid data in the air. These data also compared to other external sensor from real-time data in accuweather website at an hour interval. From both results can be seen that the sensor of this device can work as it should because the value is not much different from the real temperature value of 31–32° and real humidity that is 49%. Testing of this tool also involves a distance test that was previously conducted in the urban city of one of the members and then continued in the field. Distance test results' data can be seen in Table 2.

Test the distance using a voltage of 3.3 V, so that it can be calculated the consumption or power needed. As you can see in Table 2, the lower the current consumed, the

Fig. 4. Distance testing

Table 1. Temperature and humidity data

Time	Temperature sensor (°C)	Real temperature (°C)	Humidity sensor (%)	Real humidity (%)
16.00	32	32	51	49
16.10	31,9	32	51	49
16.20	32,3	32	48	49
16.30	32,1	31	47	49
16.40	32	31	47	49
16.50	31,8	31	48	49
17.00	31,7	31	48	49

Table 2. Distance test.

Current (mA)	Range (City) (m)	Range (Field) (m)	Power (w)
20	52	81	0,066
80	280	440	0,264

weaker the range. Also, when the device operated in urban area, the distance decreased dramatically mainly caused by interference with other wireless devices where as in the field there is little interference. Because the data are sent every 10 min, and the device will only turn on for 2 s to process and send the data, it can work for quite a long time because it does not consume much battery power.

The third test conducted in the field with four different points in a certain time interval. Obtained data include nitrogen, phosphor, potassium in mg/kg unit, also sampling generic data like humidity, temperature. Besides that, RSSI data also included in order to validate the varying distance from the main post. Data are displayed in the following Table 3.

Table 3. Field test data.

Point	Time (GMT + 7)	N (mg/kg)	P (mg/kg)	K (mg/kg)	Humidity (%)	Temperature (°C)	RSSI
Point 1	10:12:05	82	26	35	50	33,5	−63
	10:12:14	83	26	35	50	33,4	−58
	10:12:23	81	25	34	51	33,4	−60
Point 2	10:15:26	52	100	136	56	34,6	−97
	10:15:37	43	88	118	56	34,7	−97
	10:18:16	62	38	52	50	33,9	−102
	10:18:25	62	38	52	50	33,9	−113,75
Point 3	10:18:43	116	38	52	55	33,9	−104
	10:18:52	118	36	50	55	33,9	−107
	10:19:01	113	35	47	54	33,8	−104
	10:19:20	110	33	45	54	33,8	−106
	10:19:28	105	33	45	54	33,8	−105,5
Point 4	10:22:18	91	29	39	57	33,8	−120,75
	10:22:27	92	29	39	57	32,7	−117,5
	10:22:36	92	29	39	56	32,6	−110,25
	10:22:45	92	28	39	56	32,5	−113,25
	10:22:54	92	29	39	55	32,4	−116,75
	10:23:03	91	26	36	55	32,3	−109,75
	10:23:12	88	28	38	55	32,3	−11,75
	10:23:21	89	28	38	57	32,3	−108,5

The experimental point and core post used in this field can be seen in Fig. 5.

After testing the sensor at all four points, distance test is rolled out in some points to find the farthest distance the device can send data reliably. This distance test is done by carrying the tool away from the core post (blue node). In the first experiment, the device can reach a distance up to 440 m, then this experiment is also not much different, reaching 401 m. The results can be seen in the image above. For RSSI, data from such distance tests can be seen in the following Table 4.

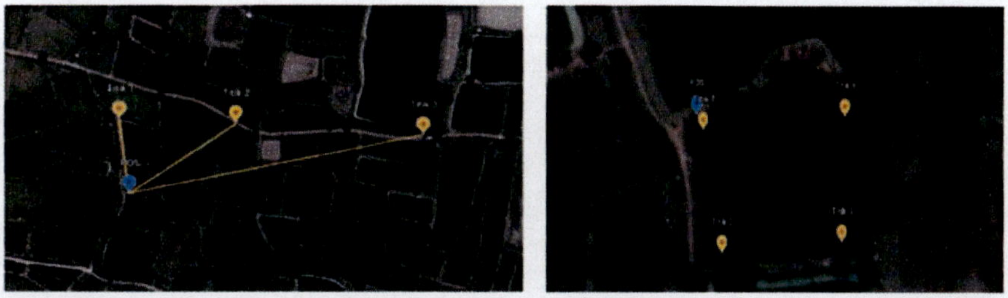

Fig. 5. Distance testing on field

Table 4. Distance test and RSSI

Distance (m)	RSSI
102,76	−114,3
168,61	−118
401,68	−133
440,1	−140

Data-Driven LoRa Precision Agriculture has the potential to be further developed both in terms of design and sensors to be used. It is also possible to provide advice to farmers through the application based on observations of farmers' fields. So that in the future, an advanced smart farming will be formed.

4 Conclusion

It can be concluded that the Data-Driven LoRa tool that has been created is capable of transmitting remote data using LoRa technology, which consumes little power. Then the sensor device uses batteries and will last quite a long time because data are sent every 10 min, and the tool will only turn on for 2 s to send the data. The farthest test distance is 440 m. The test shows that the temperature sensor has a value of 31–32°, from the humidity sensor a value of 49% is obtained. Users also do not need to monitor their land directly because it can be done through the web and applications. Furthermore, the system collects and displays data results automatically on the web. Finally, the UI display is easy to understand and easy to use.

References

1. Hashim NMZ et al (2015) Agriculture monitoring system: a study. J Teknol 77(1):55–59
2. Kusnadi N, Tinaprilla N, Susilowati SH, Susilowati SH (2011) Analisis Efisiensi Usaha Tani Padi di Beberapa Sentra Produksi Padi Di Indonesia. J Agro Ekon 29(1):25–46

3. Maintang M (2012) Pengelolaan Tanaman Terpadu dan Teknologi Pilihan Petani : Kasus Sulawesi Selatan. IPTEK Tanaman Pangan. BalaiPengkajian Teknologi Pertanian Sulawesi Selatan 7(2)
4. Semtech Homepage. https://www.semtech.com/. Accessed 13 Feb 2021

Human Gait Parameters Calculation Based on Partial Motion Data

Sandro Mihradi(✉), Ferryanto, Tatacipta Dirgantara, and Andi Isra Mahyuddin

Faculty of Mechanical and Aerospace Engineering, Institut Teknologi Bandung, Bandung, Indonesia
sandro@ftmd.itb.ac.id

Abstract. This paper presents a software development for a 3D gait analysis system based on experimental motion analysis by using partial motion data. The 3D human body model consists of eight segments that are interconnected by seven joints. To track the gait motion, seven active markers are attached to the right leg of the subject during walking. The marker configuration is based on Helen Hayes' marker arrangement with several alterations to adjust to the two cameras' image processing system. To overcome the lack of information on the orientations of the left leg segments, the seven markers representing the orientation of the right leg during walking are "mirrored" to obtain those of the left leg segments. The inverse dynamics approach is then applied in the software to calculate the human gait parameters. The program developed could successfully calculate spatio-temporal parameters, kinematics parameters, as well as kinetics parameters of the joints. To check the reliability of the system, the calculated gait parameters are compared with the parameters available in the literature. The results show that those parameters are in good agreement.

Keywords: Gait analysis · Human body model · Inverse dynamics · 3D kinematics · 3D kinetics

1 Introduction

The study of human motion is of interest and has a broad application such as in the field of physical therapies, medical diagnostics, and sport sciences [1–4]. Gait analysis is the one that studies human locomotion, such as human walking.

In gait analysis, the inverse dynamics approach is one of the fundamental techniques that is widely applied in biomechanics and gait laboratories. The objective of this technique is to determine the forces and moments, due to the movement of the body segment. In this case, the movement is represented by the displacement of markers. The opposite technique is known as the direct dynamics approach in which forces and moments applied to the segments model are known and the objective is to determine the resulting movement of the segments due to the applied forces and moments [5].

In a previous study, the authors have developed a 3D gait analysis system based on markers position data [6] by employing the inverse dynamics approach. In the system,

I. Tolj et al. (eds.), *Recent Advances in Mechanical Engineering*, Lecture Notes in Mechanical Engineering, https://doi.org/10.1007/978-981-19-0867-5_45

a program was made to calculate some gait parameters based on the marker's position placed on the lower limb of the body. In that study, 15 markers were used to represent the eight segments of the 3D human body model.

The present work is a part of the development of the previous 3D gait analysis system that requires the system to be able to process partial motion data. Such conditions may exist when there are only two cameras and limited observation space. Therefore, in this work, a program to calculate some gait parameters based on partial motion data of markers attached only to one side of the lower limb of the human body during walking is developed.

2 Method

2.1 Human Body Model

To calculate human gait parameters, a human body model should be built first. In the present work, the previous human body model is adopted [6]. The body is modeled as a multibody system with eight substantial segments and seven joints, as can be seen in Fig. 1a, b. Each joint has three rotational degrees of freedom. They are the right and left ankle, right and left knee, right and left hip, and lumbar.

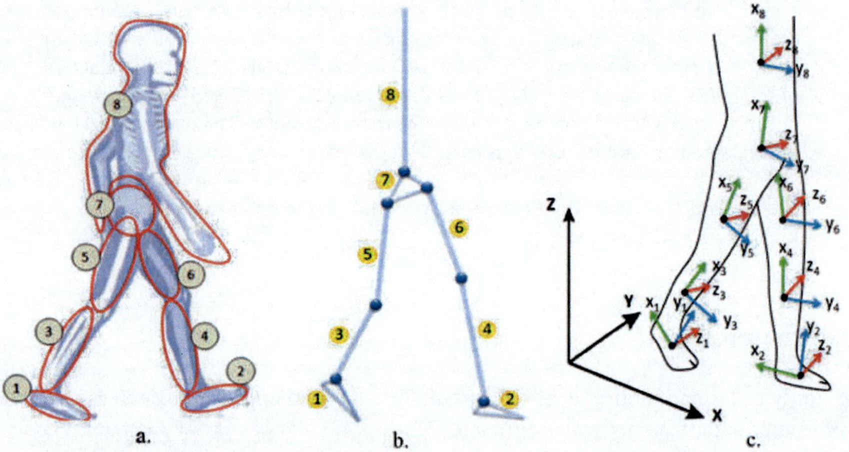

Fig. 1. **a** 3D human body model. **b** Skeleton model with rigid links and joints. **c** Reference axes in each segment [6]

For motion analysis, a system of reference axes that provide the position and orientation of each segment is needed. In the present work, the reference axes introduced by Vaughan [7] are adopted with the addition of the reference axis for the head-arm-trunk (HAT) segment. The system has a global reference system (GRS) with fixed X-Y-Z axes and a local reference system (LRS) with x-y-z axes attached to each segment as shown in Fig. 1c. This LRS is used to describe the position and orientation of each segment during motion.

2.2 Experimental Procedures

In the experiment, two video cameras (Sentech STC-CLC33A) with a speed of 90 fps are connected to the PC and are employed to record the subject's movement and all devices are in a fixed position (Fig. 2). In the experiment, 10 healthy young men participated in this study. The average age, height, mass, and BMI of subjects were 21.5 ± 1 year, 1.66 ± 0.04 m, 59.9 ± 5.65 kg, and 21.8 ± 1.5. Subjects were asked to walk at a normal pace on the walking path, as shown in Fig. 2.

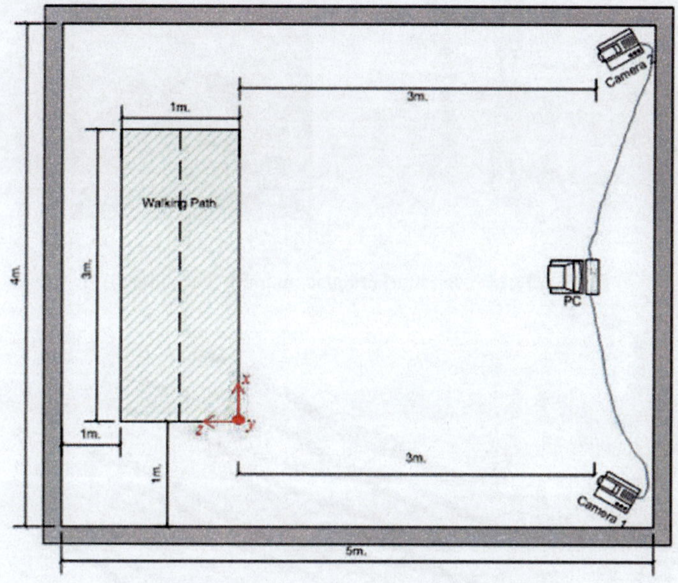

Fig. 2. Experimental setup for image recording

Figure 3 shows the location of seven markers attachment on the subject. This marker configuration is based on the Helen Hayes marker set with some alterations to adjust with the image processing system that uses only two cameras.

2.3 Mirroring Process

To perform kinetic gait analysis, motion data of all eight segments of the human body model are needed in the calculation. To overcome the lack of information on the orientations of the left leg segments, the seven markers representing the orientation of the right leg during walking are "mirrored" to obtain those of the left leg segments. It is assumed that the right and the left leg have similar dimensions and properties. The essential part of this process was to determine the initial contact (IC) and toe-off (TO) times of the right leg. The foot velocity algorithm (FVA) developed by O'Connor [8] is adopted here to determine IC and TO times. The 14 markers obtained are represented in Fig. 4, where *p8–p14* show the location of predicted left leg markers.

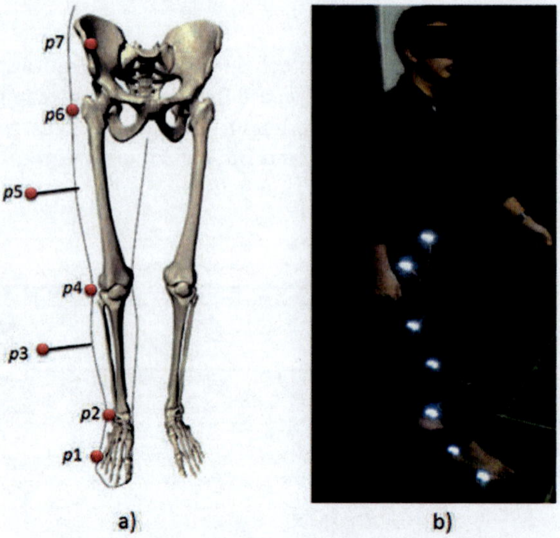

Fig. 3. **a** Seven markers placement. **b** The subject

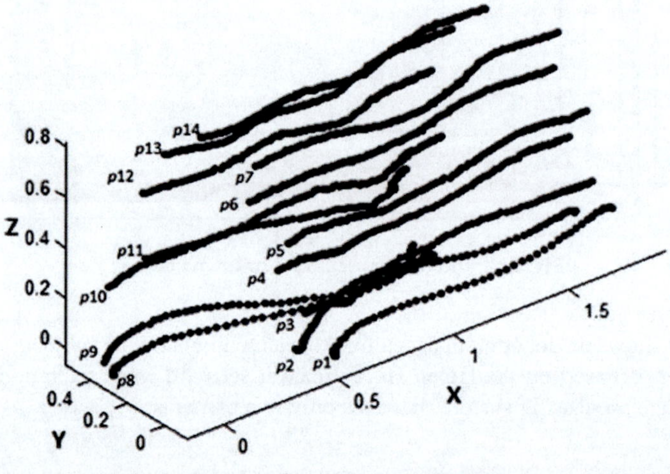

Fig. 4. The predicted left leg markers (*p8–p14*) and right leg markers (*p1–p7*)

2.4 Kinematic and Kinetics Procedures

For kinematic calculation, the orientation of one segment relative to its subsequent segment and the global reference system is calculated using the Euler angle definition. The angular velocity and acceleration are then derived using the finite-difference method. As for kinetics calculation, the mass and the center of mass (CoM) of each segment must be determined first. Subsequently, ground reaction forces (GRFs) and ground reaction moments (GRMs) at the foot are calculated during walking. Finally, the forces and moments of the other joints could be determined by considering the acceleration vector of each segment's CoM.

In the case of single support, the GRFs and GRMs during walking can be easily determined. However, for the double support condition, the calculation becomes indeterminate. To solve the problem, the ground reaction force on the trailing foot is estimated by the smooth transition assumption (STA) proposed by Ren [9], while GRFs and GRMs on the leading foot are calculated from a single support equation.

3 Results and Discussion

In this work, a program to calculate spatio-temporal, kinematics, and kinetics parameters of human walking is developed. The details of the obtained parameters are described in the following sub-chapters.

3.1 Spatio-Temporal Parameters

Before calculating spatio-temporal parameters, it is required to calculate several primary parameters such as IC and TO times and positions. After that, the spatio-temporal parameters could be obtained with the following equations and the results are presented in Table 1. The experimental results from Whittle [1] are also presented in the table for comparison. In general, the spatio-temporal results are within the range measured by Whittle [1].

$$\text{Cycle time (s)} = \text{IC}_2\text{times (s)} - \text{IC}_1\text{times (s)} \quad (1)$$

$$\begin{aligned} \text{Stride length (m)} &= \text{heel x-axis position at IC}_2\text{(m)} \\ &\quad - \text{heel x-axis position at IC}_2\text{(m)} \end{aligned} \quad (2)$$

$$\text{Cadence (steps/min)} = 60\ \text{(s)} \times 2/\text{Cycle time (s)} \quad (3)$$

Table 1. Spatio-temporal parameters of subjects

Variables	Mean	Standard deviation	Whittle [1]
Speed (m/s)	1.082	0.107	1.10–1.82
Stride length (m)	1.247	0.095	1.25–1.85
Cadence (step/min)	104.320	8.717	91–135
Gait cycle (s)	1.158	0.100	0.89–1.32

3.2 Kinematics Parameters

Linear kinematics parameters, such as joint and segment CoM positions, and their linear velocity and acceleration are successfully calculated, as well as angular kinematics parameters such as joint angles, and their angular velocity and acceleration. Due to space restriction, only joint angles in the sagittal plane will be presented in this paper. Figure 5 shows the joint angles for the ankle, knee, hip, and pelvic in sagittal planes. The results are in good agreement with the joint angular motion data available in the literature [1, 3, 10].

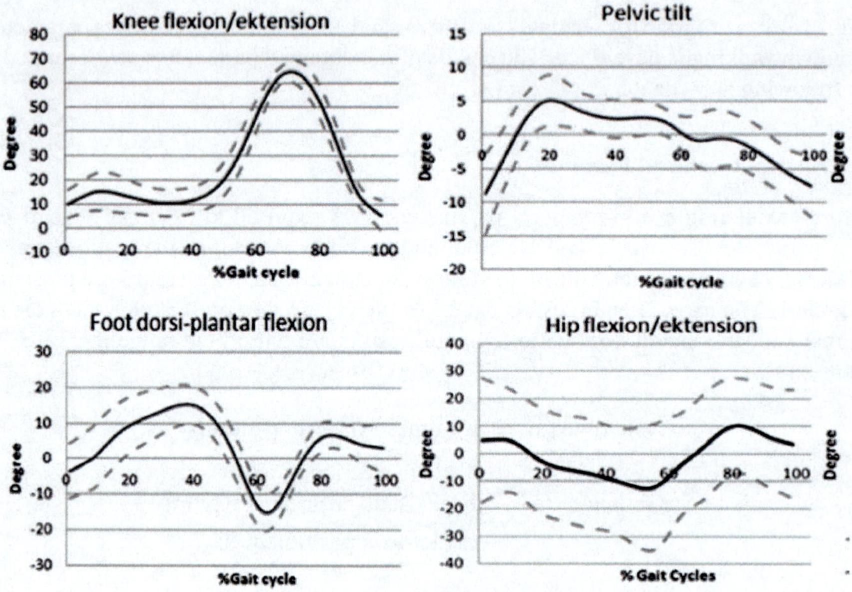

Fig. 5. Mean (solid line) and standard deviation (dash line) of joint angle in the sagittal plane

3.3 Kinetics Parameters

The kinetics parameters of human walking such as the GRFs, GRMs, joint forces, and joint moments are successfully calculated with the software. In this paper, due to space limitation, only the GRFs and joint forces are presented, as shown in Figs. 6 and 7, respectively. The calculated results show a similar pattern and are in good agreement with Ren et al. [9] and Rose et al. [10], especially for the GRFs and joint forces.

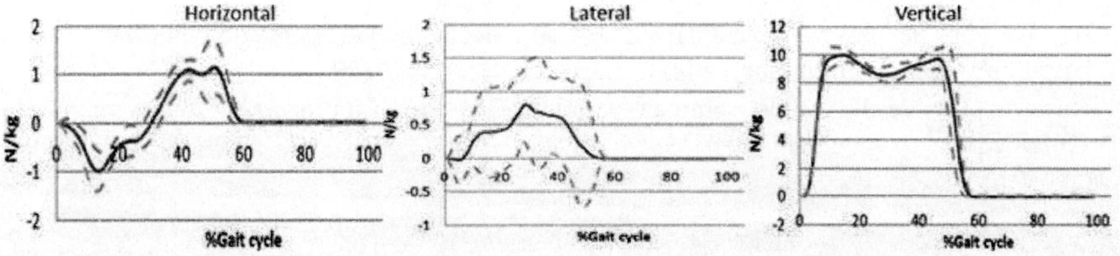

Fig. 6. Mean (solid line) and standard deviation (dash line) of ground reaction forces normalized by body mass in horizontal, lateral, and vertical directions along one gait cycle.

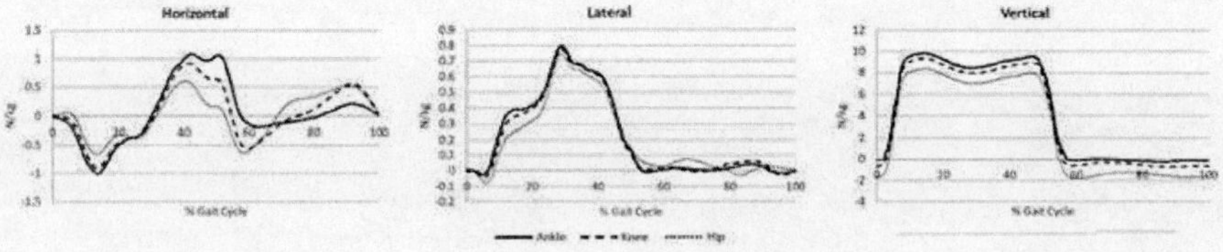

Fig. 7. Mean of the ankle, knee, and hip joint forces in horizontal, lateral, and vertical directions along one gait cycle.

4 Conclusions

In this paper, a program to calculate gait parameters based on partial motion data of markers attached only to one side of the lower limb of the human body during walking is developed. The mirroring processes were able to produce markers data for the left leg from the existing right leg data. The software could successfully calculate spatio-temporal, linear and angular kinematics parameters, and kinetics parameters of human walking based on the eight-segment human body model developed in this study.

To check the accuracy and reliability of the program, the calculated parameters are compared with the ones in the literature. The results show that the calculation is sufficiently accurate and therefore reliable.

References

1. Whittle MW (2007) Gait analysis: an introduction, 4th edn. Elsevier
2. Winter DA (2009) Biomechanics and motor control of human movement, 3rd edn. Wiley, New York
3. Perry J (1992) Gait analysis: normal and pathological function. Slack Inc., Thorofare, NJ
4. Medved V (2001) Measurement of human locomotion. CRC Press, New York
5. Otten E (2003) Inverse and forward dynamics: models of multi-body systems. Philos Trans R Soc Lond B Biol Sci 358(1437):1493–1500
6. Mihradi S, Henda AI, Dirgantara T, Mahyuddin AI (2013) Development of 3D gait analyzer software based on marker position data. ASEAN Eng. J A 3(2):4–14
7. Vaughan CL, Davis BL, O'Connor JC (1992) Dynamics of human gait, Kiboho, Cape Town
8. O'Connor CM (2007) Automatic detection of gait events using kinematic data. Gait Posture 25(3):469–474

9. Ren L, Jones R, Howard D (2008) Whole-body inverse dynamics over a complete gait cycle based only on measured kinematics. J Biomech 41:2750–2759
10. Rose J, Gamble JG (2006) Human walking, 3rd edn. Lippincott Williams & Wilkins, Philadelphia

Robust H_∞ Controller with Parametric Uncertainties for Anti-Rolling Gyro System

Unggul Wasiwitono(✉), Latifah Nurahmi, Ari Kurniawan Saputra, and Bambang Pramujati

Department of Mechanical Engineering, Institut Teknologi Sepuluh Nopember, Surabaya 60111, Indonesia
unggul@me.its.ac.id

Abstract. This paper presents the robust control of an anti-rolling gyro system based on gyroscopic precession momentum. Parameter uncertainties are considered by using the polytopic approach. The linear matrix inequalities-based (LMI) technique is used to design the robust controller. The designed controller is applied to a two-wheeled self-balancing vehicle, an unstable and sensitive system in terms of parametric uncertainties. Simulation results show that the robust controller works well for a given parameter uncertainty bound.

Keywords: Parametric uncertainty · Linear matrix inequalities · Self-balancing · H_∞ Controller

1 Introduction

Bicycles and single-track vehicles, in general, are widely known for having major stability issues. In recent years, two-wheeled self-balancing vehicles have begun to receive more interest among researchers [1–5]. It is due to the potential for energy saving and the advantages of simple structure, flexible operation, and portability. Moreover, instability, nonlinear characteristics, and uncertainties become the reason that balancing control of a single-track vehicle is a challenging topic for many researchers in dynamics control fields [3–6].

Riderless bicycles for carrying products and learning bicycles for children are potentially applying self-balancing single-track vehicles. There are three primary methods in the balancing mechanism: using flywheel [2–4], heading steering [6–8], and mass moving [9]. Quite possibly, the most well-known technique is based on the angular momentum concept. A rotating wheel is used for maintaining the angular momentum. The balancing torque in roll angle is generated by accelerating or decelerating a rotating wheel.

The idea of using gyroscopic stabilization is dated back to the early 1960s. The precession of a high-moment-of-inertia gyro rotor rotating at high speed is used to stabilize the roll of a motorcycle [10]. In [11], the use of two identical gyroscopes rotating and precessing in opposite directions has been recently proposed. Many control strategies were proposed in addition to the balancing mechanism. PID controllers and LQR, for

I. Tolj et al. (eds.), *Recent Advances in Mechanical Engineering*, Lecture Notes in Mechanical Engineering, https://doi.org/10.1007/978-981-19-0867-5_46

example, have been used to control self-balancing bicycles. The control problem is similar to the inverted pendulum problem. In [12, 13], a constrained H_∞ control scheme is used for stabilization and disturbance attenuation control of the gyroscopic inverted pendulum.

In this study, a robust H_∞ control is studied. Following the result in [12, 13], the concept of reachable sets in a state-space ellipsoid defined by a quadratic storage function is used to represent the output and control input constraints. Additionally, the effect of parameter uncertainties is modeled using the polytopic approach. The control problem is then derived using linear matrix inequality (LMI) optimization. In the literature, the LMI methodology has been considered a versatile tool for synthesizing robust control systems.

2 Dynamics Model

The two-wheeled self-balancing vehicle is shown in Fig. 1, and the orientation of the vehicle part is shown in Fig. 2. The system uses stabilizing device based on gyroscopic effect precession. As illustrated in the figures, the system consists of five parts: the main body, two gyro-flywheels, and two gimbals. Two brushless DC motors spin the flywheel and one DC motor with a transmission system to drive the gimbal. Each flywheel's angular velocity is constant and turns at the same speed and in the opposite direction. The transmission system contains belts, gears, and pulleys, so the rotational magnitude of each gimbal has the same value but in the opposite direction.

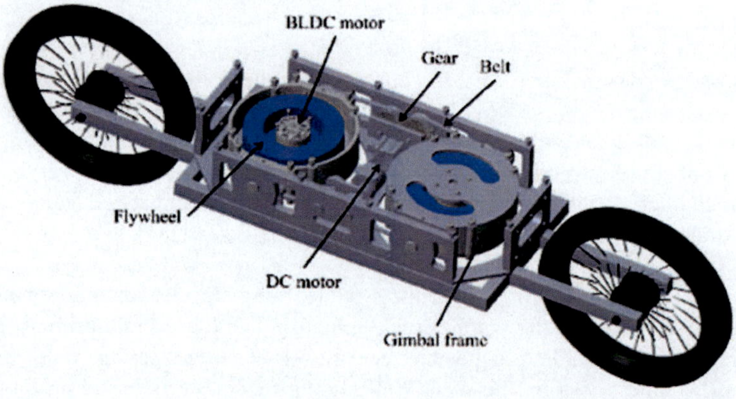

Fig. 1. The gyroscopic two-wheeled vehicle

The kinetic energy is obtained as the sum of the linear and rotational kinetic energy of each part.

$$T_B = \frac{1}{2}m_B(\dot{\phi}d_B)^2 + \frac{1}{2}I_{Bx}\dot{\phi}^2 \tag{1}$$

$$T_{Gi} = \frac{1}{2}m_{G_i}(\dot{\phi}d_{G_i})^2 + \frac{1}{2}\left[I_{Gix}(\dot{\phi}\cos\alpha_i)^2 + I_{G_{iy}}\dot{\alpha}_i^2 + I_{G_{iz}}(\dot{\phi}\sin\alpha_i)^2\right] \tag{2}$$

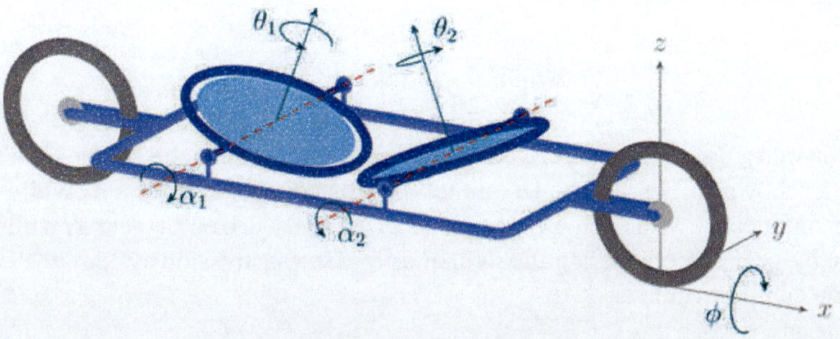

Fig. 2. Orientation of the vehicle

$$T_{F_i} = \frac{1}{2}m_{F_i}\left(\dot{\phi}d_{F_i}\right)^2 + \frac{1}{2}\left[I_{Fix}\left(\dot{\phi}\cos\alpha_i\right)^2 + I_{F_{iy}}\dot{\alpha}_i^2 + I_{F_{iz}}\left(\dot{\phi}\sin\alpha_i + \dot{\theta}_i\right)^2\right] \tag{3}$$

The total kinetic energy is

$$T = T_B + \sum_{i=1}^{2} T_{G_i} + \sum_{i=1}^{2} T_{F_i} \tag{4}$$

The total potential energy of the system comprises the potential energy of the main body, gimbal frames, and flywheels:

$$U = \left(m_B d_B + \sum_{i=1}^{2} m_{G_i} d_{G_i} + \sum_{i=1}^{2} m_{F_i} d_{F_i}\right) g\cos\phi \tag{5}$$

The mathematical model can be obtained by applying Lagrange's equations:

$$\frac{d}{dt}\left(\frac{\partial L}{\partial \dot{q}_j}\right) - \frac{\partial L}{\partial q_j} = Q_j, \ (j = 1, 2, 3) \tag{6}$$

where q_j is the j th generalized coordinate, Q_j is the generalized force corresponding to each generalized coordinate, and $L = T - U$ is the Lagrangian. Solving Lagrange's equation and linearizing around the equilibrium points, we obtain

$$\ddot{\phi} = \frac{M_V}{M_P}\phi - \frac{I_{F_{1z}}\dot{\theta}_1}{M_P}\dot{\alpha}_1 - \frac{I_{F_{2z}}\dot{\theta}_2}{M_P}\dot{\alpha}_2 + \frac{1}{M_P}d \tag{7}$$

$$\ddot{\alpha}_1 = \frac{1}{I_{F_{1y}} + I_{G_{1y}}}u_{\alpha_1} + \frac{I_{F1z}\dot{\theta}_1}{I_{F1y} + I_{G1y}}\dot{\phi} \tag{8}$$

$$\ddot{\alpha}_2 = \frac{1}{I_{F_{2y}} + I_{G_{2y}}}u_{\alpha_2} + \frac{I_{F2z}\dot{\theta}_2}{I_{F2y} + I_{G2y}}\dot{\phi} \tag{9}$$

where

$$M_p = m_B d_B^2 + I_{B_x} + \sum_{i=1}^{2} m_{G_i} d_{G_i} + \sum_{i=1}^{2} m_{F_i} d_{F_i} + \sum_{i=1}^{2}(I_{Gix} + I_{Fix})$$

$$M_v = \left(m_B d_B + \sum_{i=1}^{2} m_{G_i} d_{G_i} + \sum_{i=1}^{2} m_{F_i} d_{F_i} \right) g$$

In this study, each gimbal is driven by a single DC motor attached to the transmission system, as shown in Fig. 3. The DC motor shaft is connected to gear 1, while gears 6 and 7 are connected with each gimbal. Gear 2 is directly driven by gear 1, while gear 3 is driven by gear 2. Considering the dynamics of the transmission system, the equation of motion can be written as

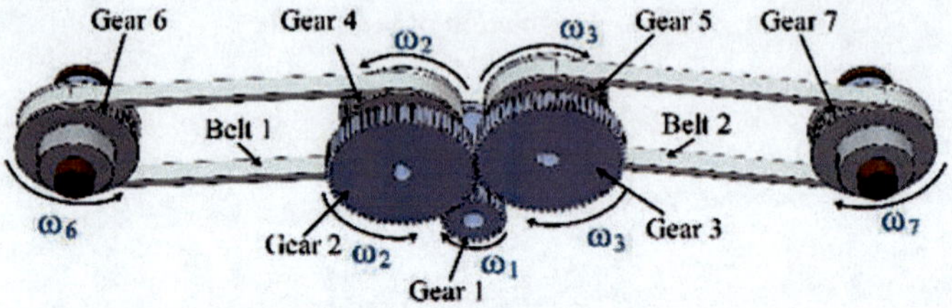

Fig. 3. Transmission system

$$\ddot{\phi} = \frac{M_V}{M_P}\phi + \left(\frac{I_{F_{1z}}\dot{\theta}_1}{M_P} - \frac{I_{F_{2z}}\dot{\theta}_2}{M_P} \right) N_t \dot{\alpha}_g + \frac{1}{M_P} d \tag{10}$$

$$\ddot{\alpha}_g = \left(\frac{N_1 - N_{J1} + N_{J2}}{M_n} \right) \tau_a + \left(\frac{N_{J3} - N_{J4}}{M_n} \right) \tag{11}$$

where

$$M_n = \left(\frac{J_2 + J_3 + J_4 + J_5}{N_1} + N_1 J_1 \right)$$

$$N_{J1} = \frac{N_2 N_t J_6}{I_{F_{1y}} + I_{G_{1y}}}, N_{J2} = \frac{N_3 N_t J_7}{I_{F_{2y}} + I_{G_{2y}}}, N_{J3} = \frac{N_2 J_6 I_{F_{1y}} \dot{\theta}_1}{I_{F_{1y}} + I_{G_{1y}}}, N_{J4} = \frac{N_3 J_7 I_{F_{2y}} \dot{\theta}_2}{I_{F_{2y}} + I_{G_{2y}}}$$

Define the following state variable $x = [x_1\, x_2\, x_3\, x_4]^T = [\phi\, \alpha_g\, \dot{\phi}\, \dot{\alpha}_g]^T$. The state-space equation can be written as follows:

$$\dot{x} = Ax + B_d d + B_u \tau_a \tag{12}$$

with

$$A = \begin{bmatrix} 0 & 0 & 1 & 0 \\ 0 & 0 & 0 & 1 \\ \frac{M_V}{M_P} & 0 & 0 & a_{34} \\ 0 & 0 & a_{43} & 0 \end{bmatrix}, B_d = \begin{bmatrix} 0 \\ 0 \\ \frac{1}{M_P} \\ 0 \end{bmatrix}, B_u = \begin{bmatrix} 0 \\ 0 \\ 0 \\ b_{41} \end{bmatrix}$$

3 Controller Synthesis

The full state-feedback control law is considered in this study, namely

$$u = Kx \tag{13}$$

The closed-loop system, including the controller, can be written as

$$\dot{x} = A_{cl}x + B_d d, z = C_{cl}x \tag{14}$$

where $A_{cl} = A + B_u K$ and $C_{cl} = C_z + D_{zu}K$

3.1 LMI Condition for H_∞

The criteria in terms of LMIs for the following H_∞ norm condition

$$\|C_{cl}(sI - A_{cl})^{-1}B_d\|_\infty < \gamma \tag{15}$$

can be stated as the following theorem.

Theorem 1 The H_∞ problem has a solution if and only if there exists a matrix W, and a symmetric positive definite matrix P, such that.

$$\begin{bmatrix} AP + PA^T + B_u W + W^T B_u^T & B_d & PC^T + W^T D_{zu}^T \\ B_d^T & -\gamma I & D_{zu}^T \\ CP + D_{zu}W & D_{zu} & -\gamma I \end{bmatrix} < 0 \tag{16}$$

In addition, the time domain constraints on control input and output are respected when the following inequalities hold:

$$\begin{bmatrix} \frac{1}{\epsilon}X & W \\ W^T & P \end{bmatrix} \geq 0 \text{ with } X_i \leq u_{i,\max}^2 \tag{17}$$

$$\begin{bmatrix} \frac{1}{\epsilon}Z & S_z C_z P \\ PC_z^T S_z^T & P \end{bmatrix} \geq 0 \text{ with } Z_i \leq z_{i,\max}^2 \tag{18}$$

Assuming the state trajectory remains in the ellipsoid if the disturbance energy is bounded

$$\Omega(P, \epsilon) = \left\{x \in R^n | V(x) \leq \epsilon\right\} \tag{19}$$

with $\epsilon = \gamma d_{max} + V(x(0))$, where $V(x)$ is a quadratic storage function.

Then a state feedback controller $K = WP^{-1}$ exists such that H_∞ from the disturbance d to the performance output z is less than γ.

3.2 The Polytopic Approach

In real applications, parameter variations may exist. The vehicle's mass is uncertain due to, for example, the loading that should be carried. The polytopic approach [14] can be used to insert such uncertainties in system matrices (12). Hence the uncertain system can be modeled as

$$\dot{x} = A(\delta)x + B_d(\delta)d + B_u(\delta)\tau_a \tag{20}$$

where

$$A(\delta) = \sum_{k=1}^{N} \delta_k A_k,\ B_d(\delta) = \sum_{k=1}^{N} \delta_k B_{dk},\ B_u(\delta) = \sum_{k=1}^{N} \delta_k B_{uk} \tag{21}$$

Being $\delta \in \Xi$, with

$$\Xi = \left\{ \delta \in R^N | \delta_i > 0, i = 1, \ldots, N, \sum_{k=1}^{N} \delta_k = 1 \right\} \tag{22}$$

As a result, the control problem with parametric uncertainty can be expressed as follows.

Theorem 2. *If there exists $P = P^T > 0$, $W > 0$, and γ such that.*

$$\begin{bmatrix} A(\delta)P + PA^T(\delta) + B_u(\delta)W + W^T B_u^T(\delta) & B_d(\delta) & PC_z^T + W^T D_{zu}^T \\ B_d^T(\delta) & -\gamma I & D_{zu}^T \\ C_z P + D_{zu} W & D_{zu} & -\gamma I \end{bmatrix} < 0 \tag{23}$$

$$\begin{bmatrix} \frac{1}{\epsilon}X & W \\ W^T & P \end{bmatrix} \geq 0 \text{ with } X_i \leq u_{i,\max}^2 \tag{24}$$

$$\begin{bmatrix} \frac{1}{\epsilon}Z & S_z C_z P \\ PC_z^T S_z^T & P \end{bmatrix} \geq 0 \text{ with } Z_i \leq z_{i,\max}^2 \tag{25}$$

are satisfied then the state feedback controller $u = Kx$ with $K = WP^{-1}$ stabilizes the uncertain system (20) for all $\delta \in \Xi$ and the H_∞ norm form the disturbance to the performance output less than γ.

4 Simulation Results

The simulation result of the robust constrained H_∞ control for anti-rolling gyro is presented. The parameters of the single-track vehicle are listed in Table 1. Figure 4 shows the system's response to an initial position of 10° for all state-variable for the case of constrained H_∞ control and robust constrained H_∞ control. The constrained H_∞ control experiences more significant overshoot and longer oscillation. However, the constrained H_∞ control is experiencing a faster settling time, as shown in Fig. 4a. The maximum gimbal angle for both controllers is still less than 90°. Figure 5 shows the pendulum angle response due to variation of parametric uncertainties. In this study, the body's mass is uncertain but bounded in the interval of 0.85–1.4% of the nominal value. It can be seen from the figure that the constrained H_∞ control show significant degradation, while the robust constrained H_∞ control only deviated slightly from the nominal case.

Table 1. Single-track vehicle parameters

Parameter	Symbol	Value	Unit
Mass of the body, gimbal, and flywheel	$\left[m_B, m_{G_{1,2}}, m_{F_{1,2}}\right]$	[5.306, 1.106, 0.71]	(kg)
Mass moment inertia of the body	$\left[I_{B_x}, I_{B_y}, I_{B_z},\right]$	[0.248, 0.260, 0.261]	(kg m^2)
Mass moment inertia of the gimbal	$\left[I_{G_x}, I_{G_y}, I_{G_z}\right]$	[0.198, 0.172, 0.332] × 10^{-3}	(kg m^2)
Mass moment inertia of the flywheel	$\left[I_{F_x}, I_{F_y}, I_{F_z}\right]$	[0.877, 0.877, 1.642] × 10^{-7}	(kg m^2)
Height of each part at COG	$\left[d_B, d_{G_{1,2}}, d_{G_{1,2}}\right]$	[0.152, 0.165, 0.171]	(m)
Angular velocity of the flywheel	$\left[\dot{\theta}_1, \dot{\theta}_2\right]$	[628.31, −628.31]	(rad/s)
Mass moment inertia of each gear	$\left[J_1, J_{2,3}, J_{6,7}\right]$	[0.025, 0.539, 0.266] × 10^{-5}	(kg m^2)

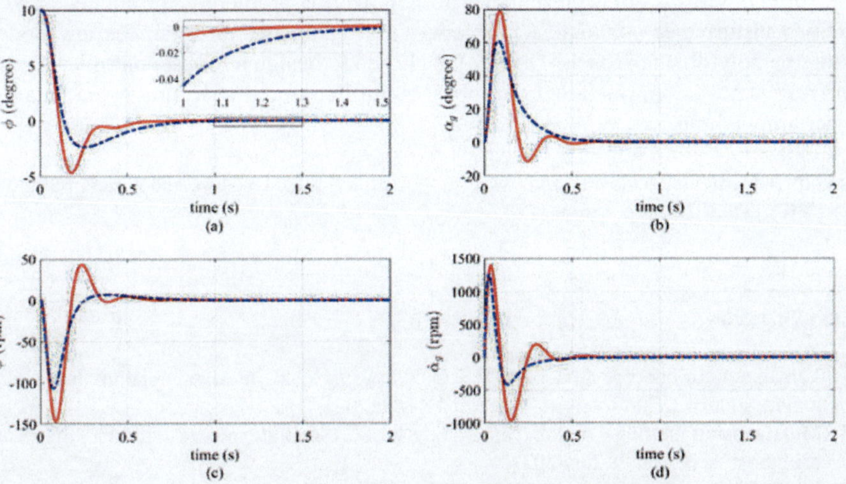

Fig. 4. Response due to initial condition of 10°. **a** Pendulum angle, **b** gimbal angle, **c** pendulum velocity, and **d** gimbal velocity (solid line: H_∞; dash-dotted line: robust H_∞)

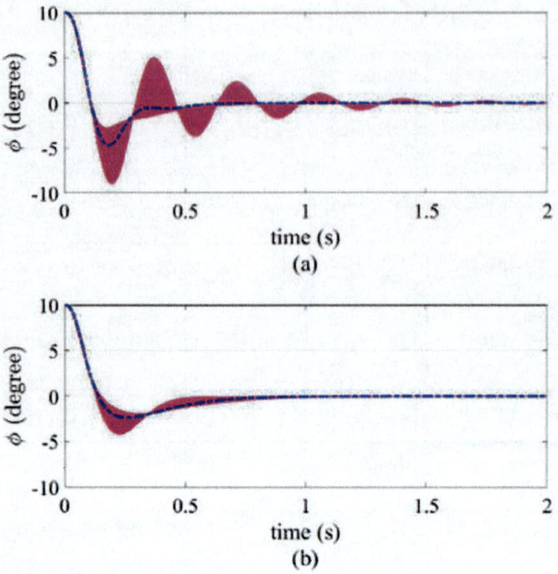

Fig. 5. Response due to parameter uncertainty, **a** H_∞ and **b** robust H_∞

5 Conclusion

The application of robust constrained H_∞ control with parametric uncertainties for the anti-rolling gyro system is presented. The stabilization problem is formulated as a constrained disturbance attenuation problem, while parameter uncertainties are modeled using the polytopic approach. The LMI is chosen to design a robust controller. Simulation results have confirmed that the robust controller works well for a given parameter uncertainty bound.

Acknowledgements. This research was supported by the ITS Research Grant (Grant no.: 1294/PKS/ITS/2021).

References

1. Yamaha motor concept model MOTOROiD. https://global.yamahamotor.com/designtechnology/design/concept/motoroid/. Accessed 2 Mar 2021
2. Murata manufacturing Co., Ltd. Murata boy. https://corporate.murata.com/more_murata/robots/mboy. Accessed 28 July 2021
3. Lot R, Fleming J (2020) Gyroscopic stabilizers for powered two-wheeled vehicles. Veh Syst Dyn 1–26
4. Tian J, Ding J, Tai Y, Ma Z (2020) Control of different-axis two-wheeled self-balancing vehicles. IEEE Access 8
5. Dai F, Gao X, Jiang S, Guo W, Liu Y (2015) A two-wheeled inverted pendulum robot with friction compensation. Mechatronics 116–125

6. Seekhao P, Tungpimolrut K, Parnichkun M (2020) Development and control of a bicycle robot based on steering and pendulum balancing. Mechatronics 69
7. Tanaka Y, Murakami T (2004) Self-sustaining bicycle robot with steering controller. In: Proceedings of the 8th IEEE international workshop on advanced motion control, pp 193–197
8. Vatanashevanopakorn S, Parnichkun M (2011) Steering control-based balancing of a bicycle robot. In: Proceedings of the 2011 IEEE international conference on robotics and biomimetics, pp 2169–2174
9. Yamakita M, Utano A (2005) Automatic control of bicycles with a balancer. In: Proceedings of the 2005 IEEE/ASME international conference on advanced intelligent mechatronics, pp 1245–1250
10. Summer TO (1968) Gyro stabilized vehicle. United States patent US 3410357
11. Klews M (2015) Method and device for the driving stabilization of a motorized two-wheeled vehicle using a double-gyroscope device. United States patent US 2015/0353151 Al
12. Wasiwitono U, Wahjudi A, Saputra AK, Constrained control application to inverted pendulum with control moment gyroscope. In: Proceeding of 2019 international conference on advanced mechatronics, intelligent manufacture and industrial automation (ICAMIMIA)
13. Wasiwitono U, Wahjudi A, Saputra AK, Yohanes (2021) Stabilization and disturbance attenuation control of the gyroscopic inverted pendulum. Int J Vib Control 27:415–425
14. Apkarian P, Tuan HD (2000) Parameterized LMIs in control theory. SIAM J Control Optim 1241–1264